3.50

AF768731

Modeling the Fate of Chemicals In the Aquatic Environment

Edited by

Kenneth L. Dickson
Alan W. Maki
John Cairns, Jr.

ANN ARBOR SCIENCE
THE BUTTERWORTH GROUP

Copyright © 1982 by Ann Arbor Science Publishers
230 Collingwood, P.O. Box 1425, Ann Arbor, Michigan 48106

Library of Congress Catalog Card Number 82-71527
ISBN 0-250-40552-0

Manufactured in the United States of America
All Rights Reserved

Butterworths, Ltd., Borough Green
Sevenoaks, Kent TN15 8PH, England

PREFACE

This book is the product of the fourth Pellston Environmental Workshop, held August 16–21, 1981. This workshop series was initiated to contribute to the need that existed for guidance in assessing the hazards of chemicals in the aquatic environment. Support for the workshop series has been provided by a variety of sources, including industry, the Environmental Protection Agency and trade organizations. Each workshop has been attended by professionals who were selected for their expertise and come from industry, regulatory agencies and universities. By serving as a forum where professionals can exchange ideas and debate issues, these workshops have attempted to provide guidance to scientists, engineers and decision-makers.

A concept that has evolved from the Pellston Environmental Workshop Series is that hazard assessment requires an understanding of the relationship between the concentration of chemicals causing an adverse effect on aquatic life and environmental exposure concentrations. This book and the workshop from which it emanates assess the role that mathematical environmental fate models can play in predicting environmental exposure concentrations of chemicals. A variety of models are examined in the book in regard to their construct, use and degree of validation. The book presents a state-of-the-art overview of the use of mathematical environmental fate models as tools for decision-makers in assessing the hazards of chemicals in the aquatic environment.

K. L. Dickson
A. W. Maki
J. Cairns, Jr.

ACKNOWLEDGMENTS

The 1981 Pellston Workshop and this publication were made possible by grants from the U.S. Environmental Protection Agency and the Chemical Manufacturers Association to North Texas State University.

The co-chairmen wish to thank the staff of the University of Michigan Biological Station at Pellston, Michigan, for their hospitality and assistance.

The workshop participants are commended for their commitment to the goals of the workshop and their cooperation in the production of this book.

Gratitude is also expressed to Bonnie C. Yates for her editorial assistance and to Elaine Curry for typing the manuscripts.

Dickson **Maki** **Cairns**

Kenneth L. Dickson is Director of the Institute of Applied Sciences at North Texas State University. Dr. Dickson received his BS and MS from North Texas State University and his PhD from Virginia Polytechnic Institute and State University. Before coming to North Texas State University in 1978, he was Associate Director of the Center for Environmental Studies at Virginia Tech. As an environmental scientist, Dr. Dickson has published extensively in the areas of aquatic biology, aquatic toxicology and environmental assessment. His specialty is pollution biology and water quality. He is the author of 17 books and monographs, 125 technical articles and numerous reports. Dr. Dickson is a consultant to a number of industries on water pollution problems and serves as an advisor to the National Research Council, the National Academy of Science and the U.S. Environmental Protection Agency (EPA). He has received a number of professional awards for his technical writings, including the Best Technical Paper Award in 1973 from the American Water Works Association Water Resources Division, and the Dudley Award in 1977 for Outstanding Publication of the Year from the American Society for Testing and Materials (ASTM). He is a member of the American Fisheries Society, ASTM and the Society of Environmental Toxicology and Chemistry.

Alan W. Maki is an environmental toxicologist with the Research and Environmental Health Division of the Exxon Corporation in East Millstone, New Jersey. He received a BS in fisheries biology from the University of Massachusetts, an MS in aquatic biology from North Texas State University in Denton and a PhD in aquatic biology from Michigan State University in East Lansing. For the past seven years he has been associated with the U.S. chemical industry and with the Procter & Gamble Co. He has done research on the fate and effects of chemicals and introduced substances on the productive capacity and physiology of aquatic species, microbiology of wastewater treatment and biodegradation methodology, model ecosystem testing, and computer applications to ecological problems. He is the author of more than 50 scientific articles concerning aquatic toxicology and risk assessment of chemicals in the environment.

John Cairns, Jr., is University Distinguished Professor and Director of the Center for Environmental Studies, Virginia Polytechnic Institute and State University, Blacksburg, Virginia. He has been concerned with the ecology of polluted and unpolluted freshwater since 1947. He received his AB from Swarthmore College in 1947 and his MS and PhD from the University of Pennsylvania in 1949 and 1953, respectively. Recent publications include *Artificial Substrates* (Ann Arbor Science Publishers), *Biological Monitoring* and the chapter "Restoration of Damaged Ecosystems" (in *Research on Fish and Wildlife Habitat,* EPA). Dr. Cairns has served on committees for the National Academy of Sciences and the National Research Council, and the Ohio River Sanitation Committee. He is a member of the governing board of the American Instiute of Biological Sciences and of the editorial board of the American Water Resources Association.

CONTENTS

Section 2
Modeling Processes

Section 3
State-of-the-Art and Case Study Examples

Section 4
Validation and Application Testing

INTRODUCTION

A. W. Maki

Exxon Corporation
East Millstone, New Jersey 08873

Kenneth L. Dickson

North Texas State University
Denton, Texas 76203

J. Cairns, Jr.

Virginia Polytechnic Institute and State University
Blacksburg, Virginia 24061

The development of the science of ecotoxicology and the overall discipline of environmental risk assessment have identified the need for sound, predictive safety-testing methods for chemicals and effluents potentially discharged into the environment. Regulatory requirements set in motion under the Toxic Substances Control Act (TSCA), the Clean Water Act and amendments, the Resource Conservation and Recovery Act (RCRA), the Federal Insecticide, Fungicide and Rodenticide Act (FIFRA), and, internationally, the Organization for Economic and Cooperative Development (OECD) Hazard Assessment Project, as well as several others, have further underscored the need for methodology with direct application to the predictive risk-assessment process. This book reports the proceedings from the fourth of what have come to be known as the Pellston Environmental Workshops.

This workshop series, begun in 1977, has primarily focused on the development of consensus philosophy and methodology for environmental risk assessment. We feel these workshops have been extremely useful for defining an overall framework of methods for measuring risk to the environment associated with the development, refining, use and disposal of chemical substances. The risk-assessment process is now recognized as an objective and probabilistic exercise based on empirical data and scientific judgment. Results from the integrated risk-assessment procedure can be used as a uniform approach to provide a means for estimating the limiting concentrations of chemical substances or effluents that will produce no observable effects on organisms or communities potentially exposed.

The first of the Pellston workshops was held June 13–17, 1977, at the University of Michigan Biological Station on Douglas Lake in Pellston, Michigan. The workshop was entitled "Application of Toxicity Testing Methods as Predictive Tools for Aquatic Safety Evaluation" and was designed to address the state-of-the-art of aquatic toxicology appropriate for determining the effects of chemicals on aquatic life. The workshop participants helped to formulate the concept that the relative hazard of a chemical or effluent is a function of (1) the measured concentration causing biological effects as determined through the conduct of laboratory toxicity tests and (2) the predicted or estimated environmental exposure concentration [e.g., hazard = f(toxicity, exposure)]. The proceedings from the workshop were published by the American Society for Testing and Materials [1].

Recognizing that the environmental questions attendant to the safety evaluation of new chemicals were truly an international problem, a subsequent workshop was organized to examine evolving risk-assessment procedures in this country, as well as several from Europe and Japan. This workshop was held August 13–18, 1978, at Waterville Valley, New Hampshire, and included representatives from industry, universities and governmental agencies to analyze and assemble a degree of international consensus for environmental risk assessment of new chemicals. The results of this workshop [2] identified the need for refined methods to assess and predict the environmental fate of chemicals for use in risk-assessment procedures.

This emphasis on the estimate of environmental exposure concentration as an essential part of the risk-assessment procedure in turn spawned the need for reliable methods of measuring and predicting those chemical, physical and biological processes that determine fate. The processes identified by the 1978 Waterville Valley workshop as important in determining the fate of a chemical included chemical degradation, photochemical degradation, evaporation, sediment binding, uptake and depuration by biota, and microbiological transformation. As an attempt to guide these efforts to develop process-oriented methods, the third workshop in the series was held August 14–18, 1979, again at Pellston, Michigan. The proceedings were published by the American Society for Microbiology [3].

With a consideration of the diverse environmental processes influencing the fate of chemicals now in hand, it was felt that a distinct need existed to examine the methods and approaches for integration of these processes into a conceptualized modeling context to make effective use of process data and to yield accurate predictions of resultant environmental concentrations of chemicals. Several environmental fate models are either in direct use or in various stages of development, and a definite need exists to bring these modelers together to exchange ideas and approaches to predictive fate modeling. Thus, the fourth workshop was conducted August 16–21, 1981, at the University of Michigan Biological Station in Pellston. The proceedings of this workshop constitute the contents of this book.

BACKGROUND, SCOPE AND OBJECTIVES

To effectively design a timely and relevant workshop program, a planning committee was assembled to meet in Atlanta, Georgia, on December 13, 1979. The members of that planning committee were as follows:

- George Baughman, U.S. Environmental Protection Agency (EPA), Athens, Georgia;
- Ken Dickson, North Texas State University, Denton, Texas;
- Bob Jamieson, Proctor & Gamble Co., Cincinnati, Ohio;
- Ray Lassiter, U.S. EPA, Athens Georgia;
- Al Maki, Exxon Corporation, East Millstone, New Jersey; and
- Fred Zeigler, ERM-Southeast, Inc., Brentwood, Tennessee.

The planning committee was responsible for the identification of specific program topics for the workshop sessions and recommended the list of participants given in Table I.

WORKSHOP CONDUCT

Approximately one month before the workshop, all participants were provided with a notebook containing copies of the discussion-initiation papers to be presented at the workshop. This created the opportunity to study, in advance, the substance of the presentations and to formulate

Table I. Participants Recommended by the Planning Committee

Dr. Robert Bailey, Dow Chemical Co., Midland, MI
Dr. Phil Bedient, Rice University, Houston, TX
Dr. Harold Bergman, University of Wyoming, Laramie, WY
Dr. Stuart Brown, Battelle – Pacific Northwest Laboratories, Richland, WA
Dr. Larry Burns, U.S. EPA, Athens, GA
Dr. John Cairns, Jr., Virginia Polytechnic Institute and State University, Blacksburg, VA
Dr. J. J. Conner, Massachusetts Institute of Technology, Cambridge, MA
Dr. Bill Cooper, Michigan State University, East Lansing, MI
Dr. Norman Crawford, Hydrocomp International, Inc., Mt. View, CA
Dr. Charles Delos, U.S. EPA, Washington, DC
Dr. Kenneth Dickson, North Texas State University, Denton, TX
Dr. Dominic Di Toro, HydroQual, Mahwah, NJ
Dr. Tony Donigian, Anderson Nickles Co., Palo Alto, CA
Dr. Larry Games, Procter & Gamble Co., Cincinnati, OH
Dr. Steve Hern, U.S. EPA, Las Vegas, NV
Dr. A. W. Klein, Umweltbundesamt, Berlin, Germany
Dr. Ray Lassiter, U.S. EPA, Athens, GA
Dr. Chris Lee, Unilever, Merseyside, England
Dr. R. M. Li, Colorado State University, Ft. Collins, CO
Dr. Gordon Loewengart, Allied Corporation, Morristown, NJ
Dr. Alan Maki, Exxon Corporation, East Millstone, NJ
Dr. Yuzuru Matsuoka, Kyoto University, Kyoto, Japan
Dr. Lee Mulkey, U.S. EPA, Athens, GA
Dr. Brock Neely, Dow Chemical Co., Midland, MI
Dr. Donald O'Connor, HydroQual, Mahwah, NJ
Dr. Carolyn Offutt, U.S. EPA, Washington, DC
Dr. John Rodgers, North Texas State University, Denton, TX
Dr. Jerald Schnoor, University of Iowa, Iowa City, IA
Dr. Mike Slimak, U.S. EPA, Washington, DC
Mr. Charles Staples, North Texas State University, Denton, TX
Dr. Fran Werner, Monsanto Chemical Co., St. Louis, MO
Dr. Bill Wood, U.S. EPA, Washington, DC
Dr. George Yeh, Oak Ridge National Laboratories, Oak Ridge, TN
Dr. Fred Ziegler, ERM – Southeast, Inc., Brentwood, TN

questions for the subsequent discussion session. The following workshop objectives were presented to the participants during the opening session:

- analyze the state-of-the-art of environmental fate modeling, considering both development and application;
- critically evaluate existing models for predicting environmental exposure concentrations of chemicals in various aquatic systems;
- critically examine the utility of various models as decision-making aids for their specific applications;

- position the role of environmental fate modeling for aquatic hazard assessment, considering both regulatory applications and new product development; and
- develop recommendations for future research needs in environmental fate modeling.

Additionally, a specific statement of charge to participants was presented to facilitate completion of the above objectives:

- Participants must recognize their multiple responsibilities, not only to respective employers, but also to the environment and user groups such as the public, government and industry.
- Participants should engage in critical discussions of topics introduced in presentations.
- Participants should assist in the preparation of summary statements and recommendations for their assigned sessions.
- Rapid completion of final drafts and revisions is necessary to ensure timely preparation of the proceedings.

Each workshop session was begun with one or more discussion-initiation paper solicited from a specific author or group of authors named by the program planning committee. These papers were presented during plenary sessions of the workshop and were followed by a discussion period open to all participants under the leadership of previously appointed chairmen. It was the responsibility of these chairmen and their committees to summarize each session and to identify the major consensus conclusions resulting from each of the sessions. Each committee presented its findings and conclusions to the plenary session of all workshop participants on the final day of the workshop. The workshop sessions and general content are outlined below.

Session 1: Introduction

K. L. Dickson, A. W. Maki and J. Cairns, Jr. held a discussion covering the organizational background of the workshop and its relationship to the previous Pellston Environmental Workshops. Participants were presented with the statement of objectives of the workshop and charge to participants.

Session 2: Modeling Overview: Use and Needs

The session chairman was S. Brown; the session summary committee consisted of D. Di Toro, G. Yeh, C. Offutt, M. Slimak and A. W. Klein.

The major emphasis of this session was an overview of what modeling is all about from the standpoint of use and application of modeling data. Presentations included several examples of regulatory use of and need for predictive models. Also considered were the philosophical and scientific bases for contemporary modeling and the general use of models in environmental science programs.

The discussion-initiation papers were:

1. D. O'Conner: An Assessment of the Philosophical and Scientific Base for Modeling (presented by Di Toro);
2. G. Yeh: Modeling the Distribution of Chemicals;
3. W. Wood: Modeling and TSCA;
4. C. Offutt: Modeling and Pesticide Programs;
5. M. Slimak: Modeling and Water Programs; and
6. A. W. Klein: Modeling Applications for OECD.

Session 3: Modeling Processes

The session chairman was R. Bailey; the session summary committee consisted of L. Burns, J. Schnoor and B. Neely. This session focused on the physical, chemical and biological environmental processes that lend themselves to description by modeling and how these individual processes can be coupled to yield predictive environmental fate models.

The discussion-initiation papers were:

1. L. Burns: Identification and Evaluation of Fundamental Modeling Processes;
2. B. Neeley and D. Mackay: Evaluative Model for Estimating Environmental Fate;
3. J. Schnoor: Fate and Transport Modeling for Toxic Substances; and
4. D. Di Toro: Modeling the Fate of Partitioning Chemicals.

Session 4: State-of-the-Art and Case Study Examples

The session chairman was G. Loewengart; the session summary committee consisted of L. Mulkey, N. Crawford, P. Bedient and B. Cooper. A number of case study examples of contemporary environmental fate models were examined. Discussions included development and use of the model, state of validation, and specific limitations.

The discussion-initiation papers were:

1. N. Crawford: Runoff Models;
2. P. Bedient: Groundwater; and
3. H. Bergman: Complex Effluents.

Session 5: Validation and Application Testing

The session chairman was D. Di Toro; the session summary committee consisted of R. Lassiter, L. Games, T. Donigian and Y. Matsuoka. This workshop session examined critically the elements of model validation. How does one run controls on modeling predictions from laboratory data to test the accuracy of resultant predictions? What types of data are most useful for model validation? Examples of field research programs designed to test model prediction were included. The practical elements of legal validation and defense of modeling data were also discussed.

The discussion-initiation papers were:

1. R. Lassiter: Testing Environmental Fate Models;
2. T. Donigian: Field Validation and Error Analysis;
3. L. Games: Field Validation in Streams;
4. W. Cooper: Validation via Experimental Stream Ecosystem; and
5. F. Ziegler: Legal Validation.

Session 6: Workshop Summary

The session chairman was C. Lee; the session summary committee consisted of W. Wood, J. Rodgers and H. Bergman. The final day of the workshop was devoted to committee presentations and a synthesis of the session papers. An overall workshop summary, including consensus conclusions and an identification of areas for future research needs, was presented by the summary committee.

The sections of this book correspond to the workshop sessions. Each section except this introduction and the final workshop summary, contains one or more discussion-initiation papers and a short summary report prepared by the session summary committee. Complete drafts of all text were prepared by the final day of the workshop, and final drafts of each of the summaries were completed via subsequent mailings. However, no substantial changes were permitted after the workshop was finished. Every attempt has been made to ensure that editorial review has been carried out in the spirit of the workshop.

KEY TO MODELS

Some of the models shown in Table II are frequently mentioned in the text by their acronyms. The key in Table II is furnished as an aid to the reader.

Table II. Key to Models Discussed in This Book

Model Name	Acronym	Reference
Agriculture Runoff Model	ARM	4
Chemical Manufacturing Association—HydroQual Inc. Model	CMA-HydroQual	5
Environmental Partitioning Model	ENPART	a
Exposure Analysis Modeling System	EXAMS	6
Fates of Aromatics Model	FOAM	7
Hydrocomp Simulation Programming	HSP	8
Hydrologic Simulation Program—Fortran	HSPF	9,10
Mackay Level I, II, III		11,12
Model of Watershed Hydrology	USDA HL−74	13
Oxidation/Reduction Equilibrium Model	REDEQL	14
Seasonal Soil Model	SESOIL	b
Sediment Radionuclide Transport Model	SERATRA	15
Stanford Research International (SRI, Inc.)	SRI	16
Stanford Watershed Model IV		17
Storage, Treatment, Overflow, Runoff Model	STORM	18
Unified Transport Model	UTM	19

[a] Developed by EPA Office of Toxic Substances.
[b] To be published by EPA.

REFERENCES

1. Cairns, J., Jr., K. L. Dickson and A. W. Maki, Eds. *Estimating the Hazard of Chemical Substances to Aquatic Life,* ASTM Technical Publication 657 (Philadelphia: American Society for Testing and Materials, 1978).
2. Dickson, K. L., A. W. Maki and J. Cairns, Jr., Eds. *Analyzing the Hazard Evaluation Process* (Bethesda, MD: American Fisheries Society, 1979).
3. Maki, A. W., K. L. Dickson and J. Cairns, Jr., Eds. *Biotransformation and Fate of Chemicals in the Aquatic Environment* (Washington, DC: American Society for Microbiology, 1980).
4. Donigian, A. S., and H. Davis. "User's Manual for Agriculture Runoff Management (ARM) Model," U.S. EPA Report 600/3-78-080 (1978).
5. O'Connor, D. J., and J. P. St. John. "An Assessment of Modeling the Fate of Chemicals in the Aquatic Environment," Chapter 1, this volume.

6. Burns, L. A., D. M. Cline and R. R. Lassiter. "Exposure Analysis Modeling System (EXAMS): User Manual and System Documentation," U.S. EPA Environmental Research Laboratory, Athens GA (1981).
7. Bartell, S. M., P. F. Landrum, J. P. Giesy and G. Leversee. "Simulated Transport of Polycyclic Aromatic Hydrocarbons in Artificial Streams," paper presented at the International Symposium on Energy and Ecological Modeling, Louisville, KY, April 20–23, 1981.
8. Hydrocomp Simulation Programming—Operations Manual," Hydrocomp International, Inc., Palo Alto, CA (1972).
9. Johanson, R. C., J. C. Imhoff, H. H. Davis, J. L. Kittle and A. S. Donigian. "User's Manual for the Hydrologic Simulation Program—FORTRAN (HSPF): Release 7.0," U.S. EPA Environmental Research Laboratory, Athens, GA (1981).
10. Hydrocomp International, Inc. "User's Manual for the Hydrological Simulation Program—FORTRAN (HSPF)," U.S. EPA Environmental Research Laboratory, Athens, GA (1979).
11. Mackay, D. "Finding Fugacity Feasible," *Environ. Sci. Technol.* 13:1218–1223 (1979).
12. Mackay, D., and S. Patterson. "Calculating Fugacity," *Environ. Sci. Technol.* 15:1006 (1981).
13. Holtan, H. N., G. J. Stittner, W. H. Henson and N. C. Lopez. "USDAHL-74 Model of Watershed Hydrology," USDA, Agricultural Research Service Technical Bulletin No. 1518 (1975).
14. Morel, F., and J. Morgan. "A Numerical Method for Computing Equilibria in Aqueous Chemical Systems," *Environ. Sci. Technol.* 6:58–67 (1972).
15. Onishi, Y., and S. W. Wise. "Mathematical Model SERATRA for Sediment and Pesticides Transport in Rivers and Its Application to Pesticide Transport in Four Mile and Wolf Creeks in Iowa," prepared for U.S. EPA by Battelle—Pacific Northwest Laboratories under Contract No. 68032613 (1979).
16. Smith, J. H., W. R. Mabey, N. Bohonos, B. R. Holt, S. S. Lee, T.-H. Chou, D. C. Bomberger and T. Mill. "Environmental Pathways of Selected Chemicals in Freshwater Systems, Part II, Laboratory Studies," U.S. EPA Report 600/7-78-074 (1978).
17. Crawford, N. H., and R. K. Linsley. "Digital Simulation in Hydrology: Stanford Watershed Model IV," Stanford University, Department of Civil Engineering Technical Report No. 39 (1966).
18. "Storage, Treatment, Overflow, Runoff Model (STORM)," U.S. Army Corps of Engineers, Hydrologic Engineering Center, Computer Program 723-58-L7520 (1976).
19. Munro, J. K., Jr., et al. "Application of the Unified Transport Model (UTM) to the Movement of Pb, Cd, Zn, Cu, S Through the Crooked Creek Watersheds," ORNL/NSF/EATC-28 (1976).

SECTION 1

MODELING OVERVIEW: USE AND NEEDS

CHAPTER 1

ASSESSMENT OF MODELING THE FATE OF CHEMICALS IN THE AQUATIC ENVIRONMENT

Donald J. O'Connor

Civil Engineering Department
Manhattan College
Riverdale, New York 10471
and HydroQual, Inc.
Mahwah, New Jersey 07430

John P. St. John

HydroQual, Inc.
Mahwah, New Jersey 07430

Organic chemicals and heavy metals are present in varying degrees in all phases of the environment—air, water and land. They are transferred between and among these media, undergo transformation within each, and accumulate in viable and nonviable constituents. The magnitude and significance of the problem has become increasingly evident, particularly in the accumulation of toxic substances in food chains and in contamination of water supplies due to the release of these substances from disposal areas. In both of these respects, the terrestrial and aquatic environments are affected.

The benefits derived from the use of these substances are evident in many facets of our society, particularly with respect to increased food production. The demand for these materials and, more specifically, the benefits derived from their use continuously increase. A balance must therefore be sought between the extreme positions—complete ban and no control. Such a balance leads to the use of certain chemicals, which may safely be assimilated in the environment at levels that yield the benefits without deleterious effects. This goal necessitates the development of assessment methods that permit evaluation and ultimately prediction of environmental concentrations. An understanding is therefore required of transport, transformation and exchange of toxic substances in and between the various media of the environment. This chapter addresses the present state-of-the-art in this respect, with emphasis on the areas about which our knowledge and understanding are minimal and inadequate.

The various phenomena that affect transport, transformation and accumulation of organic chemicals and heavy metals in the various phases of the aquatic environment are addressed. Taken into account is the exchange of these materials between water and the other phases of the environment (air and land). However, the primary emphasis is directed to aquatic systems, specifically to the spatial and temporal distribution of these materials in the various types of natural water bodies, i.e., rivers, lakes, estuaries and the coastal zone. The approach taken is similar to that defining the distribution of substances that are natural components of biochemical and ecological cycles of oxygen, nutrients, minerals, dissolved and suspended solids, and the basic elements of the food chain (bacteria and phytoplankton).

The first sections of this chapter briefly describe and present an overview of the modeling framework. Then, a series of observations and questions are presented concerning relevant factors about which, in the opinion of the authors, the theoretical understanding and laboratory and field data are inadequate. Finally, some comments are made concerning validation procedures in view of these questions.

BASIC PRINCIPLES

Mass Balance

The equations describing the spatial and temporal distribution of organic chemicals are developed using the principle of mass conservation, including the inputs with the transport, transfer and reactions components. The general expression for the mass-balance equation about a specified volume, V is:

$$V \frac{dc_i}{dt} = J_i + \Sigma R_i + \Sigma T_I + \Sigma W \qquad (1)$$

where c_i = concentration of the chemical in compartment i
J = transport through the system

R = reactions within the system
T = transfer from one phase to another
W = inputs

Equation 1 describes the mass rate of change of the substance because of the net effect of the various fluxes and transformations.

The general term "compartment" refers to each phase of the physiochemical regime, viz., dissolved and particulate solids in the water, atmosphere and bed, as well as to each element of the food chain (phytoplankton, zooplankton, fish and detrital material). The transport, reaction and transfer terms may be positive or negative, depending on the direction of kinetic routes between the chemical in compartment i and its concentration in the other compartments with which it reacts or exchanges. The pathways are determined by the hydrodynamic, geophysical, chemical and biological characteristics of both the system and the chemical. The hydrodynamic components transport material from one spatial location to another by dispersion and advection. The physical factors transfer from one phase to another, such as exchange with the atmosphere, adsorption to and desorption from the suspended and bed solids, and settling and scouring of these solids. The chemical factors transform the substance by processes such as photooxidation, hydrolysis and oxidation/reduction reactions. The biological phenomena affect transference and transformation: the latter primarily by microorganisms that may metabolize the chemical, and the former by assimilation and excretion by the various aquatic organisms. Accumulation in the food chain is brought about by ingestion of the chemical from the water and by predation on contaminated prey.

Consider the concentration c_d to be the dissolved component of the chemical in the water. It interacts with the particulate concentration c_p. The interaction may be an adsorption/desorption process with the solids or an assimilation/depuration process with the aquatic organisms. In either case the particulate concentration is defined as:

$$c_{pi} = r_i m_i \tag{2}$$

where c_p = particulate concentration in compartment i (M/L^3)
r_i = mass of chemical/unit of interacting mass (M/M)
m_i = concentration of the interacting species (M/L^3)

The mass-balance equation for the particulate component, similar to Equation 1, is then:

$$\frac{dc_{pi}}{dt} = r_i \frac{dm_i}{dt} + m_i \frac{dr_i}{dt} = J + \Sigma R + \Sigma T + \Sigma W \tag{3}$$

It is apparent from Equation 3 that an equivalent expression must be written for the concentration of the interacting compartment m_i. In principle, analysis of the problem requires simultaneous solution of three equations: the concentration of the chemical dissolved in the water c_d; the mass concentration of the chemical per unit mass of interacting species r; and the concentration of the species itself m. Since this compartment may be subdivided further (inorganic and organic solids, multiple species of fish), Equation 2 is more generally expressed as a summation of the individual components of the interacting substances:

$$c_{pi} = \sum_{1}^{n} r_i m_i \tag{4}$$

The specific conditions for which the analysis is performed frequently permit simplifying assumptions to be made. In laboratory batch reactors and in certain prototype situations, the rate of change of the interacting species may be zero, i.e., a constant concentration of suspended solids or biomass. Thus $dm/dt = 0$, from which an equilibrium concentration of solids or biomass follows, resulting in two simultaneous equations to be solved, instead of three.

Dynamic Equilibrium

One of the essential properties of the analysis is the interaction between the dissolved and particulate states of the constituent, which, in time, leads to a dynamic equilibrium between the two components. Consider the most simplified conditions of a batch reactor, in which the mixing is of sufficient magnitude to maintain a uniform concentration throughout the volume of fluid. Assume the concentration of absorbing solids m (M/L^3) is constant. Let c_d and c_p be the concentrations (M/L^3) of the dissolved and particulate components. If there is neither transfer nor decay of the chemical, the total concentration c_T remains constant in time and is equal to the sum of the dissolved and particulate:

$$c_T = c_d + c_p \tag{5}$$

The latter is related to the concentration of suspended solids m as shown by Equation 2: $c_p = rm$.

The equilibrium between the dissolved concentration in the water and the mass concentration of the solids is usually expressed in terms of a partition coefficient:

$$\P = \frac{r}{c_d} = \frac{c_p}{mc_d} \quad \text{or} \quad \P m = \frac{c_p}{c_d} \tag{6}$$

Equation 6 is the linear portion of the Langmuir isotherm. Although not always representative of actual conditions, it is a reasonable approximation when the solid phase concentration r is much less than the ultimate absorbing capacity of the solids. Combining Equations 5 and 6, the total concentration may be expressed as:

$$c_T = c_d + \P mc_d = \frac{c_p}{\P m} + c_p \tag{7}$$

The product $\P m$ is a convenient dimensionless parameter, characteristic of a particular system under equilibrium conditions. For a specified value of $\P m$, the equilibrium distribution between the dissolved and particulate concentrations is established by Equation 7.

The distribution between the dissolved concentration and the particulate concentration in the various levels of the food chain may be expressed in an identical fashion. Accounting for the distribution of various types of absorbing solids and various levels of the food chain, each with its characteristic partition coefficient, Equation 7 may be more generally expressed:

$$c_T = c[1 + \Sigma \P_i m_i] \tag{8}$$

The distribution may thus be categorized in accordance with the absorbing solids (organics, clays, silts and sands) or the accumulating biomass (phytoplankton, zooplankton, fish and macrophytes). Since the total biomass in most natural water systems is usually an order of magnitude less than that of the nonviable solids, the equations defining each category may be decoupled and the former may be solved independently. Under those conditions in which it may be significant, it may readily be incorporated as shown in the above equation.

MODEL COMPONENTS

Models for the analyses of organic chemicals and heavy metals are similar to those developed for constituents that are natural components of ecological cycles. The terms relating to the particulate form and its interaction with the dissolved component are the additional components to be incorporated. The reaction and transfer mechanisms are incorporated with the transport phenomena and input functions to define the temporal and spatial distribution of toxic substances in natural systems; it is also necessary to analyze the distribution of the various types of solids. Furthermore, the exchange between the suspended and the bed constituents must be taken into account. The resulting models describe the physicochemical effects in the inorganic realm in conjunction with the solids, as well as accumulation and transfer through the various elements of the aquatic food chain. Each of the components are discussed in the following sections.

Kinetic and Transfer Elements

The equilibrium described above is a result of the kinetic interaction between the dissolved and particulate phases. This property, which distinguishes the analysis from that of purely dissolved substances, leads to equations of a more complex form. To gain an insight into the nature of these interactions and an understanding of a practical simplification, consider the kinetics in a natural water system in which interaction is taking place between the dissolved and particulate components, each being transferred by physicochemical processes or transformed by chemical or biological degradation (Figure 1). The reversible reaction ($k_1 = k_2$) may represent adsorption/desorption between the dissolved component and the particulate in the inorganic phase (suspended or bed solids) or the assimilation/depuration in the organic phase (aquatic organisms). The ratio of these kinetic coefficients, which may be functions of the solids concentrations, is a measure of the distribution of partition coefficient ¶.

Assuming these processes are taking place in a batch reaction in which the exchange is occurring from the water to the other media, the kinetic equations for the dissolved and particulate components are:

$$\frac{dc_d}{dt} = -k_1 c_d + k_2 c_p - k_a c_d - k_c c_d \tag{9}$$

$$\frac{dc_p}{dt} = +k_1 c_d - k_2 c_p - k_s c_p - k_p c_p \tag{10}$$

Addition of Equations 9 and 10 demonstrates the rate of change of the total concentration:

$$\frac{dc_T}{dt} = -(k_a + k_c)c_d - (k_s + k_p)c_p$$

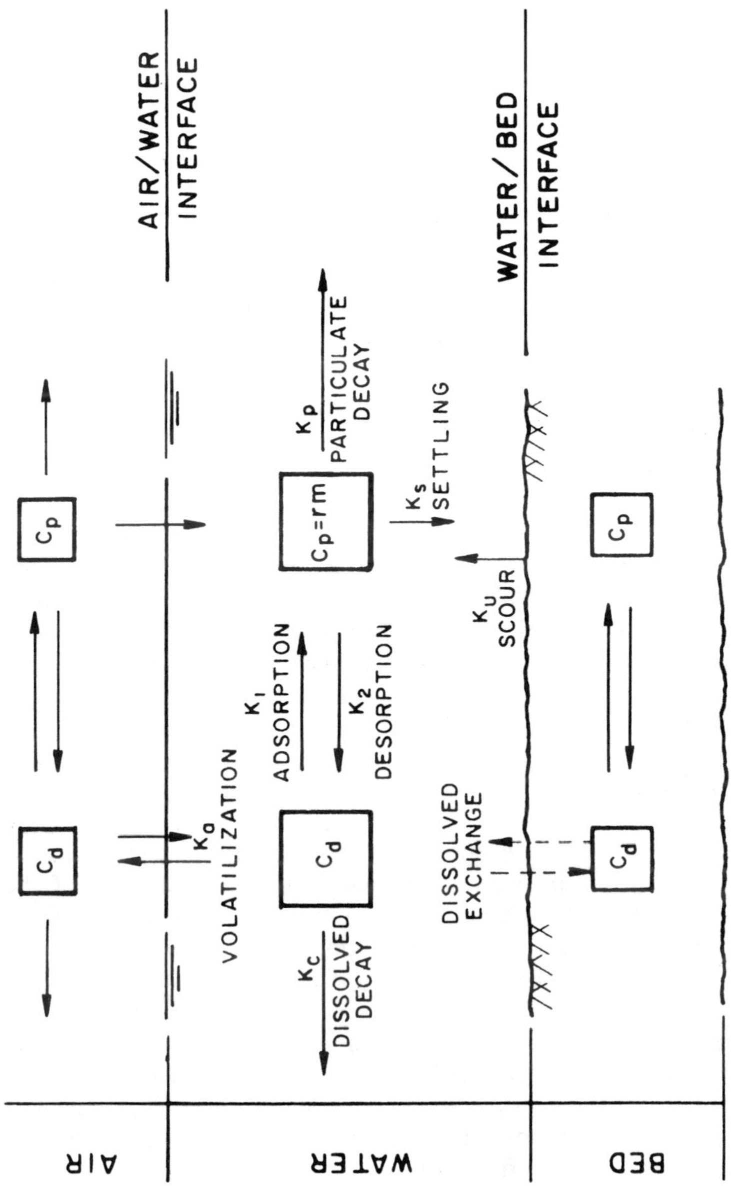

Figure 1. Kinetic and transfer factors.

Substitution of Equation 7 for the dissolved and particulate components yields:

$$\frac{dc_T}{dt} = -[f_d(k_a + k_c) + f_p(k_s + k_p)]c_T \tag{11}$$

where $f_d = 1/(1 + ¶m)$
$f_p = ¶m/(1 + ¶m)$

The assumption of "instantaneous" equilibrium as expressed by Equations 9 through 12 is a valid representation, or model, of kinetic reactions, provided the time to equilibrium, determined by k_1 and k_2, is rapid relative to the other phenomena that affect the substance k_c and k_p. This condition is generally applicable to the adsorption/desorption process, since its equilibrium time is usually much shorter (in minutes and hours) than that of other kinetic effects (which may be on the order of days, months or years).

Although various mechanisms have been described in accordance with first-order kinetics in Equation 12, some reactions are dictated by nonlinear kinetics and depend on the concentration of the interacting substance or organism. In many cases, however, these may be approximated by simpler expressions. This assumption leads to equations for which analytical solutions are available, and thus provides insight and understanding that would not otherwise be possible with more complex formulations.

Transport Regimes

Equation 1 is the most general expression to define the distribution of a chemical in a natural water body. Given the characteristics of the drainage area and water system, and the nature of the substance, it takes on a more definitive form. The distribution between the dissolved and particulate components of the chemical, and the kinetic interactions are the essential factors that are common to all types of models. What distinguishes the various models are the transport components of a specific water system and the characteristics of the bed with which it interacts. Thus, the basis of the classification lies, to some degree, in the transport regimes of the general types of water systems (lakes, streams and estuaries), but rests more significantly on the transport characteristics of the bed itself and the magnitude of the water/bed interactions.

The transform and transfer routes are shown in Figure 2. For conservative, nonvolatile chemicals, these transform routes are negligible, but the settling/resuspension transfers are potentially important for any substance, regardless of its other characteristics. These are the characteristics of the system and the substance that essentially determine the complexity of the analysis and the structure of the model.

Each of the general types of natural water systems may be classified in

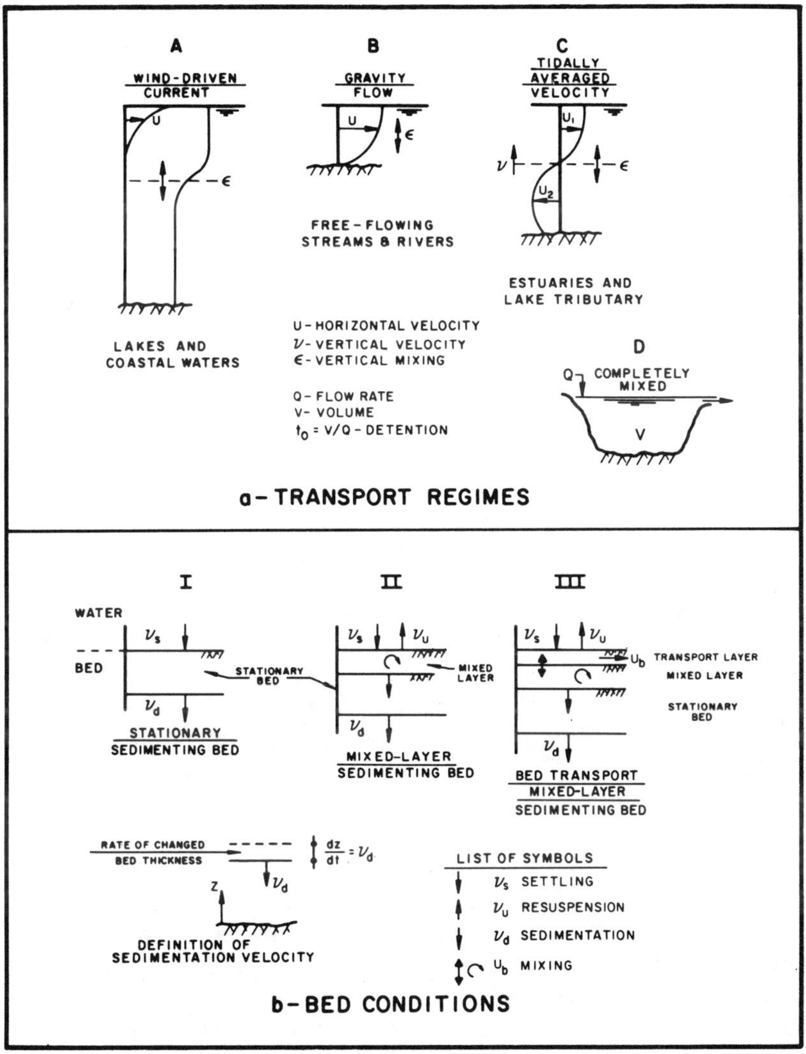

Figure 2. Transport regimes and bed conditions.

accordance with characteristic fluid transport regime and the interaction of the water with the bed (as shown in Figure 2). The components of the transport field are the advective (U) and the dispersive (E) elements, which in general are expressed in three-dimensional space. Each of the natural systems, i.e., lakes, streams, estuaries and coastal waters, is usually characterized by a predominating component in one or more dimensions. As shown in Figure 2a, the transport in lakes may be frequently approximated by one or two dimensions (A), in which the vertical is the major component; in streams by a one-dimensional longitudinal analysis (B); and in estuaries by a two-dimensional scheme (C) both longitudinal and vertical. A spatially uniform condition (completely mixed) is type (D), the transport coefficient of which is the detention time t_0.

Bed Characteristics

The bed conditions that are relevant to the analysis are shown in Figure 2b. They may be classified as inactive or stationary, and active or exchanging. The latter may be further subdivided as with and without horizontal transport. A further characteristic of bed conditions relates to the phenomenon of sedimentation. All natural water bodies accumulate, in varying degrees, materials that settle from the water column. In freshwater systems, reservoirs and lakes are repositories of much of the suspended solids that are discharged by the tributary streams from upstream drainage. In marine systems, estuaries and embayments accumulate solids in similar fashion, and the coastal zones do so to a lesser degree. In flowing freshwater streams and tidal rivers, suspended solids may settle or scour, depending on the magnitude of the velocity and shear associated with the flow. Bed conditions in these systems are therefore subject to seasonal and daily variations; however, the beds of estuaries and lakes, which are also subject to such variations, tend to accumulate material over long time scales. The increase in bed depth and concentration is expressed in terms of a sedimentation velocity, which is many orders less than the settling velocity of the various solids in suspension.

A stationary bed is basically characterized by zero-to-negligible horizontal motion. This condition is most commonly encountered in lakes and reservoirs of relatively great depth, with minimal winds. It also occurs in freshwater streams under low-flow conditions and in marine systems with little tidal mixing. It may therefore be associated with any one of the three transport systems. The essential characteristic of this type of system is a relatively low degree of vertical mixing in the fluid. The hydrodynamic environment is one that permits the gravitational force to predominate and suspended particles of density greater than that of water to settle. The accumulation of this material in the bed causes an increase in the thickness of the benthic layer; the rate of increase is sedimentation velocity. The bed is also characterized by minimal or zero particle mixing in the layer in contact with the water.

An exchanging bed, which is probably more common, is characterized by some degree of mixing in the active layer of the bed. The mixing may be the result of either physical or biological factors, such as increased levels of shear associated with horizontal or vertical velocities and gradients, or bioturbation attributable to the feeding behavior of benthic organisms. It exists, therefore, in lakes where wind effects extend to the bottom, and in streams and rivers under moderate flow conditions. In each of these cases, the shear exerted on the bed is sufficient to bring about mixing in the interfacial layer, but not sufficient to cause significant erosion and bed motion. The net flux of material to the bed is the difference between the settling flux and that returned by the exchange due to the mixing. Thus, the bed thickness may increase or decrease, and the sedimentation velocity may be positive or negative. This type of bed condition may also be associated with any of the three fluid transport types, but is more usually associated with type B (Figure 2a), and in the littoral zones of lakes where the water depth is sufficiently shallow to permit wind effects to be transmitted to the bed.

Finally, the bed may be characterized by both mixing and advective properties. The shearing stress exerted by the fluid is of sufficient intensity to cause erosion and resuspension of the bed, and the fluid velocity is of sufficient magnitude to induce horizontal motion of resuspended material and/or the interfacial bed layer. This phenomenon involves the complex field of sediment transport, which has been developed greatly in streams, but much less in estuaries and lakes. The bed system may now be envisioned as three distinct segments: a moving interfacial layer, a mixed zone and a stationary bed beneath. There is vertical exchange between the moving and mixed layers, and the vertical transport in the bed is characterized by the sedimentation velocity. This type of bed regime is associated only with types B and C fluid transport systems. The direction of horizontal motion of the bed is in accordance with the velocity vector of the fluid in contact with the bed surface. In freshwater streams and rivers, the bed transport is downstream in the direction of the flow; in estuaries, the net bed transport is upstream in the saline zone as a result of tidally averaged motion.

Classification of Models

A suggested classification is based on the types of bed conditions shown in Figure 2b, in conjunction with one of the three types of fluid transport shown in Figure 2a. The three general types are enumerated in a progressive fashion from the simplest to the most complex, as presented in Table I and described above. The final form of the equations is based essentially on one of the three types, in conjunction with the kinetic interactions shown in Figure 1.

INPUT FUNCTIONS

Mass rates of discharge of chemicals from point and nonpoint sources constitute the input to the model. They are of primary importance because they are forcing functions, to which the concentration of a relevant constituent in a natural water body responds. Whereas much is known (or at least can be measured) about mass input rates from point sources (such as municipal and industrial effluents, and agricultural and irrigation drains), knowledge of the spatial and temporal distribution of nonpoint sources is meager. Drainage from urban and rural areas, combined sewer overflows and groundwater seepages are distinct areas about which there is a dearth of information. Specification of inputs is essential to calibration and validation of the models. Although a discussion of inputs is not a specific part of this evaluation, it is emphasized that these factors are critical—not only with respect to calibration and validation procedures, but also with respect to model application and remedial actions.

FLUID TRANSPORT AND BED CHARACTERISTICS

Although much is known about fluid transport in most natural water systems, the understanding of bed transport is minimal. It is appreciated that whereas much work has been done in both the laboratory and field in elucidating the factors affecting the transport of sands and gravel, comparatively little has been accomplished concerning clays, silts and detrital material. Because of its affinity to adsorb many constituents, detrital material is of major importance in the analysis of chemicals in natural systems. Theoretical formulations, as well as reliable field data, are particularly lacking on estuaries and the littoral zones of lakes and coastal waters.

Accepting as a framework of analysis the schematic representation

Table I. Classification of Models: Bed Conditions

	I Stationary Bed	II Mixed Layer with Bed	III Sediment Transport with Layer
U	0	0	>0
E	>0	>0	>0
A: Lakes, Reservoirs Coastal Zone	Deep, minimal winds	Moderate depth, winds	Littoral zones
B: Streams, Rivers	Low flow	Moderate flow	High flow
C: Estuaries, Embayments	Enclosed bays, minimal tides	Littoral zones, main channel, low flow	Main channel, moderate flows
D: Completely Mixed	Applicable to all types, particularly to lakes, reservoirs and embayments		

shown in Figure 2b, the questions of paramount importance are related to the concentration, depth, velocity and dispersion of the mixed and moving bed layers. The problem is particularly acute in estuaries, which are subject to upstream and downstream inputs, changes in freshwater flows, variable tidal forces, and salinity intrusion. The beds of most estuaries consist of relatively high fractions of clays, silts and organic matter. There is a significant lack of fundamental understanding and reliable data.

In addition to the motion and characteristics of the bed material, additional features of importance are also uncertain. Rates of scour and entrainment are only now being addressed. The effect of agglomeration and flocculation on settling and resuspension of these solids must be evaluated to a greater degree than presently exists.

The above considerations relate primarily to the particulate component of the chemical. In addition, the characteristics of the bed that affect the dissolved component must also be addressed more fully. Diffusion of the dissolved chemical through the interstitial waters and interaction with other constituents, both dissolved and particulate, require further study. This last factor is particularly important with respect to ionic species of heavy metals.

CHEMICAL TRANSFER MECHANISMS

Once chemical materials are input to and transported in natural water systems, a number of transfers occur within the water column and bed sediment compartments. Such transfers include sorption/desorption processes between dissolved material and suspended sediment; volatilization of the dissolved fraction; settling of sorbed material to the bed with particulates and subsequent resuspension through scour; diffusive exchange between the water column and bed; and possibly deep burial and loss. All of these processes must be quantified in an appropriate manner for inclusion in the aquatic fate modeling framework.

Sorption/Desorption

Because certain chemical substances have a strong affinity for sorption to particulate materials, and the subsequent transport of such particulate chemical forms may be markedly different than for the soluble chemical phase, this process is fundamentally important for chemical fate modeling. It is known that soluble materials will be sorbed onto particulate surfaces in a reversible reaction at some rate until an equilibrium is achieved. Once equilibrium is established, the relationship between sorbed and soluble material can be described by various isotherms, such as Langmuir or Freundlich and others. In linear portions of these functions, the relationship between particulate and soluble forms of the material can be accounted for in a simple manner by use of a solid/liquid partition coefficient.

These relationships are well known in principle and are all subject to various quantitative treatments. However, for any chemical substance, a number of issues must be addressed for model application; for example:

- What are the kinetic characteristics and rates that govern the sorption/desorption process?
- Can the kinetic processes be characterized by first-order reactions or are more complicated kinetic functions required?
- How does the time to equilibrium compare with the overall time scale of the problem?
- In this regard, can it be assumed that equilibrium occurs almost instantaneously, which will greatly simplify the mathematical treatment?
- To what degree is the process reversible?
- Is substantial material sorbed onto particulates and not readily desorbed at lower soluble concentrations?
- Assuming that instantaneous equilibrium is a valid assumption for the problem at hand, will the practical aspects of the problem (soluble concentrations, etc.) allow the use of solid/liquid partition coefficient relationships?
- Further, if the process for a given chemical is only partly reversible, which partition coefficients are to be used?
- To what degree does the observation that partition coefficients may vary for certain substances depending on suspended sediment concentration influence the analysis?
- In all of the foregoing, how do different size classes of particulates and nature of the material (sand, silt or clay) influence these considerations?
- How reliable are results from laboratory experiments conducted with standardized sorbents and laboratory water as compared to field samples?

The answers to these questions will materially influence the sorption/desorption transfer structure that must be provided for a realistic aquatic fate model for a particular substance.

Volatilization

An additional transfer factor that must be considered in the establishment of the dissolved phase of a chemical substance is volatilization.

Many materials volatilize readily, so this process becomes an important loss mechanism in the overall materials balance. For substances of a highly refractory nature, volatilization may be low, but it may be the only active loss mechanism and is therefore important.

Water/air transfer is governed by the partial pressure of the substance in the atmosphere in accordance with Henry's law. Besides Henry's constant, the transfer process is controlled by resistances at the liquid/gas interface. The two-film theory has been used to characterize this process, whereby a thin liquid and/or gas film mass transfer coefficient may control the transfer.

Mass transfer rates can be determined by laboratory experiments in a convenient manner. Laboratory procedures are also available to measure water evaporation and oxygen transfer so that the experimental apparatus can be calibrated and scaled up to field conditions. However, the laboratory environment is decidedly different from most natural water bodies. How good are available methods to scale up laboratory results to actual field conditions? What are the relative merits of the available techniques to estimate volatilization losses in streams and rivers of various flow, velocity, slope and depth characteristics? What about estuaries? In impoundments with large surface areas, how do atmospheric, wind, surface turbulence and water motion affect water/air transfer? What happens in highly stratified estuaries and lakes? Although answers to many of these questions are reasonably well established for commonplace gases and substances, how good are available data and information for many chemicals of interest?

Settling, Resuspension and Sediment Mechanics

Transfer mechanisms for chemical substances bound to particles may involve analysis of settling and resuspension mechanisms, and perhaps bed sediment transport and deep burial. These transfer processes become more or less important depending on the receiving-water context as discussed above. In lakes and coastal waters, the principal particulate chemical transport mechanisms are likely to be settling and subsequent burial. In free-flowing streams, rivers, certain estuaries and shallow lakes, settling and resuspension may be the particular sediment transport mechanisms of concern. Sediment burial may or may not be important, depending on the magnitude, frequency and duration of the scouring mechanism. Flow velocity and shearing stresses may not be sufficient to induce horizontal transport of the interfacial bed layer itself. In high-flow streams and moderate-flow estuaries, all of the foregoing factors, including horizontal bed sediment transport, may be pertinent.

Field-scale studies on the fate of chemicals in the water environment have indicated the importance of sediment, both suspended and in the bed, as reservoirs of material. Environmental transfer of chemicals is therefore associated with transfer of particulates to strongly sorbing substances. Hence, the questions outlined above in the section above in the section on fluid and bed transport are directly applicable to these forms of chemical transfers.

Sediment/Water Diffusive Exchange

Another basic transfer mechanism is diffusive exchange of dissolved substances between the water column and sediment. To some degree, the process is analogous to the water/air exchange discussed above. Once a chemical substance is introduced into the water column of a water body, a diffusive exchange will take place to transfer dissolved substances to the interstitial water in bottom sediment. The process will continue until an equilibrium is achieved between dissolved material in the water column and sediment interstitial waters as affected by all transport, transfer and kinetic processes. If the equilibrium is disturbed in any manner, for example by continual deposition of particles with a sorbed chemical that then tends to desorb on the bottom, diffusive exchange may continue, with the soluble chemical transforming from sediment-based interstitial waters to the overlying water column. This process is controlled by concentration gradients that develop in thin films at the sediment/water interface. The rate at which transfer occurs is a function of diffusivity, the thickness of the films and geometrical characteristics of the water column and sediment. Quantification of this transfer process, therefore, requires a value for the molecular diffusivity of the chemical and an estimate of the film thickness. Each factor may vary by an order of magnitude, so the overall process may have a range of two orders of magnitude. The ability to measure or estimate these values is the key question regarding this transfer mechanism.

CHEMICAL KINETIC TRANSFORMATIONS

A number of kinetic transformations also affect the environmental distribution of chemical substances. The principal transformations are photolysis, hydrolysis, free-radical oxidation and biodegradation. Because these processes may be the major ultimate sinks of reacting chemicals in the water environment, it is necessary that they be determined properly and included in the analytical framework.

Photolysis

Photolysis is the degradation process by which chemical bonds of a substance are broken as the result of transfer of light or radiant energy to these bonds. Insofar as the energy transfer is direct or indirect, the process is referred to as direct photolysis or indirect (sensitized) photolysis. The direct energy transfer is by absorption of light of a certain wavelength by the substance, occurring probably as a result of resonance. In the indirect transfer, the energy is transferred to the compound through other intermediary substances, which may or may not actively be photosensitized. In view of the fact that a substance can exist in both aqueous and particulate phases, it is probable that overall photolytic decomposition is a result of both direct and sensitized photolysis.

The rate of a photochemical reaction is usually expressed in first-order kinetics. The rate coefficient is a function of the quantum yield of the reaction, the available light intensity and the presence of intermediary compounds that contribute to sensitization. The available light intensity is variable at any location in the water column and diminishes with depth because of absorption and scattering. These processes in turn are affected by the nature of dissolved and particulate substances in the aqueous environment, especially organic and inorganic solids. Questions therefore involve the transferability of laboratory-derived photolysis rates to field conditions. Has the effect of light attenuation been accounted for properly in the laboratory apparatus? If so, can light attenuation in the field be estimated properly without direct measurement? Even if measurements are obtained, how does light penetration vary spatially and temporally as particulates vary with flow and season? Of what importance are these to the analysis as a whole?

Similar concerns can be expressed regarding those materials that sensitize the photolytic reactions. Are they properly accounted for in laboratory tests? How do they vary and of what significance is the variation under field-scale conditions?

Hydrolysis

Hydrolysis is defined as the cleavage of a chemical bond and the formation of two new bonds, each one having either the hydrogen or the hydroxyl component of the water molecule. It is a common process, and it is usually mediated by an acid or base, or, less commonly, occurs in a neutral environment. In biochemical systems it is usually enzyme-mediated (enzymatic hydrolysis) and leads to decomposition of complex substances. Depending on the origin of the enzymes, it may more properly be considered a biological reaction.

Chemical hydrolysis is fundamentally a second-order process, containing the product of the hydrolyzed substance and the concentration of hydrogen or hydroxyl or water mediators, depending also on whether it is acid, alkaline or neutral hydrolysis. The two major influences on the rate of the reaction are pH and temperature. The effect of pH is that the rate of the reaction undergoes drastic changes within 1–1.5 pH units, often around pH 7–9. The precise pH value at which this change occurs, as well as the type of the change (increase or decrease), is experimentally determined. The effect of the temperature or the reaction rate is given by the Arrhenius equation relating the reaction rate to the activation energy of the compound.

Questions related to chemical transformation by hydrolysis may involve determination of the practical range of hydrolysis rates under field conditions in accordance with the natural variability of pH. Substantial background pH variability may occur as a result of activity of aquatic plants. This may have a substantial effect on the rate of hydrolysis. Reaction rates have also been observed to increase markedly in soil, perhaps by an order of magnitude, possibly from biological activity. This may affect the effective hydrolysis rate in bed sediments. To what degree can this be effectively accounted for in laboratory determinations?

Free-Radical Oxidation

Free-radical oxidation is a chemical degradation reaction that occurs in many steps. The degraded substance is usually a complex organic compound, represented as R–H. The overall reaction can be thought of as a sequence of initiation, propagation and termination segments. During initiation, an initiator substance absorbs energy heat or light, and fragments. The high-energy fragment transfers the energy to the R–H bond of the organic compounds, fragments the bond, and leaves R in a high-energy state. In case of lack of initiator, there is a slow oxidation of the R–H bond by dissolved oxygen, and the R component is left at a high-energy state. Thus in either case, the result of the initiation step is the breaking of the R–H bond and the energizing of the R fragment. In the propagation stage, the energetic R fragment is oxidized to a peroxide, which in turn uses another R–H molecule, oxidizes it, recoils an energetic R fragment, and the process repeats in a cyclic fashion. These sequences are sustained, provided that the intermediate products, $ROO^\bullet$, $R^\bullet$, do not follow their own degradation pathways. If that occurs, the reaction terminates.

Because of complexity and the fact that these reactions, as most of the efficient cyclic pathways, are mediated by living systems, the degree to which free-radical oxidation can be considered independent of biodegradation is still to be evaluated. Consequently, as with other transformations, how is this process to be measured effectively in the laboratory and modified to reflect field conditions?

Biodegradation

In solutions that contain living organisms, such as bacteria and fungi, it is difficult to speak of any degenerative process as either biological or chemical. For whereas the chemical composition of the substance is altered by a definite chemical (e.g., ring cleavage oxidation and hydrolysis), it may also be true that the presence of the microbe, with its complex enzyme system, is indispensable for the observed change of the substance. It could also be that the substance is used as food or an energy source by the microbe or that the observed changes of the substance result from the substance's proximity to the microbe and its metabolic activities (cometabolism).

In the case in which the microbe utilizes the substance in its metabolic process, its degradation can be described by the Michaelis-Menten or Monod kinetics. These nonlinear kinetics are usually reducible to first-order reactions, provided the simplifications are justified and reflect the prevailing condition.

In all of the microbially mediated reactions, their rates are strongly affected by conditions that affect species population and composition, and by factors that affect the efficiency of enzyme systems such as pH, temperature and salinity. However, no theory exists for these reactions, and the degradation rate is usually evaluated experimentally. How well these experimental procedures simulate the wide variety of field conditions that may prevail is therefore the pertinent question.

ASPECTS OF APPLICATIONS

Various modeling frameworks may be postulated to incorporate and synthesize the foregoing considerations into a systematic quantitative procedure. All such models will contain certain approximations, simplifications and an amount of uncertainty that is related to our ability to formulate answers to the foregoing questions. The final test of any modeling structure, however, lies in its demonstrated ability to reproduce, within acceptable limits, observed conditions. The degree to which various models can do this depends on available data, calibration/validation procedures and the required accuracy as indicated by water quality criteria/standards.

Available Data

Application of any modeling framework requires a certain amount of input data and data against which to compare model results. Input data consist of information on chemical loadings, transport processes, transfer mechanisms and kinetic transformations, as discussed above. Modeling frameworks will therefore have limitations comparable to the limitations in measuring or estimating these factors. How good, therefore, is the available data base for various chemical substances of interest? Is the ability to understand and quantify the various processes limited by the available data base, or are modeling frameworks much more complex than justified by existing data? What is the proper balance?

In terms of model testing, what is the role of microcosm and macrocosm data against which to compare model results? Have proper experimental procedures been developed so that these tests can be conducted properly and yield meaningful results? Can they be developed to substitute for field-scale data in terms of model testing? Is there a sufficient body of field-scale data available at present by which to test existing model frameworks adequately?

Calibration/Validation Procedures

Assuming that sufficient laboratory- and field-scale data are or become available, questions involving calibration/validation procedures gain relevance. To what degree have guidelines for calibration procedures been established? Are the principles developed for more commonplace water quality variables applicable to chemicals? How many sets of data are required for adequate calibration/validation? What is the relative reliability of input data on transfer and kinetic coefficients as guidance for parameter adjustment during calibration? What are the appropriate measures of model-data reproducibility? Do an adequate data base and modeling analysis require seasonal information? In terms of model projections for waste-load allocations, is the precedent of low flow/high temperature as established for conventional pollutants applicable to chemicals? Or may high flow and low temperature be critical and more appropriate?

Criteria and Standards

Other considerations by which to judge the requirements for model calibration and accuracy are water quality criteria and standards. These values are established to protect some aspect of beneficial water use. These values will be used as targets in the waste-load allocation process. The degree of model accuracy should therefore be related in some degree to criteria/standards. If the criterion for a particular chemical is based on biological effects, is the same number applicable to every waterway, given the diversity inherent in resident biosystems? How reliable are biological-effects testing procedures? Is a large factor of safety included in the criterion/standard? If the modeling analysis in any specific case is not accurate within acceptable limits, what risks to the ecosystem will result?

SUMMARY

The foregoing represents an attempt at a brief categorization of the state-of-the-art of modeling the fate of chemicals in the aquatic environment and resulting questions of pertinence. The ability to calculate the environmental impact of chemical substances and heavy metals has been advanced greatly in the recent past, but some work remains to be done. It will be through continuing dialog and effort on the part of the scientific, engineering, industrial and regulatory community that the state-of-the-art will gradually but inevitably be advanced to the benefit of society.

CHAPTER 2

MODELING THE DISTRIBUTION OF CHEMICALS IN AQUATIC ENVIRONMENTS

G. T. Yeh

Environmental Sciences Division
Oak Ridge National Laboratory
Oak Ridge, Tennessee 37830

Many chemical products are known or suspected to be hazardous or to have an unknown degree of hazard. Thus, among the critical issues of chemical production are the safe manufacture, processing, transporting, use and disposal of these chemical products. Comprehensive assessments are needed to quantify the effects and safety of chemical perturbations. The 1976 Toxic Substances Control Act (TSCA) provided the legal impetus to undertake assessment of the potential risk resulting from the manufacture and use of new chemicals.

The ultimate purpose of chemical assessment in the aquatic environment is to arrive at the best possible evaluation of hazard and/or risk to individual species, and ideally to the ecosystem, which would permit reasoned decisions about the consequence. Hazard here means an effect that would decrease or increase the population of animals and plants and one that would change the well-being and health of man. Whether the hazard will in fact take place depends, of course, on the level of chemical concentration and the population distribution. It is not inconceivable to assume that when biota are exposed to chemicals there will be no hazard, even though the level of chemical concentration may be high. If there is a hazard, the degree of severity will depend on (1) the dose-response relationship (effects) for individual species and (2) the extent of exposure that results from overlaying the population distribution to chemical concentrations (Figure 1). Knowing the extent of exposure and the dose-response relationship is therefore essential to formulating an evaluation of the hazard (Figure 1). However, the hazard will itself change the population distribution because of mortality, which in turn will decrease or increase the extent of exposure (Figure 1). In the meantime, the level of chemical concentration may affect the movement of some volitional (motile) biota, resulting for example, in a different population distribution, which in turn will increase or decrease the chemical concentration by uptake.

Thus, to make a complete hazard and/or risk assessment (probability of hazard) of chemical perturbation, this complex iterative interaction must be dealt with. The simplest approach is to decouple the prediction of chemical concentration from analysis of population distribution and hazard assessment. This can be achieved by assuming that the chemical concentration will not influence the movement of biota and that mortality will not alter chemical concentration. This assumption is valid as long as the chemical concentration is not so intolerably high that it changes the habitat distribution (affects the predetermined exposure) and yields severe mortality (greatly reduces the number of predetermined uptaking biota). When decoupling the system fails to be a reasonable approximation, one must deal with the iterative system in its entirety. Unfortunately, models for the coupled system do not exist.

When prediction of chemical concentration can be separated from analyses of population distribution and mortality determination, the sequence of hazard and/or risk assessment is as shown in Figure 2. First, hydrodynamic state variables, including velocity ($\vec{V}$), pressure (p), density (ρ), salinity (S) and temperature (T), are obtained by hydrodynamic models (HYDM) in terms of the model parameters, sources, boundary conditions (BC) and initial conditions (IC). Using these hydrodynamic variables, one may obtain the distribution of suspended sediment (D_{SS}), bed sediment (D_{BS}) and drifting biota (D_{DB}) with sediment distribution models (SDM) and biota distribution models (BDM). In this sequential modeling approach, the most important assumption is that concentration of suspended sediment, change in bed sediment and abundance of drifting biota do not alter the flow of water. Second, population distribution models (PDM) may be employed to predict the distribution of volitional biota (D_{VB}). In the meantime, the distribution of sedentary biota (D_{SB}) may be obtained by observational models (OBM). Third, the chemical concentrations of various species in dissolved and absorbed phases (C_W, D_{BS}, C_{SS}, C_{DB}, C_{VB} and C_{SB}) may be obtained by chemical distribution models (CDM). Finally, ecosystem models (ESM) may be used to predict the mortality of each or all species considered, using the

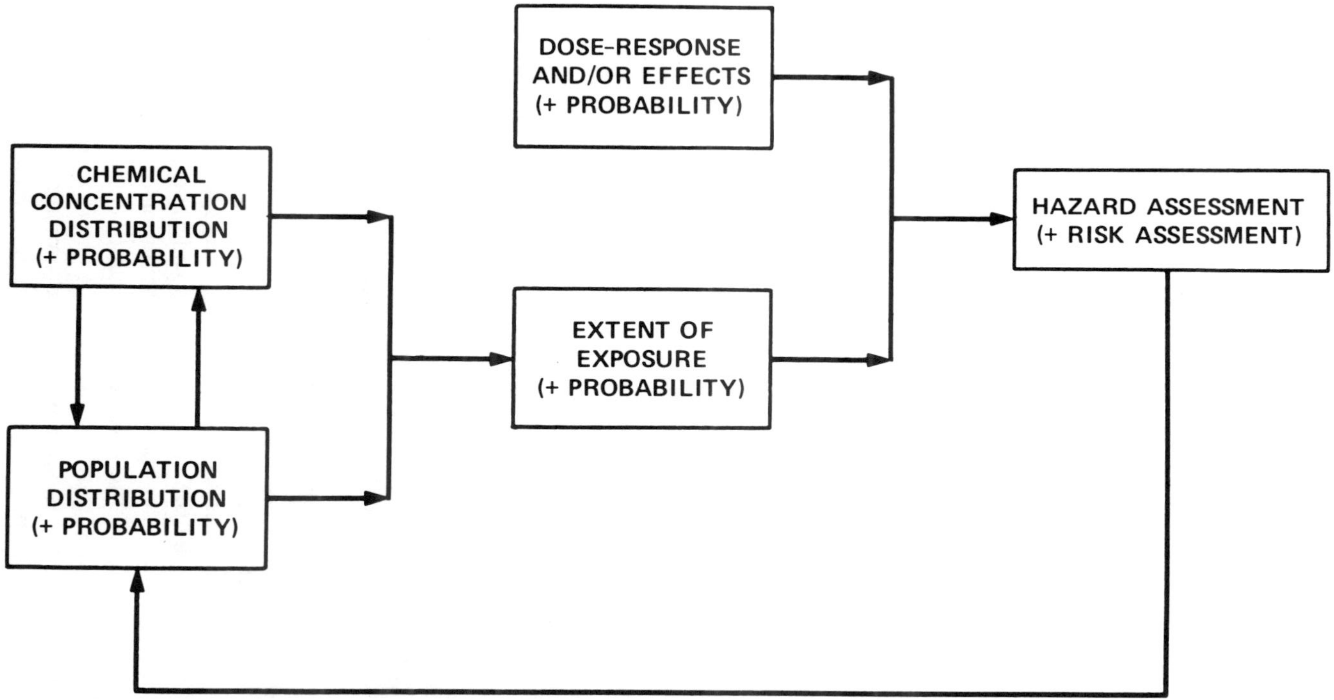

Figure 1. Interactive interaction among chemical concentration, population distribution, extent of exposure, dose-response and hazard assessment.

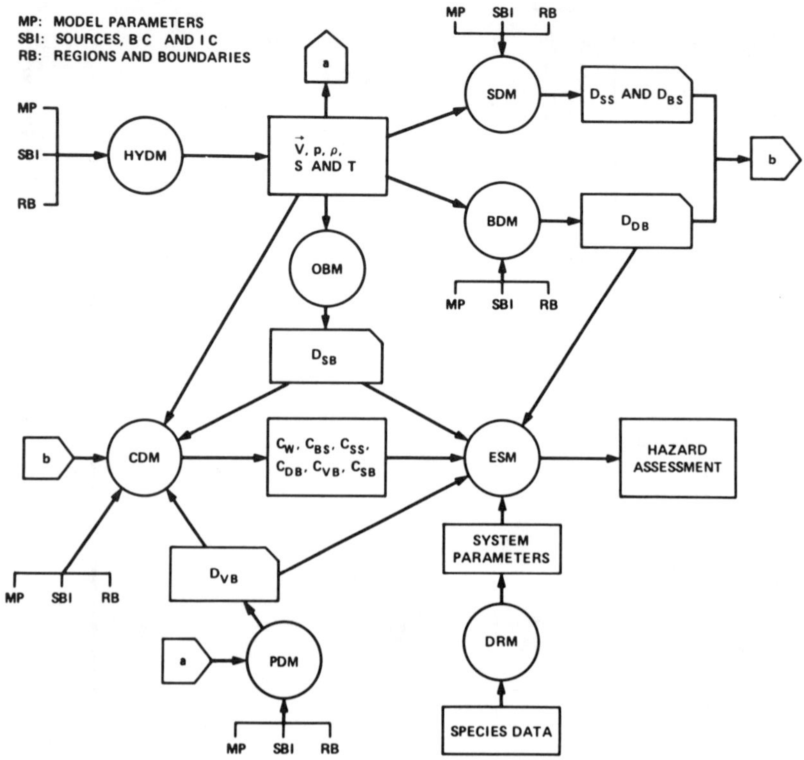

Figure 2. Procedure of hazard and risk assessment of chemicals in the aquatic environment.

system parameters that are derived by the dose-response models (DRM) in conjunction with the spatial/temporal distribution of chemicals. The output from ecosystem models is used for hazard assessment. This chapter concerns itself mainly with analysis of chemical distribution.

Dose-response is often the least understood component of the three that determine ecosystem change (Figure 1). Extensive laboratory tests and further research are required from health physicists, physiologists, biologists and medical personnel to meet this need. Population distribution prediction may be the component that will have the most uncertainty. The difficulty lies in obtaining the distribution of volitional biota such as fishes [1–3]. Of the three components discussed thus far, the ability to predict concentration appears to be the most promising, but also, unfortunately, the most complex. The simultaneous presence of numerous interactive processes—physical and chemical as well as biological—makes it difficult to depict in advance the possible spatial/temporal distribution of chemicals in a given aquatic environment. Although models are admittedly deficient, imperfect and limited, they provide perhaps the best available tool to forecast the response of the aquatic environment to chemical perturbations. This chapter discusses modeling approaches in perspective. Emphasis is put on the models derived from physical, chemical and biological principles and on the need for continuing model development and improvement. To accommodate the data availability and the detail of information desired, it is necessary to provide models with several levels of resolution.

PRINCIPLES-BASED MATHEMATICAL MODELS

Chemical distribution models can be classified in several ways [4,5]. A possible classification [6] is briefly discussed here, consisting of four types of models: descriptive, physical, analogy and mathematical.

Descriptive models are those expressed in one's "native language" [7]. The reliability of descriptive models depends on the expertise and also, unfortunately, on the reputation of modelers. These types of models are limited by their qualitative rather than quantitative predictions. Thus, different experts may easily reach different conclusions, because model predictions are generally very subjective.

Physical models represent scaled-down versions of the real world. Considerable difficulties are normally encountered in the proper scaling of all relevant factors, inevitably resulting in misrepresentation. They are, in general, site-specific and very expensive. Even though physical models can yield very detailed pictures of the phenomena if properly scaled, extrapolations of the results outside the region of study are, however, difficult and inappropriate because of the presence of unique local conditions.

Analogy models employ a transformation of a given physical system into another one that behaves in a similar manner but is easily measured. Examples are the electrical circuit analog models of the potential flow system. Because of nonlinearities in the aquatic flow system, it is doubtful that such models can be designed and constructed in the near future for simulating chemical perturbations.

Mathematical models describe the system with mathematical expressions based on obvious principles, axioms and acceptable assumptions. Generally speaking, a set of mathematical equations is used to describe the relationships between system parameters and state variables. Based on formulism, they can be classified into parametric (empirical), principles-based and stochastic approaches. "Parametric models" are mainly based on observations and experiments with some intuition and judgment. The models simulate the distribution of chemical mass and/or fluid flow by a series of empirical functions that transform inputs into outputs. The most ubiquitous example is the pioneer Stanford Watershed Model IV [8] and its subsequent derivatives [9–11] used for basinwide hydrologic simulations. Dimensional analysis can be employed to reduce the number of variables involved. This type of model does not quantify the mechanisms that link inputs and outputs. However, the predictions can be reliable because the model is tuned to site-specific data. "Stochastic models" often make mathematical assumptions and axioms concerning system phenomena and then derive the required results directly [12–14]. A classic example is the derivation of radioactive disintegration from the probabilistic viewpoint to yield a Poisson-type distribution. The problem is that the assumptions and axioms are not obvious for complex systems. "Principles-based (mechanistic) models" make use of (1) physical, chemical and biological principles; (2) the constitutive laws of fluxes; and (3) the assumptions about processes in the system. A set of governing differential or integral-differential equations is generally derived and subsequently solved using analytical or numerical techniques. Only this class of models can have both site-specific and generic applications and incorporate the effects of aquatic chemical factors in obvious and rational ways. The functional forms obtained by this approach are universal, meaning that they are applicable to any location, as long as the assumptions concerning the model derivations are satisfied. However, computer storage capacity and computational time may prohibit the application of detailed, complicated mechanistic models.

Of the different types of models discussed above, principles-based mathematical models appear to be the most promising for describing the spatial/temporal distributions of chemical perturbations to the aquatic environment as completely and accurately as possible. The physical principles employed to formulate the problem include conservations of mass, momentum and energy. The chemical and biological principles often involve the basic formulation of chemical reactions and exchange among various carriers. Before applying these principles to the formulation of problems, one should concern oneself with the processes that control the chemical distribution. After all, mathematical models are no more than the computations of assumed processes.

Any chemical perturbation in an aquatic environment will be subject at least to transport, transfer and transformation. Transport is the physical process of the flowing water that moves along with it and spreads suspended sediment, drifting biota and dissolved chemicals (Figure 3). It

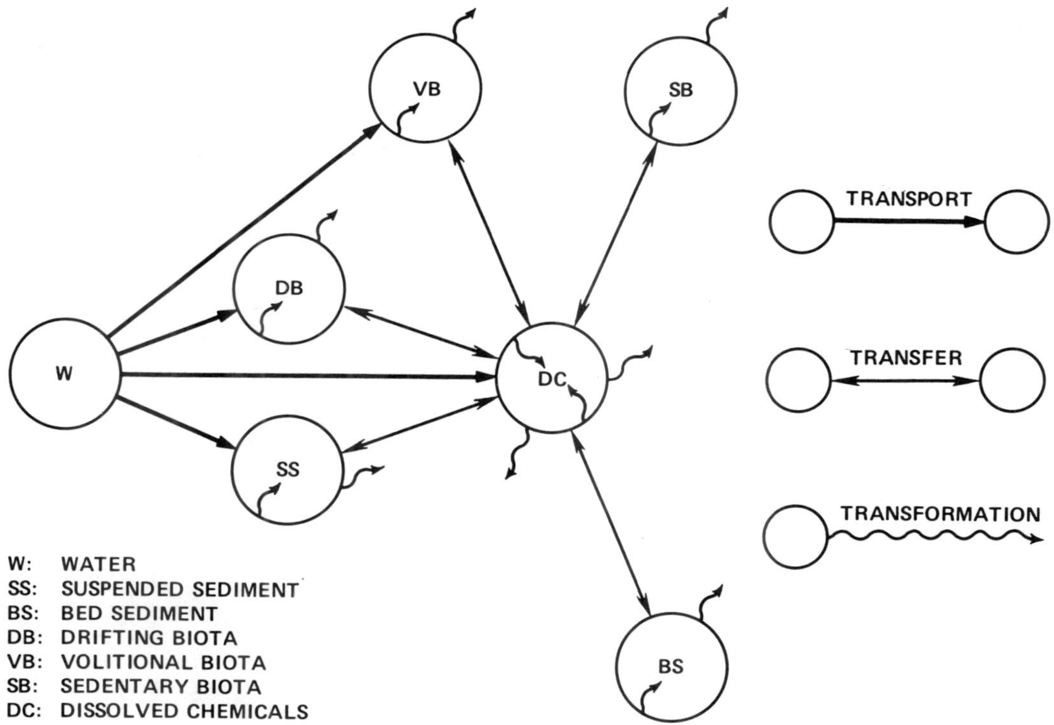

Figure 3. Interrelationship among transport, transfer and transformation mechanisms of various carriers in the aquatic environment.

includes advection and convection resulting from hydrodynamic and buoyancy forces, respectively, and dispersion and diffusion due to turbulence and molecular diffusion, respectively. Advection and convection move passive materials to the statistically mean position. They are characterized by the fluid velocity as functions of space and time. Dispersion and diffusion induce the spreading of these passive materials about their mean position. These physical mechanisms do not directly contribute anything to the attenuation of chemicals in the dissolved phase. Transfer is the mechanism that partitions chemicals among various carriers, such as water, sediment and biota, in the aquatic system (Figure 3) or exchanges the chemicals with other environmental regimes such as atmosphere and lithosphere. Transformation here means the mechanisms that decompose the chemicals or cause interaction among them to form new chemicals. Transformation may be achieved through dissolved, sediment-adsorbed and biota-adsorbed phases (Figure 3).

All models incorporating the above processes to predict the spatial/temporal distribution of chemicals can be derived by the conservation statement:

$$\int \frac{as_i f_i}{v\, at} dv = -\int_s q_i ds - \int_v f_i d_v - \int_v R_i d_v \qquad (1)$$

where
S_i = generalized storage coefficient for state variable i
f_i = content of the state variable i
v = environmental volume
S = surface enclosing the environmental volume
q_i = outward normal flux through S of the state variable i
F_i = rate of transfer per unit volume of state variable i from a carrier to other carriers within v
R_i = rate of transformation per unit volume of state variable i within v due to physical, chemical and biological processes, and/or artificial withdrawal and addition from and to v

This statement simply expresses the fact that the total rate of accumulation of chemicals in a carrier within v is equal to the total flux by the carrier across S minus the total rate of transfer of state variable i from the given carrier to other carriers and the total rate of transformation through the given carrier within v. Constitutive equations relating q_i, F_i and R_i to measurable variables, such as density, pressure, velocity, temperature, salinity and concentration, must be quantified based on mechanisms with acceptable approximations and obvious hypotheses.

For example, the continuity equation in the hydrodynamic models may be obtained by taking $S_i = 1$ and $f_i = p =$ density of water mixture with the constitutive equations given as follows:

$$q_i = \vec{n} \cdot (\rho \vec{V}) \tag{2}$$

where $\vec{V}$ = velocity vector
$\vec{n}$ = outward unit vector normal to the enclosing surfaces
F_i = 0
R_i = 0

This continuity equation is the most desirable among all principles-based derived equations because it does not contain any other empirical system parameters. All of the variables in the equation are state variables. Thus, there is no empiricism to be dealt with. We will not have such luxuries for all others.

For example, if we consider the state variable to be the dissolved chemical concentration, we may then take $f_i = C_W$ = mass of chemicals per unit liquid volume. If a unit environmental volume is occupied mostly by the liquid-phase carrier, we may assume $S_i \simeq 1$. The question is how to relate the normal flux q_i to C_W and other state variables. As we have mentioned earlier, the flux may be due to advection/convection and dispersion/diffusion. For the advection/convection flux q_{ia} we may reasonably approximate it by

$$q_{ia} = \vec{n} \cdot (\vec{V} C_W) \tag{3}$$

For dispersion/diffusion flux q_{id} the classical approach is to assume that it is proportional to the gradient of concentration to result in

$$q_{id} = \vec{n} \cdot (-\bar{\bar{D}}\, \vec{\nabla} C_W) \tag{4}$$

The main problem in this assumption is that the dispersion/diffusion tensor, $\bar{\bar{D}}$, is not universal. Similarly, one must propose or derive constitutive equations to relate R_W and F_W to the state variables. The state-of-the-art models all seem to contain, more or less, some empirical coefficients [15–17].

From the above discussion, it is obvious that further model development and improvement must be made to yield constitutive equations for relating q_{id}, R_i and F_i to state variables with less-empirical parameters. We must strive for the day when we will have the constitutive equations free of empiricism. Although one may pressure empirical relationships to be a close representation of reality, empirical data reflect many uncertainties of sampling bias and errors, inherent variability of nature, and artifacts of data manipulation and summarization. "Real-world" data may, in fact, reflect the underlying mechanisms that operate. The only possibility, however remote, that we may ever achieve the goal of having principles-based models that are free of empiricism is through intensive research on the mechanisms and processes based on physical, chemical and biological principles. Genuinely new laws may be proposed and revised based on experimental evidence.

LEVEL OF ANALYSIS

Depending on the complexity of the problem being analyzed, any level of resolution may be adopted to describe the chemical concentration distribution for a given physical system. The factors that determine the level of analysis for a given field problem may be as follows:

1. the number of processes and the complexities of the corresponding constitutive equations;
2. the spatial/temporal variability and nonlinearity of model parameters;
3. the number of state variables that are considered as given parameters rather than simulated with models;
4. the degree of complexity of boundaries and regions;
5. the availability of data and the level of detail desired;
6. time and budget constraints; and
7. the method of solving the resulting system of equations.

These factors will be addressed below.

The formulation of the advection/convection processes of transporting any chemical species is universally given without much dispute or controversy as

$$q_{ai} = \vec{n} \cdot (\vec{V} f_i) \tag{5}$$

The constitutive equation of the Fickian type that is used to represent dispersion/diffusion processes has become a well accepted formulism. Although this approach has met with some success in practical applications, the controversies and disputes over the value of the dispersion/diffusion tensor to measurable physical and dynamic characteristics of the system are definitely needed. In the meantime, more fundamental studies on turbulence must be conducted to enable the proposition of more appropriate constitutive equations for this transport process that could contain less-controversial parameters. The transfer process between dissolved phases and sediment- and biota-adsorbed phases has been formulated both with equilibrium and with nonequilibrium equations [16,17]. These equations have been proposed mainly based on laboratory experiments. How well these formulas actually represent the real-world system remains to be seen.

Transformation processes for a chemical may result from, among other things, precipitation of chemicals, oxidation, hydrolysis, photolysis, biolysis and volatilization [15,16]. There are many formulations to describe these processes [15–17]. If a linear kinetic approach is accepted as an adequate first approximation to the problem, the transformation processes may be characterized by three combined rate constants: one for the decomposition in the dissolved phase and the other two to represent the decomposition through sediment- and biota-adsorbed phases, respectively. When more than one chemical is of concern, the appropriate chemical-reaction equations must be used to establish the constitutive equations for values of R_i. This may result in a system of complex nonlinear equations with state-variable–dependent coefficients.

Once the constitutive equations are set up, they will inevitably be represented by some parameters. These parameters may vary with location and time because of the changing environment. Of course, the best constitutive equations will be the ones with universal constants only, which are not possible given the present state-of-the-art of modeling efforts. Sometimes the state variables may be considered as input parameters. For example, the velocity field required for the advection/convection and dispersion/diffusion transport of chemicals may be given as input if one can afford continuous measurements over the whole region of interest. Under such circumstances, hydrodynamic models will not be needed, thus yielding a great degree of simplification. Another problem in deciding the level of analysis will be the size of the region and the complexity of the boundaries.

Depending on the number of processes included in the governing equations, variability of parameters, complexity of boundaries, availability of data and detail of information desired, and budget and time constraints, different types of solutions may be obtained for the governing system of equations. These may range from phenomenological estimation and analytical techniques to realistic numerical simulations with finite differences and finite elements or other special methods. Phenomenological results may be used to estimate the arrival time of chemicals at specific locations and for screening purposes. Analytical computations often involve numerous assumptions for simplification. These include constant parameters, simple invariant-flow field, and regular or infinite extent of region. Therefore, analytical computations are best applied to preliminary assessments. Numerical simulations, on the other hand, are required when realistic and detailed studies of spatial/temporal distribution of chemical concentrations are needed for comprehensive assessment of various chemical perturbations.

A three-level resolution scheme was suggested to analyze radionuclide transport in saturated/unsaturated porous media [18]. Two examples were used to illustrate that scheme using hypothetical parameters [18]. The examples were constructed to represent drainage from the bottom of a water body containing liquid wastes and leaching from trench burials, respectively. The drainage problem indicated that three levels of analysis yielded answers that differed by two orders of magnitude. Furthermore, the leaching problem exemplified the case where simple levels of resolution are neither adequate nor possible because the hydrodynamic state variables cannot be determined by any rational approach other than the level 3 (most detailed) resolution. Thus, only the highest level of analysis could be used to achieve the goal of determining chemical arrival time and its detailed distribution as a function of time and space. The same line of reasoning may be applicable to the prediction of chemical distribution in aquatic environments.

CONCLUSION

Models to deal with the coupled system of dose-response, chemical distribution, population pattern and mortality do not exist. Therefore, efforts should be made to develop such models, since decoupling is rarely achieved in nature. Until this can be done, modeling the chemical distribution among various carriers in an aquatic environment is a prerequisite for assessing the hazard to the ecosystem and to the well-being and health of man. Because of the complexity of various processes in transporting, transferring and transforming chemicals, principles-based (mechanistic) mathematical models offer the most promising means for describing the spatial/temporal distribution of chemical perturbations to aquatic environments as completely and accurately as possible. Many levels of resolution can be applied to a problem, depending on (1) the number of processes and the complexity of the corresponding constitutive equations; (2) the variability and nonlinearity of model parameters; (3) the number of state variables modeled; (3) the existence of complex boundaries and regions; (4) the availability of data and detail of information desired; (5) time and budget constraints; and (6) the method of solving the resulting system of governing equations. Different levels of resolution often yield very different answers, with lower-level resolution giving more conservative solutions. For many situations, lower-level analyses are neither adequate nor possible, because hydrodynamic state variables cannot be determined by any rational approach except by the highest level of simulation. Thus, realistic, numerical mechanistic models must be made available in anticipation of such occasions. Continuing model development and improvement must be made to enable the proposition of constitutive equations with less-empirical parameters. It is also important to bear in mind that a comparable degree of sophistication for each process should be made for any model.

ACKNOWLEDGMENTS

This work is supported by the Office of Pesticides and Toxic Substances, U.S. Environmental Protection Agency, under Interagency Agreement No. EPA-78-D-X0387. Many of the topics and ideas in this work have been drawn from weekly discussions with S. Bartell, J. E. Breck, C. C. Coutant, S. W. Christensen, R. H. Gardner, R. V. O'Neill and M. J. Sale, of the Environmental Sciences Division, and M. R. Patterson, of the Computer Science Division, Oak Ridge National Laboratory.

REFERENCES

1. DeAngelis, D. L. "A Model for the Movement and Distribution of Fish in a Body of Water," ORNL/TM-6310 Oak Ridge National Laboratory, Oak Ridge, TN (1978).
2. Coutant, C. C., and D. S. Carroll. "Temperature Selection by Ten Ultrasonic-Tagged Striped Bass in Freshwater Lakes," *Trans. Am. Fish. Soc.* 109(2):195–202 (1980).
3. Hackney, P. A., T. A. McDonough, D. L. DeAngelis and M. E. Cochran. "A Partial Differential Equation Model of Fish Population Dynamics and Its Applications in Impingement Impact Analysis," U.S. EPA Report 600/7-80-068 (1980).
4. Fishman, G. S. *Concepts and Methods in Discrete Event Digital Simulation* (New York: John Wiley & Sons, Inc., 1973).
5. Oren, T. I. "Concepts for Advanced Computer Assisted Modeling," in *Proceedings of the Symposium on Methodology in System Modelling and Simulation,* B. P. Zeigler et al., Eds. (Rehovot, Israel: Weizman Institute of Science, 1978).
6. van Genuchten, T. M. "Simulation Models and Their Applications to Landfill Disposal Siting: A Review of Current Technology," paper presented at Land Disposal of Hazardous Wastes, Fourth Annual Hazardous Waste Management Symposium, Southwest Research Institute and U.S. EPA, San Antonio, TX, March 6–8, 1978.
7. Emshoff, J. R., and R. L. Sisson. *Design and Use of Computer Simulation Models* (New York: The MacMillan Co., 1970).

8. Crawford, N. H., and R. K. Linsley. "Digital Simulation in Hydrology: Stanford Watershed Model IV," Stanford University, Department of Civil Engineering Technical Report No. 39 (1966).
9. Crawford, N. H., and A. S. Donigian. "Pesticide Transport and Runoff Model for Agricultural Lands," U.S. EPA Report 660/2-74-014 (1973).
10. Donigian, A. S., and N. H. Crawford. "Simulation of Agricultural Runoff," U.S. EPA Report 600/9-76-016 (1976).
11. Leytham, K. M., and R. C. Johanson. "Watershed Erosion and Sediment Transport Model," U.S. EPA Report 600/3-79-028 (1979).
12. Carter, L. L., and E. D. Cashwell. "Particle-Transport Simulation with Monte Carlo Method," U.S. Energy Research and Development Administration Report-TID-26607 (1975).
13. Ahlstrom, S. W., H. P. Arnett, R. C. Cole and R. J. Serne. "Multicomponent Mass Transport Model: Theory and Numerical Implementation," BNWL-2127, Battelle—Pacific Northwest Laboratories, Richland, WA (1977).
14. Alsmiller, F. S., R. G. Alsmiller, Jr., H. W. Bertini and C. L. Begovich. "Monte Carlo Simulation of Turbulent Atmosphere Transport and Comparisons with Experimental Data," ORNL/TM-6291, Oak Ridge National Laboratory, Oak Ridge, TN (1978).
15. Lassiter, R. R., G. L. Baughman and L. A. Burns. "Fate of Toxic Organic Substances in Aquatic Environment," *State-of-the-Art Ecol. Model.* 7:219-246 (1979).
16. Burns, L. A., and D. M. Cline. "Exposure Analysis Modeling Systems (EXAMS): User Manual and System Documentation," U.S. EPA Environmental Research Laboratory, Athens, GA (1981).
17. Browman, M. G., M. R. Patterson and T. J. Sworki. "Formulations of the Physicochemical Processes in the ORNL Multi-media Model (MMM)," ORNL/TM- Oak Ridge National Laboratory, Oak Ridge, TN (1981).
18. Yeh, G. T., and T. Tamura. "Geohydrological Considerations in Land Disposal of LLW," ANS/ED/TP-4, American Nuclear Society, La Grange Park, IL (1981).

CHAPTER 3

ROLE OF ENVIRONMENTAL TRANSPORT AND FATE MODELS IN IMPLEMENTING THE TOXIC SUBSTANCES CONTROL ACT

William P. Wood

Office of Toxic Substances
U.S. Environmental Protection Agency
Washington, DC 20460

As the issues regarding environmental protection management become more and more complex and intertwined, increasing reliance is being placed on the use of models to support decision-making. Environmental transport and fate models are being utilized in deriving strategies directed at meeting air and water quality standards, in evaluating the suitability of proposed radioactive and toxic waste disposal sites, and in selecting among alternative energy sources, to name just a few examples. The use of models in implementing one recent piece of environmental legislation—the Toxic Substances Control Act (TSCA)—is the topic of this paper.

TSCA was enacted in 1976 out of a growing realization and concern that there is little knowledge about what effects many chemicals may have on human health and the environment. It is not the first federal law to address the serious health and environmental problems associated with toxic chemicals. Unlike TSCA, however, the other laws largely enable the government to take action only after widespread exposure and perhaps serious harm have occurred.

Under TSCA, the U.S. Environmental Protection Agency (EPA) has been given a broad mandate to protect public health and the environment from unreasonable risks associated with chemicals in commerce. This law directs EPA to gather information on chemicals, to identify harmful substances and to control those substances whose risks outweigh their benefits to society and the economy.

TSCA is not only a new regulatory tool, it marks a watershed in environmental legislation. It focuses for the first time on chemicals as products, not as pollutants, and represents a preventive approach to protecting the public health and the environment. It accomplishes this by including the universe of existing chemicals, as well as new chemicals, under its purview, and by providing EPA with the authority to gather information on chemicals for which few data exist on their effects and to review new chemicals before introduction into commerce. Finally it is among the few environmental laws that involve a decision-making process that takes both risks and benefits into account.

EPA's implementation of TSCA has centered around four major activities:

1. collection of basic information on chemicals where such information does not now exist;
2. identification of potentially harmful substances and the requirement that industry provide information on how these chemicals affect human health and the environment;
3. review of all new substances produced domestically for commercial purpose or imported into the United States prior to their introduction, and
4. control of commercial chemical substances that pose an unreasonable risk.

USE OF MODELS IN IMPLEMENTING TSCA

TSCA is a powerful and far-reaching piece of legislation, encompassing the tens of thousands of chemical substances manufactured for commercial purposes. It authorizes EPA to evaluate the risks associated with the entire life cycle of the substance, i.e., from its production, through its use, to its eventual disposal. This legislation places a heavy burden on EPA in terms of the number of chemical substances that need to be evaluated, as well as the number of possible exposure situations that must be assessed.

From its inception, the Office of Toxic Substances (OTS) has actively been pursuing the development of various analytical capabilities to support its risk-assessment activities. There are several reasons for this, among which is the need to process efficiently a large number of chemicals and rapidly to screen out substances that pose no significant risk.

Associated with this, and even more important, is the need to have a systematic and reproducible approach to risk assessment that results in a consistent and equitable evaluation of chemicals under the purview of TSCA. Decisions must be documented and defensible.

TSCA has unique information-gathering authority; however, the burden is on EPA to decide what kinds of information are needed to support risk assessments. It is important that EPA be able to explain why the data are needed and how EPA will use these data before requiring the chemical industry to bear the cost of testing. Environmental fate and transport models can help provide a rationale for testing decisions based on exposure concerns by projecting levels, frequencies and durations of exposure, as well as helping to identify populations at risk.

For example, in the case of a new chemical that has not yet entered into commerce, environmental transport and fate models can play an important role in the risk-assessment process. Obviously, for these substances, monitoring data on exposure levels do not exist, and EPA is faced with having to make projections of exposure based on intended uses, projected production volumes and physiochemical properties of the substance. Environmental transport and fate models provide a framework for making these projections. Considerable effort is being expended by OTS in developing a modeling system to support Premanufacturing Notice (PMN) reviews. This system, which will be discussed in more detail below, involves not only the use of models, but also the use of automated techniques for estimating many of the physiochemical properties of the substance required by the models. This system is being designed to assist in the initial screening of PMN to efficiently eliminate some chemicals from further review if they pose no significant risks to human health or the environment. Likewise when hazard concerns are raised during the initial review of the PMN, models are used in a more detailed assessment of who is exposed to the chemical, under what conditions and to what extent.

In conjunction with this, EPA is working closely with Organization for Economic Cooperation and Development (OECD) countries to develop internationally consistent approaches to the assessment of new chemicals to avoid the establishment of nontariff trade barriers between member countries. The use of environmental transport and fate models is one method being examined by members of the OECD Expert Working Group on Exposure Assessment as a means for reaching international agreement on the use of the Minimum Premarketing Data set proposed by the OECD.

Along with review of new chemicals, environmental transport and fate models have an important role to play in evaluating the risks associated with existing chemicals in commerce and in making decisions regarding the need to develop data on how these chemicals affect human health and the environment. Although there are monitoring data indicating the levels of exposure for many of the chemicals of concern under TSCA, these data sets are rarely complete. In other words, there are few chemicals for which multimedia data exist. It is often unclear as to the representativeness of the data that have usually been collected at only a few sites and over a limited period of time, and the data sets are rarely identified with particular sources of the chemical. Modeling can be employed to help fill in the information gaps in these situations and to aid in rounding out the exposure profile. Based on that profile, decisions can then be made on the need for testing related to the hazard of the substance.

In addition to the information-gathering authorities under TSCA, EPA has the authority to control chemicals in commerce if the risks of a substance outweigh its benefits. These actions may range (as appropriate to the circumstances) from labeling, to limited regulations of handling and use, to outright bans on manufacturing and processing. In light of the number of possibilities available to EPA for mitigating the risks associated with chemicals in commerce, EPA's efforts to implement the law must be well focused and deliberate. Under the law, EPA is bound to implement TSCA in a "reasonable and prudent manner" and must "consider the environmental, economic, and social impact of any action taken under the Act." Besides their role in predicting the levels of ambient exposure, environmental fate and transport models have use in exploring control options available to EPA and in selecting those options for implementation that result in the greatest risk reduction at the least cost.

OTS MODELING ACTIVITIES

Current activities within OTS are directed at developing the analytical capability of modeling for the implementation of TSCA. These activities fall into two broad categories. The first deals with specific models and their implementation requirements, and the second deals with development of an automated system for performing integrated exposure assessments incorporating models, as well as other analytical components.

OTS is engaged in the following activities regarding the use of specific models:

- identification,
- evaluation,

- modification and/or development,
- implementation, and
- validation.

Since OTS must deal with issues that range from groundwater contamination to depletion of the stratospheric ozone layer, a library of models had to be compiled and evaluated. One of the first tasks was to establish criteria for use, against which the appropriateness of existing models could be judged. These criteria include such considerations as:

- model input data requirements and accessibility of data;
- existence of computer code and technical support documentation;
- prior use in regulatory decision-making;
- use by other EPA program offices;
- skill level and level of effort required in exercising the model; and
- existence of or plans for validation studies.

Analysis has led to the implementation by OTS of a large number of models (with more being added) for single-medium and multimedia analyses. In the area of multimedia models, OTS has undertaken a modest development program ranging from design of a simple-media partitioning model [Environmental Partitioning Model (ENPART)] for use in the initial review of PMN, to a more elaborate model [Unified Transport Model (UTM)] for use in detailed risk assessments. This multimedia modeling capability will be of particular value to EPA in light of the current emphasis within EPA of dealing with toxics assessment in an integrated multimedia fashion.

In addition to implementing and applying specific models to support TSCA risk assessments, EPA is implementing an automated system that will facilitate the use of these models and their interfacing with associated data bases and additional analytical capabilities. The OTS Exposure Modeling System is planned to help analysts develop a variety of information for integrated exposure analyses. The major functions of the system involve its user interface, data management, statistical analysis, model operations and information display capabilities. Compartments of the system are illustrated in Figure 1.

The great variation in types of OTS exposure assessments requires an easy-to-use system that is also extremely flexible for specifying a wide variety of data to be assembled, analyzed and displayed. In addition, the system must be designed to readily accept additional models, data sets and programmed capabilities to meet future needs that cannot be anticipated.

The user interface guides users through each state of an operation.

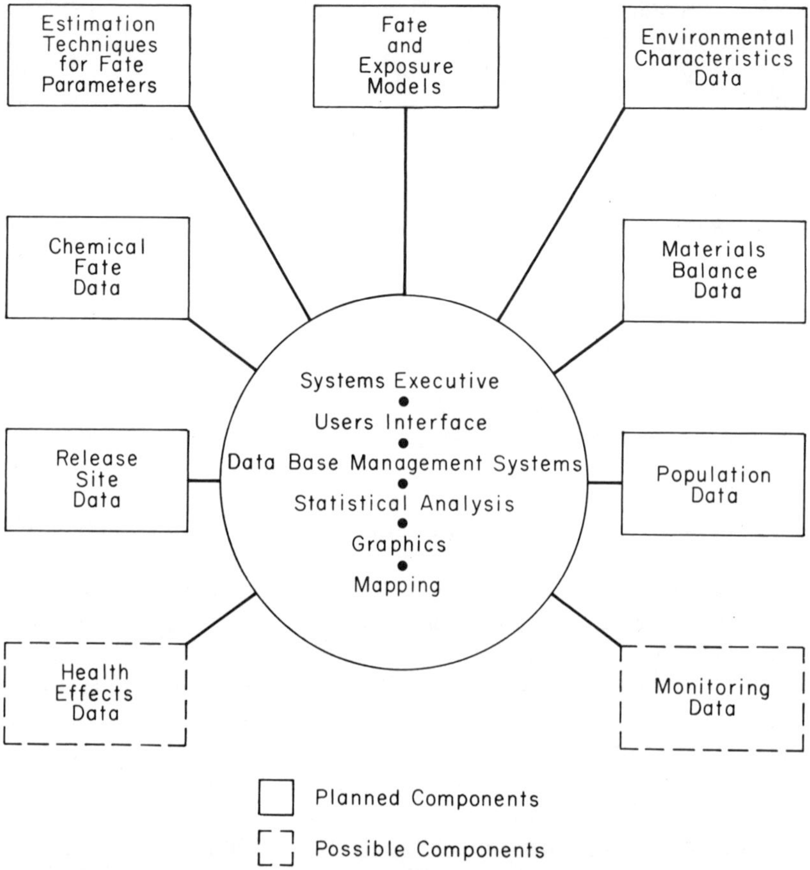

Figure 1. Proposed components of OTS Exposure Modeling System.

Numerous help messages are provided to explain what options are available at each stage. The system can accept four styles of command instructions from users, so that the degree of assistance provided is appropriate for users ranging from those inexperienced with a computer system to those who are familiar with its many features. The system includes numerous provisions for error recovery, so that any mistakes in instructions can easily be rectified.

The system is planned to have a wide variety of data management features. The various analyses performed to gain insight concerning potential exposure pathways may often involve novel ways of organizing disparate data. The OTS Exposure Modeling System will allow users to draw together readily a variety of information types and organize them as necessary. The system will support a number of specified data set formats such as those required by the various statistical methods and by the models. Manual data entry into the system will greatly be facilitated by prompting messages and terminal screen displays that mirror written forms, allowing even inexperienced users to follow a "fill-in-the-blank" approach.

The statistical analysis capabilities planned for the system include the functions of the International Mathematical and Statistical Library (IMSL) program. IMSL can perform a wide variety of mathematical and statistical analyses efficiently, and when linked with the system's user interface, will be easy to use. Examples of the studies that could be performed using IMSL include (1) analyses of the geographic distribution of population and sources of potential exposure to indicate areas of possible concern, and (2) analyses of the geographic variation in environmental characteristics such as soil types or rainfall to indicate areas in which environmental conditions may facilitate persistence of a chemical.

One of the most significant and unique aspects of the OTS modeling system is its support for model operations. Each of the transport and fate models being used by OTS will be linked to a common routine for assembling the necessary input information. Users will be able to choose to run a model using a previously compiled set of data, modify or assemble such an input data set from the data base using the system's data-management capabilities, or interactively build an input data set by typing in the data as the system prompts for it.

An example of how the user of the system can build an input data set is illustrated by the Fate Estimation Techniques module (Figure 2). Data on the physiochemical and degradation properties of the chemical being assessed are often not known (this is especially true for new chemicals); in this situation, the analyst has the option to estimate these properties via a software package being developed by OTS. This package allows the user to enter the structure of the molecule (by drawing it on the terminal screen) as well as any known properties and then to choose from a menu those properties for which values need to be estimated. Figure 2 illustrates the macrostructure of the Fate Estimation Techniques module and the interrelationships of many of these properties. Having developed these estimates on-line, they would then be inserted by the system into the model that is being used. Likewise, the environmental characteristics data required by the model can be assembled on-line and inserted into the model.

The models would then employ these data to produce their predictions while the user waits for results, or they could be set to run in the back-

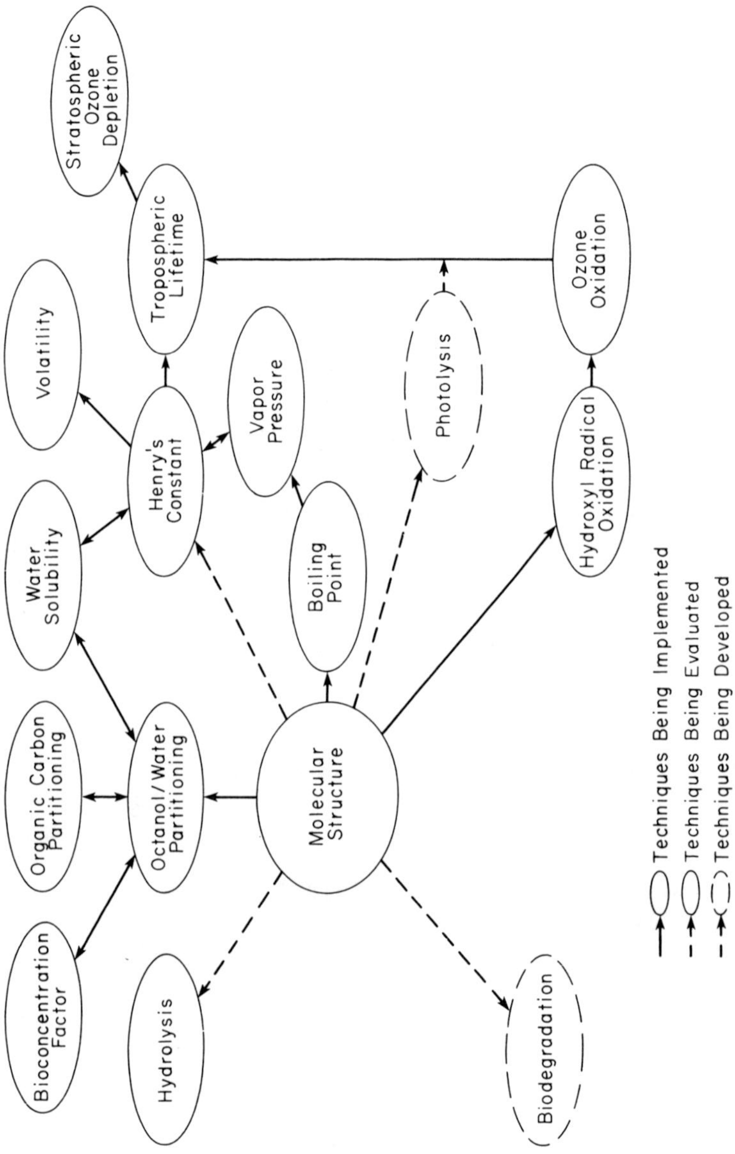

Figure 2. Fate Estimation Techniques module of OTS Exposure Modeling System.

ground while a user performs other analyses. Modeling results could be displayed at the user's terminal or directed to a high-speed printer.

Other features planned for the system include graphic and mapping information displays. Analytical results could be displayed as histograms or other frequency-distribution plots, linear regression graphs, or a variety of other line or bar charts. The mapping features are intended for use in displaying results of fate modeling analyses for particular sites and to show the geographic distribution of some parameter from the data base. For example, calculated airborne ambient concentrations at specified distances from a release site might be shown with the distribution of population near the site. County-level variations in environmental characteristics might also be displayed for various regions of the country.

WORKSHOP OBJECTIVES

Considerable effort within OTS has been directed toward making environmental transport and fate models, as well as other analytical capabilities, readily available for use by the program's scientific staff. These tools, however, will be no more than mathematical curiosities unless the modeler can communicate to the decision-maker the basic assumptions underlying the model, its range of applicability, the sensitivity of the results to uncertainties in input parameters and, perhaps more importantly, what has been done to validate the model. The decision-maker must have some proof that the model works. The purpose of developing and applying models is to minimize the chance of making wrong decisions. The use of untested, unvalidated models leaves the decision-maker in a precarious position.

In light of this, the last major area of activity has been that of model validation. OTS has requested that the EPA Office of Research and Development devise guidelines for validation of aquatic models via field studies, with the intended purpose of sponsoring studies to evaluate the robustness of those models most commonly used by OTS. In addition, OTS has helped to sponsor this workshop because it believes it important to have the benefit of other scientists' experiences and insights.

The following issues require attention:

1. What is the definition of "validation"? This term represents different aspects of model-testing to different people. A consistent definition can go a long way in helping communication between modelers and decision-makers.

2. Can criteria be established that address the question of adequate validation? These criteria might reflect some statistical measure such as the correlation between measured and predicted values.
3. Which mechanisms and models have been validated and which need to be?
4. To what extent can models be evaluated at one site and generalized to other sites?
5. What situations, ranges of conditions and time frames need to be covered in the validation exercise?
6. How reliably do laboratory data simulate field conditions?

Answers to these questions will go a long way in formulating a future course of action for those engaged in the development and/or application of models for environmental protection.

CHAPTER 4

PREDICTIVE FATE MODELS: THEIR ROLE IN THE U.S. ENVIRONMENTAL PROTECTION AGENCY'S WATER PROGRAM

Michael W. Slimak and Charles Delos
U.S. Environmental Protection Agency
Office of Water Regulations and Standards
Monitoring and Data Support Division
Washington, DC 20460

Models, in general, are an important and often-used tool in the U.S. Environmental Protection Agency (EPA) Water Office. This administrative office consists of a number of major programs that are designed to study and, when necessary, regulate chemicals in the aquatic environment. These programs include:

- Office of Drinking Water, responsible for regulating drinking water;
- Office of Water Regulations and Standards (OWRS), responsible for controlling point-source discharge of chemicals to surface waters;
- Office of Water Programs, responsible for managing the appropriate design, construction and maintenance of sewage treatment plants and evaluating the role of nonpoint-source pollution; and
- Office of Water Enforcement, responsible for enforcing the regulations emanating from these other water offices.

Without doubt, these offices rely on predictive models—ranging from simple models that predict dissolved oxygen (DO) concentrations to complex models that predict the concentration of toxic chemicals in various components of the aquatic ecosystem—to understand the role the environment plays in affecting the form and quantity of the modeled substance. Although a dilution computation for a specific outfall (i.e., taking a loading of a substance and, based on stream flow, diluting it to a concentration) is simple in its basic construction, it is an effective tool in making regulatory decisions. When applied to a frequency distribution of dilution available to an entire industrial category, it has been used for assessing effluent guidelines.

This leads into one of the most important points regarding the application of predictive models: provided that a model is based on defensible principles, the simpler it is, the more widespread its use. Although sophisticated models may predict more accurately actual conditions, oversophistication or overcomplication tend to limit a model's use in the decision-making process, since the means of arriving at each formulation and input parameter value must be supportably documented. The most complex models are perhaps better suited to research, an open-ended pursuit, than to assessment, which requires an unambiguous final result.

Chemical fate models are used for two general purposes: (1) to predict the general behavior of substances, and (2) to predict concentrations. The models used by these offices are tools for understanding complex phenomena and their interactions. The equations form a simplified, symbolic representation of the interrelationships among components of a particular system. Predictions obtained by changing certain inputs are used for setting priorities or guiding decisions.

MODELS AND OWRS

Water quality models have been important tools in the water program for many years, as shown by the use of DO models to predict oxygen sags in evaluating the need for advanced waste treatment plants. The acceptability of using models to predict the level of treatment needed to attain ambient water quality standards has thus been proven. As our understanding of fate and transport properties of conventional pollutants was building, so was our knowledge of how environmental parameters (e.g., pH, sunlight and stream velocity) affected the concentrations of conventional pollutants. This stage represents, to many within EPA, the first phase of model development and implementation—implementation that was judged successful by the widespread application of models by state-operated water quality management programs. The second phase of model development and implementation began in early 1977 as a result of the 1976 settlement agreement that resulted in our focus on toxic chemicals. As a result of this agreement, a program was established to evaluate exposure and subsequent risk from the 129 priority pollutants.

Effective regulatory action for toxic chemicals requires an understanding of the human and environmental risks associated with the manufacture, use and disposal of the chemical. Assessment of risk requires a scientific judgment about the probability of harm to the environment resulting from known or potential environmental concentrations. The risk-assessment process integrates toxicity data with information on exposure. The goal of an exposure assessment is an exposure profile that identifies subpopulations (geographic, demographic, etc.) and associates how much, and in what form, a chemical comes in contact with each subpopulation. Ideally, this exposure profile can be synthesized by matching the location and habits of various subpopulations with the location and form of the chemical. If the characteristics of environmental release are well defined (as to location and quantity), environmental fate processes can be used to determine the ultimate location of the chemical. Thus, prediction of environmental concentrations is a function of (1) amount of the chemical released, (2) location, (3) prior accumulation, (4) time period, and (5) behavior of the chemical. For predicting chemical behavior, the model applications described below have ranged from simple partitioning equations to computer models of intermediate complexity.

Partitioning Between Environmental Media

The simplest predictions of chemical behavior in the environment involve partitioning between phases. Such analyses are useful for identifying the medium most likely to become contaminated, for estimating the importance of cross-media transfers, and for quickly judging, without monitoring data, the probable relative importance of different exposure routes.

The partition coefficients governing these considerations are: (1) air/water, the Henry's law constant, a measure of the ratio of vapor pressure and solubility; (2) solids/water, including soil/water, sediment/water and biota/water (bioconcentration factor), all of which may be measured directly or estimated through the octanol/water partition coefficient; (3) air/solids, e.g., the air/soil partition coefficient, estimated as the ratio of the abovementioned air/water coefficient and the soil/water coefficient. The simple integration of all three classes of partition coefficients corresponds to what Mackay [1] calls Level I of fugacity calculations.

As an example of the rapid utility of such analyses, consider human exposure to tetrachloroethylene (TCE) a substance detectable in most untreated municipal wastewaters and some ambient waters due to its widespread use as a solvent in dry cleaning and metal degreasing. Based on its measured vapor pressure of 14 torr (at 20°C) and water solubility of 150–200 mg/l, an air/water partition coefficient (H) of about 0.75 (dimensionless, wt/vol basis) would be calculated. The bioconcentration factor (BCF) for edible fish has been measured at 31 liter/kg [2]. For a human exposed to an equilibrated environment, the ratio ER of exposure via inhalation vs exposure via consumption of water and fish (assuming equivalent absorption efficiencies) would be given by:

$$ER = HQ_a/(Q_w + BCFQ_f)$$

where Q_a = inhalation rate (20,000 liter/day)
Q_w = water consumption (2 liter/day)
Q_f = fish consumption (0.0065 kg/day, excluding ocean seafood) for a standard person

For equilibrated TCE, ER is calculated to be 6800; that is, exposure via air is 6800-fold greater than exposure via water and contaminated fish. Consequently, waterborne exposure to this substance can be important only if the contaminated water has not had an opportunity to equilibrate with air, such as within several miles of a wastewater outfall or in groundwater.

OWRS [3] has evaluated the 114 organic compounds on the list of 129 priority pollutants originating from the 1976 Consent Decree between EPA and the Natural Resource Defense Council (NRDC) and has determined the concentration ratios in an equilibrated air/water/solids environment. Examination of the above equation and the Q values indicates that drinking water would constitute more than a minor fraction of the air exposure (i.e., $ER < 10$) whenever $H < 10^{-3}$ (dimensionless, wt/vol basis). About half of the priority pollutants meet this criterion: all phenols and phthalates, most pesticides, nitrogen-containing compounds, ethers, and some polycyclic aromatic hydrocarbons (PAH). Similarly, exposure via fish would constitute more than a minor fraction of air exposure if $BCF/H > 3 \times 10^{-5}$. On this basis, several additional priority pollutants would attain waterborne importance, including most or all of the polychlorinated biphenyls (PCB), chlorinated benzenes and PAH, among others. The importance of waterborne exposure to the remaining pollutants, which include most of the halogenated aliphatics and a few other volatile compounds from various groups, depends on equilibrium not being quickly attained.

Figure 1 shows the relative quantities of pollutants residing in air, water and soil/sediment compartments of an environment having arbitrary but reasonable relative proportions (as chosen by Lyman [3]). In this case, results are a function of the environmental compartment

1. Benzene, toluene, and mono-, di- and trichlorinated ethanes and ethenes
2. *a*-Endosulfan
3. Endosulfan sulfate
4. Phenol
5. Benzidine
6. Dimethylphthalate
7. Diethylphthalate
8. 2,4-Dichlorophenol
9. PCB-1254
10. TCDD
11. Benzo[a]pyrene, chrysene, *bis*-(2-ethylhexyl)phthalate, DDT, DDD, DDE

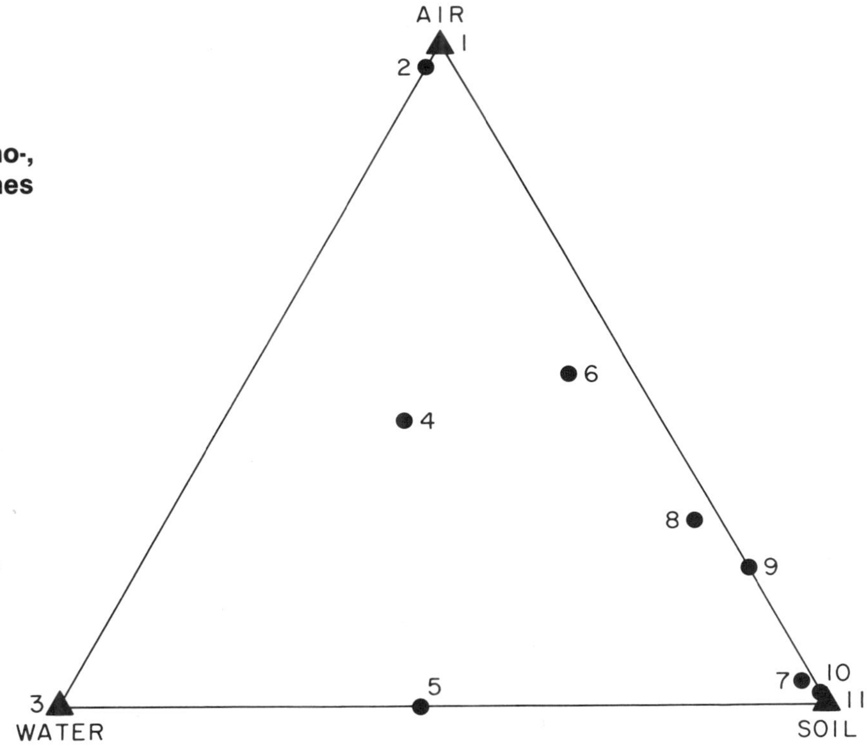

Figure 1. Equilibrium distribution of several priority pollutants between environmental media (percentages).

volumes rather than the human exposure volumes. Because the atmosphere constitutes more than 96% of the total mass of the assumed environment, a high percentage of a pollutant in air does not necesssarily imply that air hazards outweigh water hazards. It can be seen that many priority pollutants tend toward a single environmental medium, for example, air for chlorinated ethanes and ethenes, and soil/sediment for many PAH. Many may inhabit two media; very few, however, partition almost equally between all three phases, as does phenol.

Comparison of typical ambient air and water concentrations of several priority pollutants, however, suggests that the observed water concentrations generally greatly exceed the equilibrium water concentrations. Some portion of this effect may be the result of bias toward severely polluted sites in the water data base. Nevertheless, the fundamental limitation of the simple partitioning analysis is that it cannot predict, without further kinetic analysis, how quickly equilibrium will be attained.

While the Mackay [1] Level III approach allows incorporation of kinetics into the analysis, specific formulations for predicting exchange rates between media and degradation within media must be developed independently by the user. This limitation, along with the assumption that each medium is homogeneous, somewhat hampers use of this method for predictions in real environments. Rather, OWRS has utilized detailed single-medium models (described later) when absolute values (rather than ratios) of concentrations are needed or when equilibrium conditions cannot be assumed.

Metals Speciation

Another category of equilibrium models predicts speciation of metals. Given particular concentrations of organic and inorganic ligands and adsorbants, and other water quality characteristics, such as pH, models such as EPA's Oxidation/Reduction Equilibrium Model (REDEQL) can be used to predict the concentration of the metal free ion and all other species, at equilibrium [4].

Such information can suggest what the general behavior of a metal will be in the aquatic environment, particularly how long it will remain in the water column. More importantly, however, such information can provide understanding of the effect of other water quality parameters, such as pH, hardness and organic content, in controlling the toxicity of the metal to aquatic life. For example, the cupric free ion is generally regarded as being highly toxic to fish. The most prevalent inorganic complexes, hydroxide and carbonate, are of uncertain toxicity; the organic complexes, which copper readily forms, are relatively nontoxic [5]. Since EPA's water quality criteria are based on laboratory bioassays run in a standard dilution water, and since such dilution water is much lower in organic content than typical environmental waters, such water quality differences are believed to account for the observed differences in toxicity between laboratory and environmental waters [6]. That is, a given concentration of toxic free ion may be in equilibrium with a much higher concentration of total metal, complexed and adsorbed, in environmental waters than in the comparatively pure laboratory waters.

Consequently, EPA is interested in such models to predict toxicity in the field using laboratory toxicity data. Currently, however, it is not certain that these models and the supporting data have been developed sufficiently to predict correctly key interactions such as organic complexing, and organic and inorganic adsorption. Furthermore, independent of any shortcomings in the models, the identity of the toxic species, often in addition to the free ion, usually is not known with certainty. Nevertheless, it is not unreasonable to expect that, in time, effects data will be given in terms of toxic species concentrations (rather than the total metal concentration) and that such models will be applied to field situations to predict the allowable discharge of the total metal.

Organic Chemicals in Aquatic Systems

For detailed analysis of the fate of organic chemicals in water, OWRS has used Exposure Analysis Modeling Systems [7]. This model's strength is in evaluating kinetics: how rapidly equilibrium is attained for partitioning or chemical degradation. It is beneficial to determine what concentration will occur a certain distance downstream from a particular discharge, e.g., in waste-load allocation analysis.

Using the results from a particular application to generalize about behavior in the aquatic environment can lead to some difficulties. For example, again consider TCE, a substance very slowly degraded but rapidly volatilized in water. Its volatilization half-life depends on the surface-to-volume ratio (depth) of the stream, and because of its high Henry's law constant, on the turbulence of the water phase (rather than the air phase) according to the two-film theory [8,9].

In the river used as an example by the developers of EXAMS, TCE would have a volatilization half-life of 45 hr. Real streams, however, vary greatly in depth and turbulence [10], as reflected in Table I. In shallow, turbulent streams similar to those in which Owens et al. [11] made their reaeration measurements, half-life would be a mere 1.5 hr, and half-life distance only 1.6 km. On the other hand, during common low flows in the Ohio River below Cincinnati [12], half-life would be 530 hr, and half-life distance 340 km, if the impounded Markland Pool conditions were to extend that far. Thus, unlike the Mackay fugacity approach, which may be better suited to make broad generalizations about behavior than to predict local exposure, EXAMS is in some respects better suited to evaluate a specific locality than to generalize. Consequently, proper use of EXAMS in nationwide assessments appears to require that the complete distribution of environmental characteristics be represented statistically.

Table I. Effect of Stream Hydraulic Conditions on Volatilization Half-Life of Tetrachloroethylene

	Depth (m)	Velocity (m/sec)	Surface Transfer Coefficient (cm/hr) DO	Half-Life (hr)	Distance (km)
Small Stream (15)	0.25	0.3	30	1.5	1.6
Intermediate Rivers (2)	3	0.9	12	45	150
Impounded Rivers (8)	9	0.18	3	530	340

One important limitation of kinetic analyses is that one must know certain chemical properties of the pollutant. For example, errors in the photolysis rate coefficient [13] greatly affect the predicted decay of pentachlorophenol (PCP) in river and lake environments (Figure 2). Similarly, errors in biodegradation rate [13] greatly affect the predicted fate of phenol (Figure 3). However, given that biodegradation is the dominant fate process for phenol, errors in the chemical oxidation rate are of little importance. Thus, given rough order-of-magnitude estimates for various rates, such sensitivity analyses can help in identifying the critical parameters and thereby guide laboratory investigations toward determination of the most important unknowns.

Another limitation of EXAMS as an exposure assessment tool is that the analyst must know the pollutant loading to the receiving water. Consequently, it is perhaps best applied to point-source discharges, since the determination of nonpoint-source loads generally involves its own modeling.

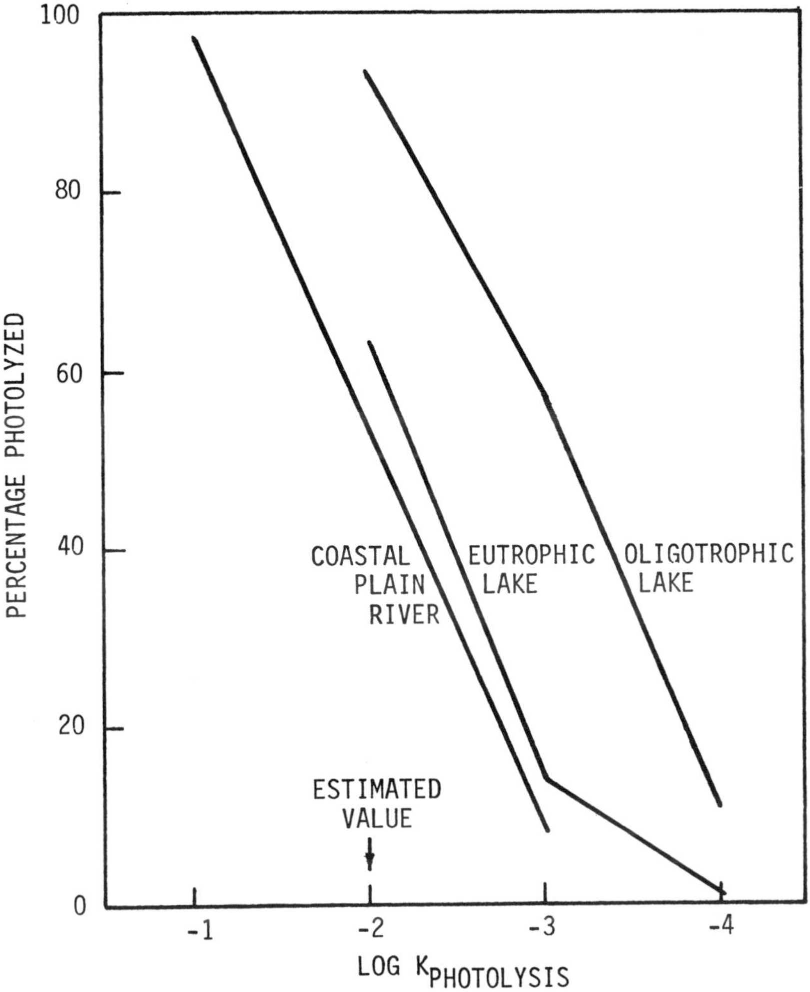

Figure 2. EXAMS sensitivity analysis: PCP.

Runoff

The EPA Water Office has, over the last few years, been involved with a variety of urban runoff models, primarily through the Water Quality Management and Construction Grants programs.

Models of pollutants in separate storm sewers have been used in assessing (1) overall importance of urban runoff as a pollutant source

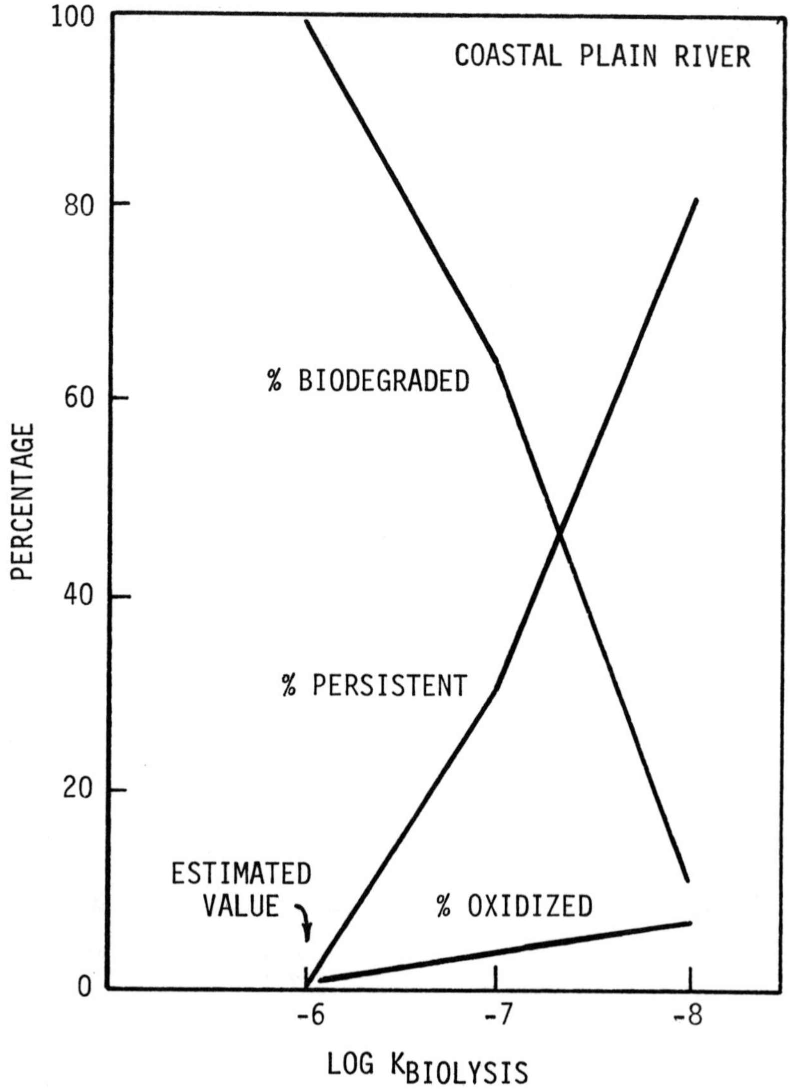

Figure 3. EXAMS sensitivity analysis: phenol.

(particularly for suspended solids, nutrients and metals), and (2) feasibility of control. The results of such analyses are being applied more to guide the direction of future federal and state efforts than to design specific projects at the local level. For combined sewer overflows, on the other hand, a variety of models [14] have been applied widely to determine the size and scope of specific local projects. Such work, however, seldom involves consideration of toxic chemicals; rather, the effects of raw sewage on biochemical oxygen demand (BOD) and coliforms dominate these concerns.

Rural runoff modeling has been undertaken much less often by the Water Office than by the pesticide program. Although complete watershed models, usually the Hydrocomp Simulation Programming (HSP) model [15], were procured a few years ago by many regional planning agencies, toxic chemicals were seldom evaluated. Significant amounts of future work in this area are not anticipated.

Solid Waste Disposal

Ideally, disposal of hazardous organic wastes should result in containment within the disposal site for a period long enough that complete breakdown of the toxicants occurs via chemical or biochemical processes. Toxic metals ideally would be as immobile in the disposal site as they were in the geologic structure from which they were originally extracted. In practical situations, however, it may be satisfactory merely if the toxicants do not at any time leave the disposal site rapidly enough to result in hazardous exposure levels (i.e., if disposal does not merely delay the occurrence of hazardous exposures). Thus, the disposal site may act partially as a destructive treatment facility and partially as a diffuser.

Disposal site models can be applied to such questions as:

- How mobile is the waste?
- Will it volatilize or leach?
- How rapidly will it degrade?
- How would containment alternatives, such as adsorption, affect degradation and mobility?

OWRS is involved with the development of Seasonal Soil Model (SESOIL), which on a seasonal basis traces the vertical movement of waste substances in a soil column until they either volatilize to the atmosphere or percolate to the depth of the groundwater table [16].

Because public perception of risk from hazardous waste dumps is rather high, this is an area of substantial regulatory interest. Consequently, considerable activity is expected in modeling the rates of migration and degradation within solid and hazardous waste dumps.

CRITICAL UNCERTAINTIES

The appropriateness of regulatory decision-making for toxicants depends on the analyst's ability to trace environmental releases to human or biotic exposure and to relate such exposure to harmful effects. In this analytical chain, the weak link may be in the determination of the harmful effect levels rather than the exposure levels. If this were the case, then efforts to sharpen the exposure analyst's accuracy would yield little benefit, since the location of the target exposure level would be the greatest uncertainty.

The difficulties of relating carcinogenic response of heavily dosed laboratory animals to cancer incidence in slightly exposed humans are well known [17]. For different reasons, the available body of aquatic toxicity tests on single species is likewise difficult to relate quantitatively, rather than merely qualitatively, to discernible ecological changes in the field. Single-species toxicity tests essentially are physical (as opposed to mathematical) models of environmental behavior. They are, however, very incomplete. Just as effect levels for whole organisms are not predictable from tests on single-cell cultures, likewise the responses of communities are not quantitatively predictable solely from measurements at the organism level of organization. Rather, quantitative measures of ecological structure (e.g., diversity) and function (e.g., energy or nutrient flow) would appear to be more directly applicable.

Consequently, when the regulatory decision-maker asks the analyst for verification that a particular industrial activity causes damage, most of the work may often involve the verification of actual effects rather than the verification of exposure. Furthermore, since cost-benefit comparisons require quantitative estimates of environmental damage, and since such comparisons may become an important element in regulatory decision-making, there is a clear need to upgrade our ability not simply to predict environmental concentrations but also to predict environmental response in quantitative terms.

REFERENCES

1. Mackay, D. "Finding Fugacity Feasible," *Environ. Sci. Technol.* 13:1218–1223 (1979).
2. "Ambient Water Quality Criteria for Tetrachloroethylene," U.S. EPA Office of Water Regulations and Standards Report 440/5–80–073 (1980).
3. Lyman, W. J. "Prediction of Chemical Partitioning in the Environment," U.S. EPA Office of Water Regulations and Standards Contract No. 68–01–5949 (1981).

4. Morel, F., and J. Morgan. "A Numerical Method for Computing Equilibria in Aqueous Chemical Systems," *Environ. Sci. Technol.* 6:58–67 (1972).
5. Perwak, J., et al. "Exposure and Risk Assessment for Copper," U.S. EPA Office of Water Regulations and Standards Contract No. 68-01-3857 (1980).
6. Geckler, J. R. "Validity of Laboratory Tests for Predicting Copper Toxicity in Streams," U.S. EPA Environmental Research Laboratory Report 600/3-76-116.PB-264867 (1976).
7. Burns, L. A., D. M. Cline and R. R. Lassiter. "Exposure Analysis Modeling System (EXAMS): User Manual and System Documentation," U.S. EPA Environmental Research Laboratory, Athens, GA (1981).
8. Canale, R. P. "Gas Transfer Processes," in *Physiochemical Processes for Water Quality Control,* W. J. Weber, Jr., Ed. (New York: Wiley-Interscience, 1972), pp. 504–508.
9. Liss, P. S., and P. G. Slater. "Flux of Gases across the Air-Sea Interface," *Nature* 247:181–184 (1974).
10. Hydroscience, Inc. "Simplified Mathematical Modeling of Water Quality," U.S. EPA, Washington, DC (1971), p. A-2.
11. Owens, M., R. W. Edwards and J. W. Gibbs. "Some Reaeration Studies in Streams," *Int. J. Air Water Poll.* 8:469–486 (1964).
12. Hydroscience, Inc. "Water Quality Analysis for the Markland Pool of the Ohio River," prepared for MSD of Greater Cincinnati (1968).
13. Mabey, W. R., et al. "Aquatic Fate Process Data for Organic Priority Pollutants," U.S. EPA Office of Water Regulations and Standards Contract No. 68-01-3867 (1981).
14. DiGiano, F. A., D. D. Adrian and P. A. Mangarella. "Short Course Proceedings: Applications of Stormwater Management Models," U.S. EPA Office of Research and Development Report 600/2-77-065 (1976).
15. "Hydrocomp Simulation Programming Operation Manual," Hydrocomp International, Inc., Palo Alto, CA (1972).
16. Bonazountas, M., J. Wagner and B. Goodwin. "Evaluation of Seasonal Soil-Groundwater Pollutant Pathways," U.S. EPA Office of Water Regulations and Standards Contract No. 68-01-5949 (1981).
17. Fishbein, L. "Overview of Some Aspects of Quantitative Risk Assessment," *J. Toxicol. Environ. Health* 6:1275–1296 (1980).

CHAPTER 5

SIGNIFICANCE AND LIMITATIONS OF ENVIRONMENTAL COMPARTMENTALIZATION MODELS IN THE CONTROL OF NEW CHEMICALS BASED ON THE ORGANIZATION FOR ECONOMIC COOPERATION AND DEVELOPMENT MINIMUM PREMARKETING SET OF DATA

Albrecht W. Klein and Friedrich Schmidt-Bleek

Umweltbundesamt
Berlin, West Germany

With increasing concern over potential and actual hazards to man and his environment through the largely uncontrolled use of commercial chemicals, a number of key chemical-producing countries have recently passed or are in the legislative process of enacting chemical control legislation. Japan moved in 1973; the United States enacted the Toxic Substances Control Act (TSCA) in 1979; and the Member States of the European Communities agreed on a common directive in 1979 [1]. Including Canada, Switzerland, Sweden, Norway and some other countries, close to 65% of world chemical production is now, or will shortly be, under general legislative control.

It is worth noticing that with this "new generation" of legal instruments, a significant shift from "curative" to "preventive" control is being instituted; furthermore, protective legislation is finally formulated to be neither medium-specific (water, air, soil) nor use-pattern–specific (drugs, agrochemicals, explosives, etc.). In this sense, "basic chemical acts" are promulgated in many countries today.

Common to all new product legislation is *notification* for chemicals before they are placed on the market. Notification of the competent authority means turning over extensive, clearly defined information about the new substance. Such information is of an economic nature (quantities to be marketed, use pattern, etc.); it includes recommendations for safe handling, packaging, labeling, disposal of the substance, etc.; and it ultimately consists of the presentation of data derived from laboratory studies.

To avoid nontariff barriers caused by different requirements on test data or by a nonmutual acceptance of these data, the Organization for Economic Cooperation and Development (OECD) has attempted the task of harmonization. For this purpose a Minimum Premarketing Set of Data (MPD), shown in the Appendix, has been designed to ascertain reliable recognition of such new substances that likely pose a considerable risk to man and the environment. Such ("red") chemicals are immediately subjected to further tests. The other ("amber" or "green") substances are called in for further studies at a later stage, normally according to a well-defined and regulated step system [2,3].

HOW CHEMICALS ENTER THE ENVIRONMENT AND THUS BECOME POTENTIALLY DANGEROUS

Figure 1 indicates various human activities within the technosphere that put substances into the environment, where they become environmental chemicals and potentially toxic substances.

Mass (natural raw materials), energy, labor, capital and human knowledge are the essential ingredients for the production of specifically designed chemical substances and their application within the technosphere, including agricultural uses.

Emissions and discharges, deliberate and accidental, occur within all areas of human commerce involving chemicals (Figures 1 and 2). The ecosphere thus receives a continuous stream of anthropogenic substances whose principal fate is decomposition back to natural chemical entities.

Air, water and soil are the dumping grounds of manmade "waste" chemicals. Once in the environment, they can virtually never be called back, collected or even influenced in their course of action or destruction. The environmental compartments differ greatly with respect to transport phenomena (e.g., mobility), storage capacity for long-lived chemicals, global uniformity, chemical reactivity, content of biotic systems (habitat) and biotic population densities. No chemical is ever truly at rest in this vast environmental system, nor is any chemical substance exclusively found in only one compartment.

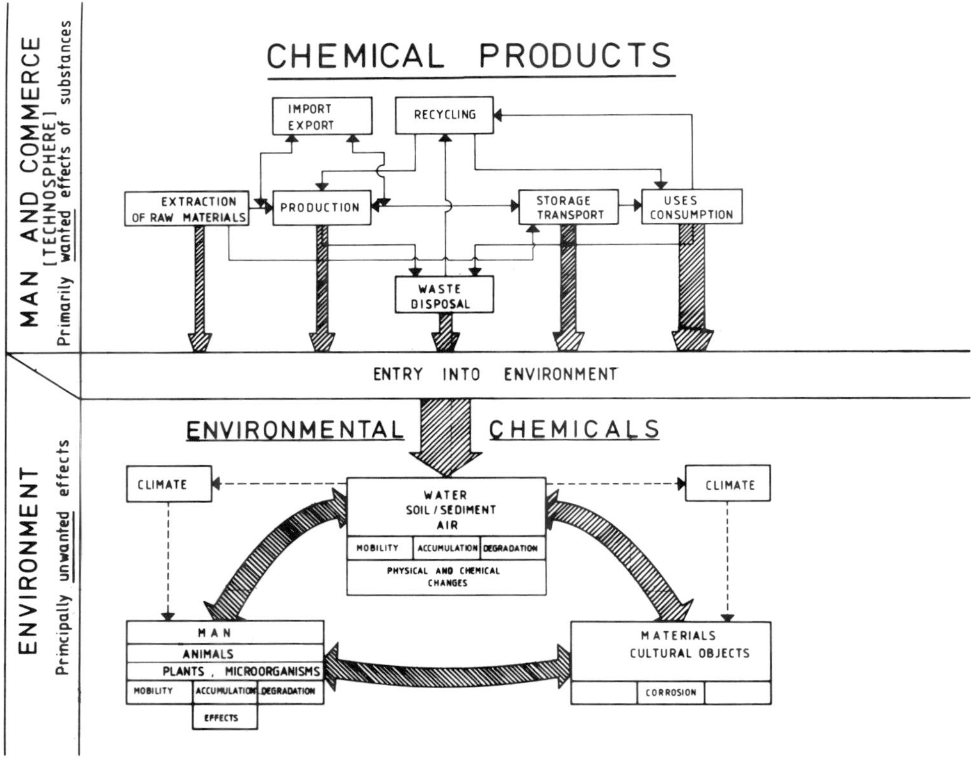

Figure 1. Human activities within the technosphere that input chemical substances.

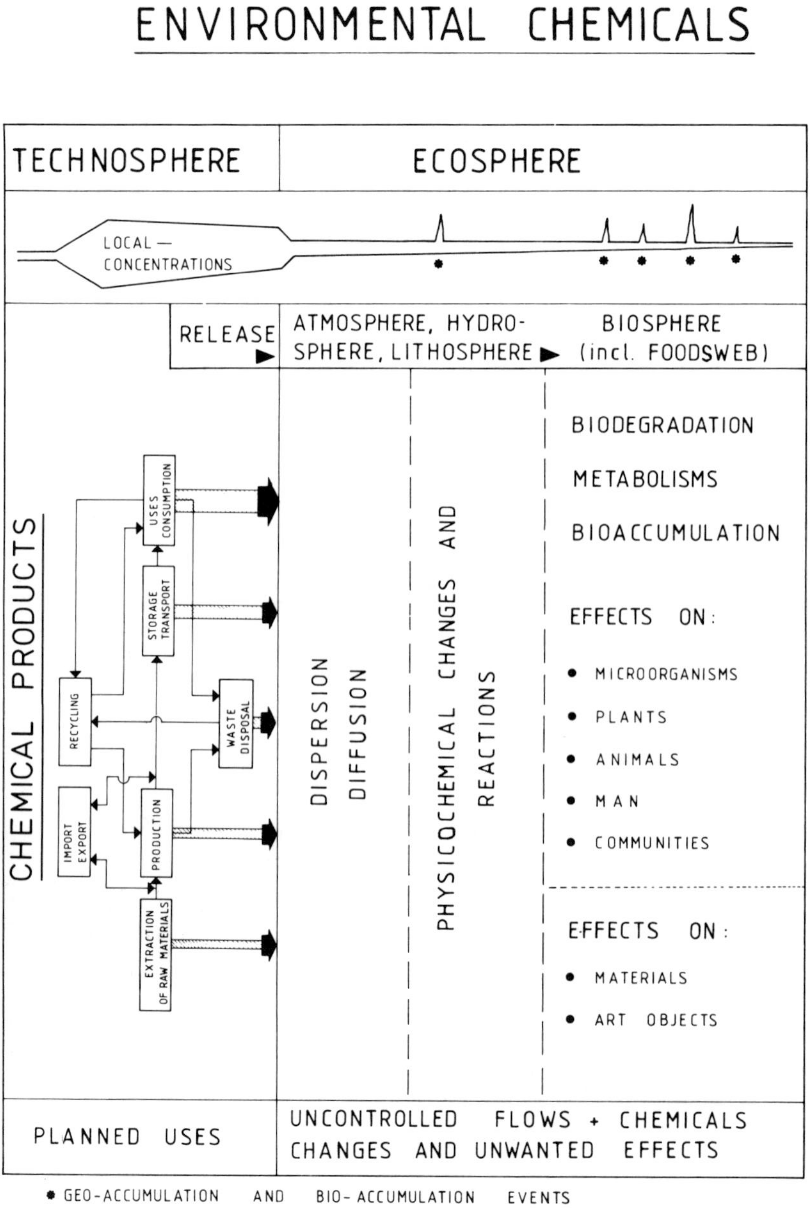

Figure 2. Geo- and bioaccumulation events.

Within the three environmental compartments, manmade chemicals encounter biological entities and may act on them, disturbing, blocking and enhancing natural activities and interdependencies. They may also (temporarily) be trapped and carried along food webs. Again, multiple-exchange mechanisms can be observed among biotic systems, as well as from and to the environmental compartments of air, water and soil.

Finally, the possibility should be recalled that environmental chemicals interact destructively with manmade building materials and objects of art. Figure 2 presents these interconnections.

It may be observed that transitions of manmade chemicals from the technosphere to the ecosphere are generally accompanied by a dramatic dilution (lowering of concentration). One must keep in mind, however, that in water and, particularly, soil, relatively high concentrations of environmental chemicals can nevertheless be obtained and that, even at low concentrations, chronic effects occur. One must furthermore consider that reconcentrations (accumulations) of occasionally astounding magnitude (factor of 10^5) can be observed in biological systems. Finally, whenever the potential hazard of an individual substance is contemplated, it should not be forgotten that hundreds or even thousands of other manmade chemicals are simultaneously behaving and acting similarly in the same environment.

Chemical substances, once released to the environment, act and re-act according to their specific physicochemical and biological activity. In this sense, they therefore have no "memory" of their source and circumstance of release from the technosphere. This in turn implies that, irrespective of the source of pollutants, experimental elucidation of their physicochemical and biological properties, and conclusions derived therefrom with respect to their potential environmental hazard are principally a function of the concentration and nature of the chemical involved in the surrounding environment.

EXPOSURE AND EFFECTS MAKE UP THE HAZARD OF A CHEMICAL

In this sense the environmental hazard (H) of a chemical may be regarded as a function of two broad considerations, the potential of the chemical to harm biological systems (T) and its potential for exposure (E):

$$H = f(E, T) \tag{1}$$

The exposure term envelops basically three considerations: (1) quantity of substance, (2) population, and (3) time (period, frequency). Although the effect term usually reflects the biological activity (toxicity) of the chemical, inorganic and physicochemical changes occasionally must be considered [e.g., destruction of building materials, objects of art, atmospheric chemical changes (fluorocarbon degradation of the ozone layer)].

STEPS TOWARD ENVIRONMENTAL HAZARD ASSESSMENT [2]

Exposure, Effects and Fate

Predictive hazard evaluations of chemicals take into account the likely exposure of environmental objects (including man) to the chemicals and the resulting effects [4–6]. It can be deduced from Figure 3 that not only are results of laboratory tests required for estimating likely environmental exposure, but the flux of a chemical from the technosphere to the environment must be estimated by the quantities, use patterns and type of application of a chemical placed on the market.

The environmental fate of a chemical once it has entered the environment is principally a function of

1. mobility and environmental compartmentalization (i.e., the theoretical proportional distribution of a chemical among the environmental media of soil, water and air, assuming a state of equilibrium);
2. degradability in the environment; and
3. potential to accumulate in specific sectors within the environment.

Abiotic effects of chemicals, as mentioned, may in some caes of high-volume products also have to be considered (e.g., fluorocarbons and solvents). Generally, however, the emphasis here must be concentrated on biotic effects.

The basis for the evaluation procedure must be the substance data prescribed within the legal frameworks. It is possible for such data to be:

1. combined with each other by the application of scientifically based equations to determine properties of a substance that cannot be tested, e.g., compartmentalization capability; and
2. combined by the application of simple algorithms in such a way that an assessment of the possibly inherent "general environmental hazard" of the chemical, which takes into account the test data for that chemical, is possible.

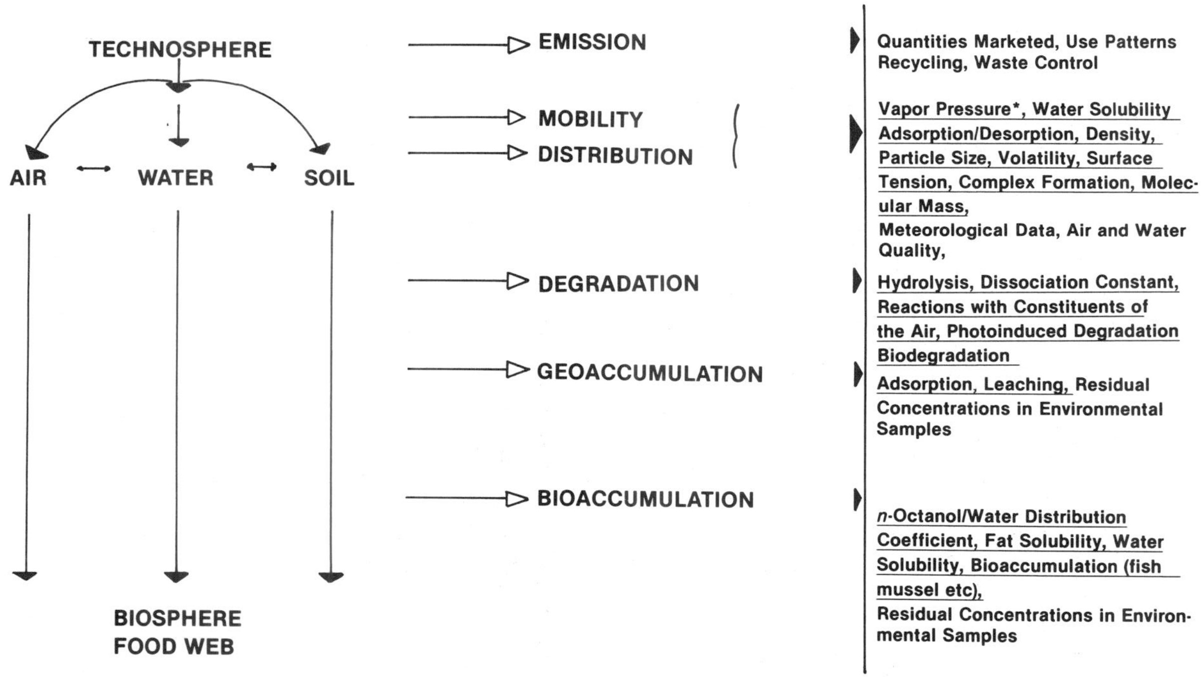

Underlined Terms: item can be determined in laboratory (field) studies (test guideline) for individual chemical under consideration.

Figure 3. Survey of information useful for determining environmental exposure.

Hazard Estimates

From a scientific and environmental-protection point of view, the evaluation must proceed stepwise, and it must be distinguished between the use of MPD information for hazard ranking (HR) of substances and environmental hazard assessment (HA) at more advanced levels of information. HR and environmental HA involve the logical integration of weighted test results and other available information, falling broadly into the categories of exposure and effects. Such hazard estimates will hardly ever be of an absolute nature, but will have to be scaled to experiences with suitable, well known "old" substances, as well as past regulatory decisions on environmental chemicals. For instance, for many compounds much of the needed information for predicting an "exact" environmental concentration is normally not available at the MPD level. So the purpose of the HR can only be to identify and preselect those chemicals that have high ("red" substances) or medium ("amber" substances) environmental hazard probabilities and therefore should receive further testing either immediately (red) or after passing a higher marketing volume ("amber"). It cannot, however, be used to make ultimate regulatory decisions, which must be left to the HA.

SIGNIFICANCE OF ENVIRONMENTAL EXPOSURE ANALYSIS WITHIN THE HR PROCESS FOR CHEMICALS [4]

HR on the MPD level is a screening exercise and no substitute for an environmental HA at later stages. A flow scheme of the process is shown in Figure 4.

To identify chemical substances that represent a potential health or environmental hazard, it first must be determined that exposure to the substance will occur. A substance to which no exposure occurs cannot be considered a hazard. Any exposure analysis must consider both the exposure occurring at the original point of entry to the environment (point source) and that resulting at some later time when the chemical has achieved, in idealized terms, an equilibrium among the several media. Each of these stages can be more finely divided for each chemical. For the point-source analysis, consideration must be made of emission quantities and use and disposal patterns. Attention should also focus on the medium of entry to the environment. To evaluate exposure, therefore, a more detailed breakdown is required. There are four major types of exposure (Figure 5):

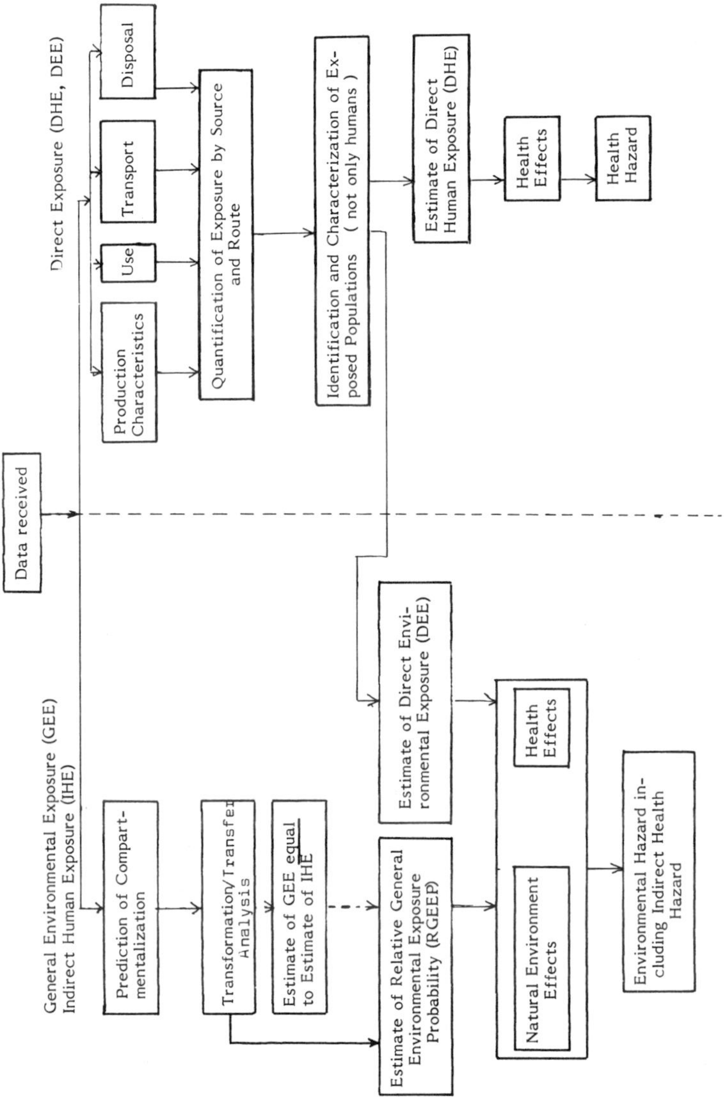

Figure 4. Hazard ranking for chemicals.

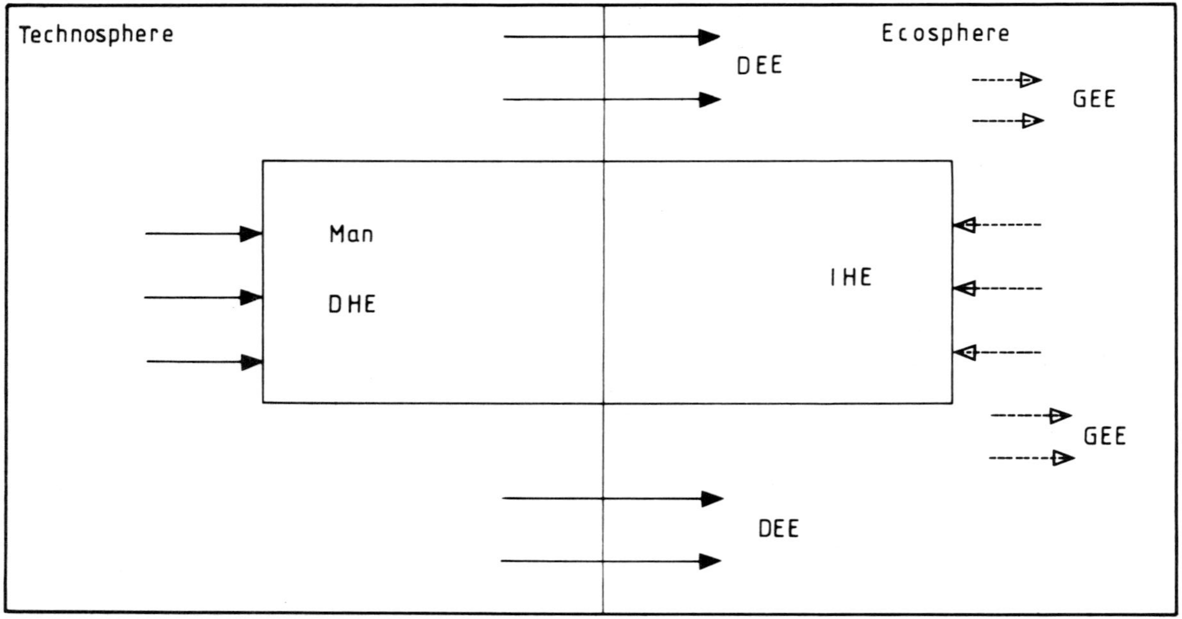

release of the chemical under consideration from a point source

impact to targets after its distribution among the environmental media and subjection to natural degradation, accumulation and transformation processes

Abbreviations are explained within the text

Figure 5. Major types of exposure.

1. Direct human exposure (DHE): the concentration of a chemical reaching human targets is sometimes a result of its release from a point source. This category includes the occupational population as an identifiable subset.
2. Direct environmental exposure (DEE): the concentration of a chemical reaching environmental (nonhuman) targets is sometimes a result of its release from a point source.
3. General environmental exposure (GEE): a chemical, once released from the technosphere into the environment, will distribute among the environmental media (air, water, soil/sediment and biota). GEE is the equilibrium concentration of the chemical after distribution and subjection to natural degradation and transformation processes.
4. Indirect human exposure (IHE): exposure of the general human population to a chemical at equilibrium distribution among the environmental media (air, water soil/sediment) may indirectly affect humans.

Since people are environmental targets identical to any other type of target in the environment, GEE is equal to IHE, and health effects data must be considered. On the other hand, DEE and DHE situations are to be assessed quite differently, since their exposures vary, depending on the chemical's use and manner of release.

High GEE can be considered an environmental hazard because chemicals, after escaping the technosphere, cannot be collected again; deleterious effects, including chemical, physical and biological changes, can occur through complex long-term mechanisms not normally observed.

At the premarketing stage, however, where the notification data are supplied to the authorities, the expected GEE is generally close to zero. Nevertheless, to fulfill the government's obligation of prospective environmental protection, relative general environmental exposure probabilities (RGEEP) can be defined that allow comparison of the relative probability for a particular compound to appear in air, water, soil, sediment and/or biota and that allow ranking of chemicals with respect to their probabilities of being present within a given medium. Once higher production/marketing volumes are reached, RGEEP are used to scale the exposure potential for each chemical at the time exposure and health effects are combined.

ENVIRONMENTAL EXPOSURE MODELING

For performing a preliminary evaluation of exposure of a new chemical, governmental scientists have to rely on the data supplied by industry as part of the premarketing notification data set. Data most relevant to exposure evaluation within the OECD's MPD are shown in Table I [7–9].

Table I. Data in the OECD MPD Most Relevant to Exposure Evaluation

Physical/Chemical Properties	Biodegration
Molecular Mass	Abiotic Degradation
Water Solubility	Hydrolysis
Vapor Pressure	Adsorption (ultraviolet/visible)
Octanol/Water Partition Coefficient	
Adsorption/desorption	

For chemicals that are marketed or placed in the technosphere (Figure 1), no point-source emission can generally be constructed, since their places of use can literally be worldwide. Thus, when the time of emitting is long enough so that "equilibrium" conditions may be assumed, GEE = IHE must be considered.

Concerning the determination of GEE, considerable progress has recently been possible [10], based on the well known publications of Neely [11] and Mackay [12]. Additional definitions are required to evaluate the models; for example, environmental compartmentalization (EC) is the estimation of the mass percentage or relative concentration of a chemical among the environmental media under the following simplifying assumptions:

1. The environment is a closed system consisting of the environmental compartments air, water, sediment/soil and biota.
2. The environment has reached thermodynamic equilibrium.
3. No degradation processes occur during distribution.

Also, predicted environmental distribution (PED) extends the EC estimated, because degradation and transfer processes are considered, and because the compartments need not be assumed to undergo rapid exchange in comparison to degradation, so that transfer processes may control the PED. An explanation of the dependencies among EC, PED and GEE (IHE) is given in Table II.

The set of regression equations developed by Neely [11] for predicting environmental distribution among air, water and soil/sediment requires only data on the three physicochemical properties: molar mass, vapor pressure and water solubility. This approach was derived from other calculations (involving a more complex dynamic model of an aquatic ecosystem) on a series of ten common chemicals ranging from toluene to DDT and exhibiting a wide range of solubilities (about 0.001–500 ppm) and vapor pressures (10^{-7} to 30 mm Hg). Although the regression equa-

Table II. Explanation of Dependencies among EC, PED and GEE (IHE)

Production/marketing figures (at base-set level close to zero)
Transfer rates
Degradability (biotic & abiotic)
Structural formula
Vapor pressure
Water solubility
Partition coefficient
Adsorption/desorption
In their combined form, these express EC, PED and GEE

tions are normalized such that the total percentage is constrained to be 100%, the value of the mass percentage in a compartment can take on any value, either positive or negative (when relationships between water solubility and vapor pressure of the chemicals under consideration do not fit the ranges used by Neely). In this case, only a qualitative indication of the location of equilibrated material can be given, taking the result as zero, if it is obtained negative, and as 100%, if it is in excess of 100%. If two compartments have values greater than 100%, it must be assumed that the chemical is split between the two compartments.

The Mackay approach [12] uses the concept of fugacity, which can be defined as the tendency for a substance to escape from a phase and relies, contrary to the Neely model, on the basic assumption of equilibrium conditions among a set of environmental compartments. Therefore, the general equation for EC values becomes:

$$EC_i = \frac{Z_i \times V_i}{\Sigma_i Z_i \times V_i} \times 100\,(\%) \tag{2}$$

where i = air, water, soil, sediment or biota components
Z = fugacity capacity
V = volumes of the different environmental compartments

The fugacity capacities (g-mol/m^3-atm) are as follows:

$$Z_{air} = 1/RT\,(\approx 40 \text{ g-mol/m}^3\text{-atm at } 25\,°C) \tag{3}$$

$$Z_{water} = 1/H \tag{4}$$

$$Z_{soil/sediment} = K_D S/H \tag{5}$$

$$Z_{biota} = K_{ow}B/H \tag{6}$$

where
R = universal gas constant
H = Henry's law constant
K_D = soil/sediment adsorption coefficient = K_{oc} (% organic carbon content of soil sediment/100)
B = mass fraction of biota times lipid part
K_{ow} = partition coefficient (*n*-octanol/water)

The physicochemical data needed include molar mass, water solubility, vapor pressure, soil sorption coefficient (K_{oc}) and partition coefficient (*n*-octanol/water).

The outputs of this model differ with regard to the assumed volumes of the environmental compartments, the assumed densities and organic carbon contents of soil and sediment, and the assumed product of mass fraction and lipid part of biota as well. To coordinate the future efforts in environmental compartmentalization modeling, it is therefore necessary to standardize certain ground rules on the media available for interaction. A first, tentative, approach was taken at the Workshop on Environmental Compartmentalization, Washington, DC March 24–25, 1981, by the description of a "generic environment" [10] shown in Figure 6. Utilizing the basic equations of Mackay's fugacity approach:

$$C_i = Z_i f_i \tag{7}$$

where
C_i, f_i = concentration and fugacity, respectively, of the chemical under consideration in the compartment i
Z_i = constant at equilibrium concentration ratios, equivalent to the respective partition coefficients K_{ij} available from laboratory measurements and appropriate assumptions on site-specific parameters

These parameters can be calculated [13] according to

$$K_{ij} = \frac{C_i}{C_j} = \frac{Z_i}{Z_j} \quad (i \neq j) \tag{8}$$

and are now independent of compartment volume.

Suitable environmental monitoring data for comparison are available for some substances [14], which are relatively stable as requested by the model. The results, based on the assumptions of the "generic environment" concerning densities and organic carbon contents, are shown in Table III. The agreement of estimated and observed concentration ratios

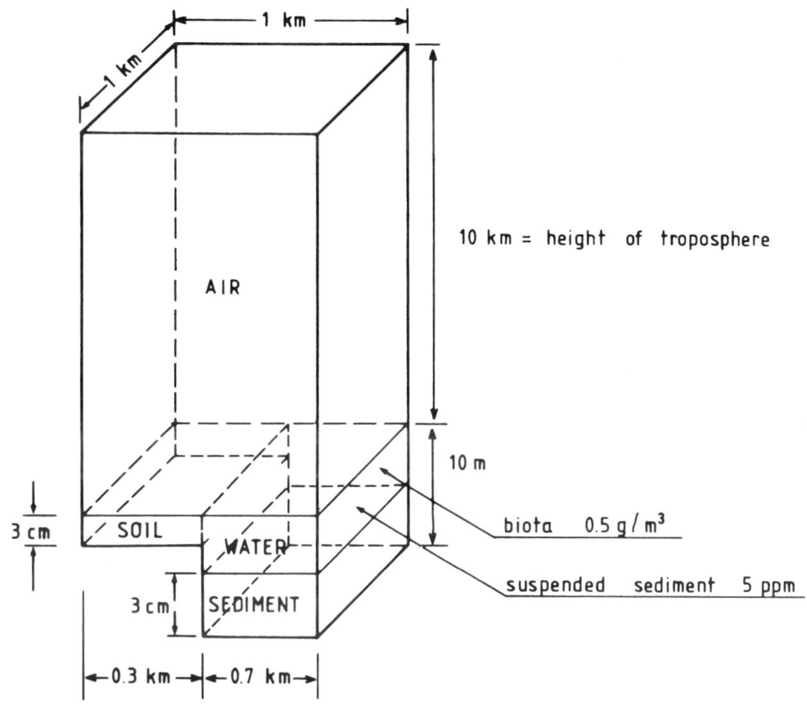

Figure 6. Generic environment.

is such that results may be considered as a basis for further adaption of the "generic environment" model to the real world, e.g., by least-squares refinement.

EC, as discussed here, only relates to the distribution of a chemical in the environment under conditions where equilibrium among the compartments is established and no degradation of the substances occurs. Of course, the real world is more complicated. Mackay has therefore expanded his basic compartmentalization model to incorporate degradation and transport, so that the model, as well as the U.S. Environmental Protection Agency (EPA) adaption by Wood [13], can explicitly include the results of respective laboratory testing.

The products of these dynamic models are the PED. For PED, however, not all of the data are available within the OECD's MPD (see Appendix), especially those used to describe reactivity in the air compartments. For example, there is a test, not currently in the MPD, that would provide as an output the photodegradation rate (Table I). The

Table III. Concentration Ratios According to the Mackay Model (Step I) and Compared with Those from Monitoring Data

	p,p' DDT[a]		PCP[b]		PCB (Aroclor 1260)[c]		γ-BHC (Lindane)[d]		HCB[e]	
	Mackay	Env.	Mackay	Env.	Mackay	Env.	Mackay	Env.	Mackay	Env.
Air/Water	6.0×10^{-4}		7.7×10^{-4}		0 . 3		8.9×10^{-5}		6.5×10^{-2}	
Soil/Water	1.9×10^{4}		1.9×10^{3}		2.6×10^{5}		36.8		6.8×10^{3}	
Sediment/Water	3.8×10^{4}		5.3×10^{4}	$0.5\text{–}0.9 \times 10^{4}$	5.2×10^{5}	$5.3\text{–}18 \times 10^{3}$	73.6	150–900	1.3×10^{4}	2.0×10^{4}
Suspended Solids/Water	1.3×10^{-1}		1.3×10^{-2}	$6\text{–}45 \times 10^{-2}$	1.7	1.0–1.5	2.5×10^{-4}	$7.5\text{–}30 \times 10^{-3}$	4.5×10^{-2}	4.8
Biota/Water	0.5		5×10^{-2}	$3\text{–}18 \times 10^{-2}$	6.9	0.7–1.4	9.5×10^{-4}	$5\text{–}750 \times 10^{-4}$	0.18	0.13–0.20
Biota/Suspended Solids	3.97		0.1–0-9							
Biota/Sediment	0.13×10^{-4}	$0.13\text{–}1.2 \times 10^{-4}$								

[a] $M = 354.49$; $P_{25} = 2.3 \times 10^{-5}$ Pa; $C_w = 5.5 \times 10^{-6}$ g/l; lg $K_{ow} = 6.00$.
[b] $M = 266.34$; $P_{25} = 0.1$ Pa; $C_w = 1.4 \times 10^{-2}$ g/l; lg $K_{ow} = 5.01$.
[c] $M = 375.70$; $P_{25} = 5.4 \times 10^{-3}$ Pa; $C_w = 2.7 \times 10^{-6}$ g/l; lg $K_{ow} = 7.14$.
[d] $M = 290.85$; $P_{25} = 7.6 \times 10^{-3}$ Pa; $C_w = 10^{-2}$ g/l; lg $K_{ow} = 3.29$.
[e] $M = 284.78$; $P_{25} = 2.8 \times 10^{-3}$ Pa; $C_w = 5 \times 10^{-6}$ g/l; lg $K_{ow} = 5.55$.

rates of •OH and O_3 radical reaction can be estimated using linear free-energy calculations [15], but no reliable estimation procedure exists for photolytic half-life.

Biodegradation test results, as required in the current MPD, only permit a determination of whether the compound is readily biodegradable, but what is needed by the PED model is a biodegradation rate constant. Therefore, to get an approximate k (observed) for biodegradation (in the case of ready biodegradability) an averaged, constant value should be used and $k = 0$; otherwise, a worst-case assumption is in order.

The obvious way to convert PED to GEE is to multiply the relative environmental distribution data by an appropriate factor as shown in Figure 7, dependent on whether the distribution has been expressed in terms of mass percentage or relative distribution. The problem here is that, if one takes the world as a dilution factor, the expected exposure to a new chemical will generally be far too low to be meaningful, since the production volume will be low.

It can be deduced from Figure 7 that the PED of a chemical is identical with the RGEEP of the chemical. For practical purposes, therefore, PED should be computed in each case and serve as a stand-in for RGEEP. This would allow (1) comparison of the relative probability for a particular compound in air, water, soil, sediment or biota; and (2) ranking of different chemicals with respect to their probability of being present within the respective compartments [2,4].

In HR, the scaled RGEEP values can then be combined with appropriate effects, persistence and bioconcentration information. This procedure will eliminate on one hand many of the chemicals from immediate further considerations and testing, since they will be transformed rapidly in the environment to nonhazardous products, will not accumulate, and will result in only low effects data. On the other hand, it will indicate which tests beyond the MPD should be undertaken, assuming some concern has been developed in the HR.

CONCLUSIONS

For the present, no specific criteria can be given for HA of substances based on information beyond the MPD. It must be assumed that testing programs at advanced levels are decided on by authorities only after extensive consultations with independent scientists and the producers/importers of such substances. At this stage, more sophisticated approaches like the EPA's Exposure Analysis Modeling Systems (EXAMS) model must be utilized to predict real environmental concentrations. What is

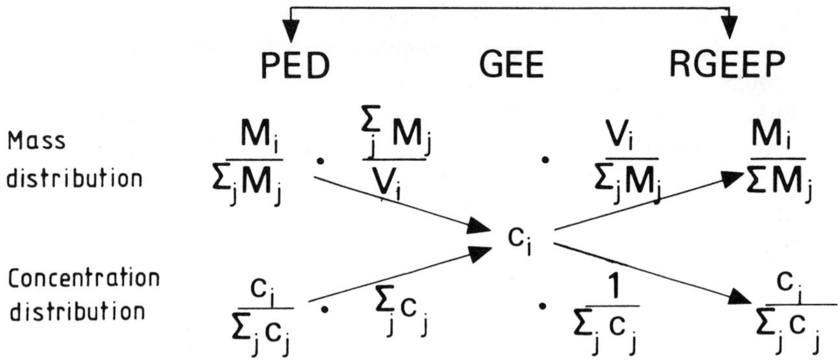

Figure 7. Correlations among PED, GEE and RGEEP.

required in the near term is (1) more thorough testing and validation of the models for predicting PED; (2) adaptation of the defined "generic environment" to real world; and (3) evaluation of the data requirements and decision processes necessary to evaluate DEE and DHE. For all these tasks, an adequate international exchange of scientific information is of key importance.

APPENDIX: OECD MINIMUM PREMARKETING DATA SET

Chemical identification:

- names according to agreed international nomenclature, e.g., International Union of Physical and Analytical Chemistry
- other names
- structural formula
- Chemical Abstracts Service Number (CASN)
- spectra ("fingerprint spectra" from purified and technical-grade product)
- degree of purity of technical-grade product
- known impurities and their percentage by weight
- essential (for the purpose of marketing) additives and stabilizers and their percentage by weight

Production/use/disposal data (exposure):

- estimated production, ton/yr
- intended uses

- suggested disposal methods
- expected mode of transportation

Recommended precautions and emergency measures

Analytical methods

Physicochemical data (exposure):

- melting point
- boiling point
- density
- vapor pressure
- water solubility
- hydrolysis (only the screening part to be done for base set)
- spectra
- adsorption/desorption
- dissociation constant
- particle size (only the screening part to be done for base set)

Degradation/accumulation data (exposure):

- biodegradation: screening-test for ready biodegradability
- bioaccumulation: screening-test (partition coefficient—*n*-octanol/water, fat solubility)

Acute toxicity data (effects):

- acute oral toxicity
- acute dermal toxicity
- acute inhalation toxicity
- skin irritation
- skin sensitization
- eye irritation

Ecotoxicity data (effects):

- fish lethal concentration (LC): at least 96-hr exposure
- *Daphnia*: reproduction 14 days
- algae growth inhibition 4 days

Repeated dose toxicity data (effects)

- 14–28 days repeated dose

Mutagenicity data (effects)

REFERENCES

1. "Council Directive 79/831/EEC of 18 September (1979), 6th Amendment of Directive 67/548/EEC on the Approximation of the Laws, Regulations, and Administrative Provisions Relating to the Classification, Packaging, and Labeling of Dangerous Substances," *Off. J. Eur. Comm. Council* L259 (October 15, 1979), p. 10.
2. Schmidt-Bleek F., W. Haberland, A. W. Klein and S. Caroli. "Steps Towards Environmental Hazard Assessment of New Chemicals (Including a Hazard Ranking Scheme, Based upon Directive 79/831/EEC," *Chemosphere* (in preparation).
3. Schmidt-Bleek, F. "Toxic Substances Control for the 80s and Beyond," *Toxic Substances J.* (in press).
4. Schmidt-Bleek, F., A. W. Klein, W. Haberland and J. Hushon. "Control of Toxic Substances—Environmental Exposure Analysis and Hazard Ranking within the European Directive 79/831/EEC," *Toxic Substances J.* (submitted).
5. Schmidt-Bleek, F. "Regulated Testing of New Substances for Environmental Exposure and Toxicity in the European Community," *Reg. Toxicol. Pharmacol.* (submitted).
6. "Common Proposal for Reference Chemicals (Yardsticks) with Respect to Their Physical/Chemical Properties in Relation to the Assessment of the Environmental Fate and Effects of Chemical Substances," CEC Document No. ENV/223/74-DE, Rev. 3 Environmental Chemicals Group, Umweltbundesamt, Berlin, West Germany (1980).
7. Klein, A. W., M. Harnisch, H. J. Poremski and F. Schmidt-Bleek. "OCED Chemicals Testing Programme, Physico-chemical Tests," *Chemosphere* 10(2):153 (1981).
8. Klein, A. W., H. J. Poremski and F. Schmidt-Bleek. "Physikalisch-chemische Laborprufung neuer Stoffe nach, dem Chemikaliengesetz," *Sicherheit Chemie Umwelt* (July 1981).
9. Poremski, H. J., A. W. Klein and F. Schmidt-Bleek. "The Predictive Value of Physical Chemical Properties," paper presented at the International Symposium on Principles for the Interpretation of the Results of Testing Procedures in Ecotoxicology, Sophia Antipolis, Valbonne (Alpes-Maritimes), 1980.
10. Hushon, J., W. Wood, J. Poremski, W. Strachan, D. Mackay, W. B. Neely, G. Parris, F. Schmidt-Bleek, W. Klopffer and A. W. Klein. "Evaluation of Current Methods for Predicting General Environmental Exposure to New Chemicals," *Environ. Sci. Technol.* (submitted).
11. Neely, W. B. "A Method for Selecting the Most Appropriate Environmental Experiments on a New Chemical," in *Dynamics, Exposure and Hazard Assessment of Toxic Chemicals,* R. Haque, Ed. (Ann Arbor, MI: Ann Arbor Science Publishers, Inc., 1980), p. 287.
12. Mackay, D. "Finding Fugacity Feasible," *Environ. Sci. Technol.* 13:1218–1223 (1979).
13. Wood, W. P. "Comparison of Environmental Compartmentalization Approaches," OECD Chemicals Group, Working Party on Exposure Analysis (EXPO), Room 80.21.
14. Strachan, W. Personal communication.
15. Henry, D. G., and R. A. Kenley. "Atmospheric Reaction Products of Organic Compounds," U.S. EPA Report-560/12-79-001.

CHAPTER 6

SYNOPSIS OF DISCUSSION SESSION: MATHEMATICAL MODELING OF CHEMICAL FATE

S. M. Brown
Session Chairman

C. Delos, A. W. Klein, W. B. Neely, C. Offutt, W. P. Wood and G. T. Yeh

Hazard assessment involves two components: (1) generation of environmental exposure concentrations, and (2) description of toxicity levels of dose-response relationship. The chapters in this section clearly indicate that mathematical models for chemical fate will play a major role in the first component.

One of the key questions facing scientists who conduct hazard assessments is "What level of modeling is required?" There are a number of models available for the prediction of chemical fate. They range from simple representations of the world (e.g., the Neely and Mackay unit world) to highly sophisticated, physics-based models.

Selection of chemical fate model(s) for hazard assessment should be based on some criteria. We propose the following:

1. the specific question(s) that will be asked of the model;
2. the level of uncertainty that can be tolerated;
3. resource availability;
4. our understanding of transfer, transformation and transport processes; and
5. the complexity of the system being analyzed.

The specific question(s) that may be confronted in a hazard assessment can be as diverse as the models available for chemical fate predictions. These questions can be broadly grouped as follows:

1. What potential problems might occur as a result of release of a chemical into the environment?
2. What must be known about a chemical to assess its environmental fate?
3. Can a given quantity of chemical be released at a specific place and time?
4. To what extent do we need to clean up the environment after it has been contaminated?

The first two questions are typical of those asked by industry in the early stages of developing a chemical substance. They are also the types of questions addressed by the U.S. Environmental Protection Agency (EPA) in the early stages of new chemical evaluations. A unit world approach or any other simple chemical fate model is appropriate to address these questions. Such models allow the user to identify quickly how a substance is likely to partition itself among the different environmental media and its residence time or persistence in each medium. Most simple chemical fate models do not account for transport processes and often use generic, rather than site-specific, representations of the environment. As a result, their value is limited in predicting exposure concentrations that can be compared with effects data.

Answers to the last two questions require a higher order of modeling sophistication, in that all important transport, transformation and transfer processes must be included, and the selected model(s) must be applicable to a specific site.

The level of uncertainty associated with predicted chemical concentrations is affected by a number of factors, one of which is the level of model sophistication. Simple models, by their very nature, ignore certain transport, transformation and transfer processes. As a result, simple models may not provide all of the information required to estimate exposure concentrations. Thus, some level of uncertainty is introduced into the overall hazard assessment. This is not to say, however, that more sophisticated models will yield proportionally more reliable results. Without adequate supporting data and model testing, sophisticated models provide nothing more than a false sense of security. The level of modeling selected for a hazard assessment should provide results with a tolerable degree of uncertainty.

Models require certain resources before they can be used properly. The availability of these resources can be a determining factor in the level of chemical fate modeling that can be accomplished. The resources affecting model selection include:

1. time constraints and funding,
2. computer requirements,

3. data availability, and
4. skill level of the user.

The last two criteria for model selection are closely related. As many of the chapters in this book indicate, the processes affecting chemical fate in aquatic systems are complex, and some are better characterized than others. The degree to which individual processes are active in a specific aquatic system or affect a specific chemical needs to be assessed before identifying which level of modeling should be used. If certain processes can be neglected, combined or simplified, a fairly simple level of modeling can be used. If, on the other hand, processes cannot be neglected, a more sophisticated level of modeling will be required. Again, it is important to note, however, that the use of more sophisticated models is justified only if the individual algorithms for each chemical fate process and the overall model have received adequate testing and if supporting data are available.

Throughout the workshop, it became clear that existing and potential model users have widely varying views regarding the benefits and limitations of mathematical models. Given the important role models they are likely to play, our group felt it was important to make several statements that might clarify some of the misconceptions that are currently held about modeling.

1. "Models are cartoons of reality" [1]. This is one very important perspective to hold on models, because the user should realize that perfect predictions of real-world behavior are not possible. There are many approximations and assumptions that have to be made to model chemical fate in the environment. Only the dominant processes affecting chemical fate can be included. Thus, there will always be some difference between model results and natural occurrences, even if the processes being modeled are precise. The extent to which science can bring the "cartoon" in line with "reality" is the extent to which valid exposure assessments can be conducted.

2. Different users will not, in general, obtain identical results when applying the same model to the same set of data. This is due to the fact that considerable judgment is required in the application of any model. Thus, the soundness of model predictions is highly user-dependent. As one of the workshop participants, Dominic Di Toro, pointed out, models are not like radios in that anyone can plug them in and get beautiful music. Rather, they are like violins—tools that require practice and experience before one can obtain beautiful music.

3. The development of sound, tested models for chemical fate in aquatic systems will be a long-term effort. We are now capable of

making good predictions of dissolved oxygen concentrations in aquatic systems. The modeling of dissolved oxygen, however, began some 50 years ago with the Streeter-Phelps equation. Through successive comparisons of predictions with field and laboratory data, it was found that certain phenomena were being ignored. This discovery led to the addition of nitrogen cycling and algal interactions to existing models. More recently, an improved description of water/sediment interactions has also improved our ability to model dissolved oxygen. The same evolution is occurring in chemical fate modeling. Hopefully, past experience will considerably shorten the time frame for chemical fate model development.

4. Models offer a unique way of asking the question "what if . . . ?" For example, because of the potential risks, it is not feasible to release a newly developed chemical substance to the environment to see if it causes any unreasonable impact. Models are an ideal way of evaluating potential impacts a priori.

5. Models help in gaining understanding. Models provide an excellent framework for sorting through massive quantities of environmental data in a logical way. The use of a model in this way can help to identify cause-and-effect relationships that are not apparent when one simply reviews data. When used in this manner, models can act as repositories for the current state-of-the-art. Another way of gaining understanding with a model invovles parameter variation, sometimes called sensitivity analysis. This procedure can help the user determine which processes dominate chemical fate and which parameters are of importance. The final way in which models improve one's understanding is in their application. Differences between observed chemical concentrations and predicted concentrations force the user to seek an improved understanding of both the model and the real world.

6. Modeling provides a means of eliminating the vagueness inherent in decision-making. That is, a model can be used as a focal point for the clarification of issues and uncertainties. This is a very subtle but powerful use for modeling, especially in the area of hazard assessment.

One final comment relative to modeling of chemical fate is that efforts need to be made to obtain both field and laboratory data for modeling. Input data to drive most chemical fate models are lacking on a site-specific basis. Estimates of transfer and transformation rates are also lacking. This is having a major impact on the use of both simple and complex chemical fate models. Sound modeling of chemical fate in aquatic systems simply cannot be accomplished without data for model input and testing.

REFERENCES

1. Chow, V. T. "An Introduction to Systems Analysis of Hydrological Problems," paper presented at the Second International Seminar of Hydrology Professors, Utah State University, Logan, UT, August 2–14, 1970.

SECTION 2

MODELING PROCESSES

CHAPTER 7

IDENTIFICATION AND EVALUATION OF FUNDAMENTAL TRANSPORT AND TRANSFORMATION PROCESS MODELS

Lawrence A. Burns

Environmental Research Laboratory
Office of Research and Development
U.S. Environmental Protection Agency
Athens, Georgia 30613

Mathematical models designed to evaluate the behavior of chemicals in aquatic environments must include descriptions of the physical, chemical and biological processes governing the transport and fate of compounds. To be fully general, and therefore useful in a wide range of applications, these process models should be constructed as a combination of properties of the chemical and as functions of independently measurable environmental driving forces. This "second-order" or "system-independent" approach makes it possible to study the fundamental chemistry of compounds in the laboratory and then, based on independent studies of the levels of driving forces in aquatic systems, evaluate the probable behavior of the compound in systems that have never been exposed to it [1].

This approach depends for its success on the development of a thorough understanding of transport and transformation processes, with a secure base in fundamental theory and physical causation. It requires that the process chemist adopt an especially sophisticated and subtle experimental approach, for in general the reaction conditions and chemical concentrations in environmental situations are inconveniently benign, making modeling of process chemistry inordinately difficult—reactions are slow; extreme analytical precision is required to trace the time-course of reactions; reaction conditions must be characterized rigorously; and a mass balance often must be maintained to confirm the integrity of the experiment.

It is sometimes plausibly argued that coarser studies can serve to delineate the dominant fate processes of chemicals and that minor and slow pathways have little effect on the outcome of environmental evaluations. Although this argument has some merit, it also has several flaws. For one thing, the behavior of synthetic chemicals can undergo significant qualitative changes when the chemical is present at high concentrations. Thus, although the behavior of synthetic chemicals at high concentrations is of interest for evaluating chemical spills, accidents and waste depositories, at the dispersed concentrations more typical of aquatic systems, such studies can exaggerate the speed of transformation and degradation of dilute residuals.

Second, the next generation of exposure models will include the production of secondary "daughter" products via transformations occurring in the environment. The possibility of production of secondary compounds of high toxicity will force attention on the kinetics of these transformation pathways even when they are relatively slow and of little importance (in a mass-balance sense) to the overall fate of the parent compound. Finally, it should be remarked that when a slow transformation pathway is the only process that degrades a chemical in the new environment, slowness cannot be equated with insignificance if we wish to estimate the persistence of the compound in the global environment.

This chapter outlines the current state of knowledge of those aspects of transformation and transport processes required for inclusion in environmental models. It includes a brief description of many of the characteristic formulations current in the literature, some discussion of the environmental information required to bring them to bear on environmental situations, and some exploration of areas of ambiguity and opportunities for further research.

PHYSICAL PROCESSES

Physical processes include hydrodynamic transport, capture of synthetic chemicals by solid phases (sorption and its kinetics), transport of solid phases containing sorbed chemicals, transport across the benthic boundary layer and volatilization (and absorption) across the air/water interface.

Hydrodynamic Transport

Turbulent and advective water motions in aquatic systems have been studied for many years. The temporal and spatial scales of these studies are very diverse, ranging from the behavior of effluent streams to overall characterizations of mean circulation patterns in large lakes and the oceans. A thorough exploration of this vast body of material is far beyond the scope of this discussion; it is sufficient to say that hydrodynamic flows carry an entrained load of dissolved materials and that, when these dissolved materials include synthetic compounds, the transport of the chemicals can be derived readily by regarding the flow as a simple carrier. In many instances, a model for synthetic chemicals can be "piggy-backed" on a hydrodynamic transport model or on a transport description derived from observational data; this strategy is not uncommon in generalized water quality models.

The study of hydrodynamics is best developed for rivers [2], estuaries [3] and flowing water systems in general. Although the hydromechanics of ponds and lakes have not been totally neglected [4], and some lakes have received fairly intensive study [5], vertical transport in stratified water bodies continues to pose difficulties for general models. Transport across the thermocline is dominated in many instances by episodic events, including wind-induced eddies, storm surges and internal seiches [6]. In some studies, these transport events have been described successfully via a summary overall dispersion coefficient [7], that is, by analogy with Fickian diffusion. This approach may ultimately be supplemented with a stochastic description based on mixing events. Significant differences in predicted chemical residuals can be envisioned, for example, from rapid photolysis of chemicals episodically injected into sunlit surface waters from the (relatively dark) hypolimnion. This possibility clearly exists, but evaluation of its potential significance for exposure modeling must await comparisons between the results of appropriate alternative whole-system models.

Sorption

The capture of neutral organic compounds by nonliving sedimentary materials is a strong function of the organic carbon content of the solid phases [8]. Equilibrium sorption of these "hydrophobic" compounds has been described successfully via a carbon-normalized partition coefficient on both a whole-sediment basis and by particle-size classes. The success of the whole-sediment approach for natural sediments apparently derives from the fact that most naturally occurring sediment organic matter falls in the "silt" or "fine" particle-size fractions. So long as the dissolved concentration of the sorbing compound does not exceed 10^{-5} M, a linear isotherm (i.e., a partition coefficient) can be used to describe the phenomenon. At higher concentrations, the capacity of the solid to accept additional compound can be exceeded, and a nonlinear Freundlich or Langmuir isotherm must be invoked. One recent study [9] suggests that partition coefficients may decrease under large sediment/water ratios. If this result represents a general phenomenon, it will require careful additional study to derive a system-independent characterization and a causal explanation of the effect.

The "bound residue" phenomenon, with its implicit sediment desorption hysteresis, is a subject of particular concern for environmental models. Karickhoff [10] has proposed a two-step sorption model that may encompass most experimental observations of "bound residues." This model suggests that sorption is composed of a labile pool in continuous equilibrium with dissolved chemical, plus a more slowly exchanging pool that may reside in interior zones of the solid phases. The rate of release of sorbed compound from the refractory pool is inversely correlated with the overall partition coefficient of the compound, with desorption half-lives in some cases approaching 40 days.

This approach has several implications for environmental models. In the case of reactive compounds [i.e., compounds with short (<10- to 40-day) free-water half-lives], desorption kinetics can become a rate-limiting step in the overall degradation of the compound in some circumstances. This phenomenon is likely to be of considerable importance in the description of transformation reactions in bottom sediments, because slow desorption half-lives occur primarily for extensively sorbed compounds, which tend to be captured by benthic subsystems. On the other hand, a finite release time for sorbed chemical simplifies the computation of volatilization rates. Because the residence time of sorbing particles in the air/water interface zone is very short, the computation of dissolved, un-ionized species (subject to volatilization) near the air/water interface can be based simply on conditions in the bulk phase of the water column.

Sorption of ionizable compounds (organic acids and bases) has yet to be described in a fully system-independent fashion. Ionization itself is quite fast and can be described readily via an equilibrium approach using the pH of the water body and chemical ionization constants (pKa, pKb). Sorption of the resulting ions involves competition with normal inorganic and organic cations and anions. Sorption of organic cations can be described fairly well via a competitive Langmuir isotherm, using cation exchange capacity (CEC) as a general capacity factor. In this approach, the (compound-specific) parameters of the isotherm are determined on a single sediment, and sorption to other sediments is estimated via sediment pH and CEC [11]. Increased precision could be achieved via a detailed characterization of the mineralogy and capacity of each particle-size class, but the level of detailed environmental data required to implement such an approach in operational exposure models may be exceedingly difficult to attain.

Given partition coefficients for neutral and ionic species, computation of the (local) equilibrium distribution of a compound in a given segment of an aquatic system can be mechanized in a straightforward way [12,13]. Nonlinear isotherms and detailed sorption kinetics, however, have yet to be incorporated in operational exposure models.

One additional aspect of sorption phenomena deserves at least passing mention. The presence of dissolved organic matter [dissolved organic carbon (DOC)] in a water sample has been observed to "solubilize" some hydrophobic compounds [14]. This phenomenon probably derives from the same hydrophobic mechanisms driving sorption of neutral organics to solid phases, and the "solubilization" phenomenon has been described successfully via a simple partition coefficient [15]. Given the dissolved organic carbon content of a water body, it should be possible to describe the equilibrium association of hydrophobic compounds with DOC using conventional equilibrium partitioning methods.

Sediment Transport

Sediment transport, like hydrodynamic transport, has been of interest to the engineering profession for many years. Many comprehensive discussions of the dynamics of fluvial sediment transport have appeared in the literature [16,17]. As with hydrodynamic transport, one possible strategy for environmental modeling is to "piggy-back" the transport of sorbed synthetic chemicals on a model or description of transport of the sediment phase itself. The success of this strategy obviously depends on the facility with which the distribution of the compound between sorbed and dissolved phases can be described, because sediment transport dynamics depend strongly on particle-size dependencies of sorption. Models combining sediment and entrained chemical transport have already begun to appear [18,19].

In nonfluvial water bodies, interactions between benthic sediments and chemical contaminants often cannot be described simply via recourse to sediment transport dynamics. The simplest situation, and a fairly common one in most developed countries, occurs in run-of-the-river reservoirs. These systems are relatively impermanent landscape features because of their tendency to trap large quantities of riverborne sediments. The "trap efficiency" of reservoirs has been of interest to engineers for many years [20], and in this special case the capture of fluvial sediments and burial of synthetic compounds in reservoir benthic systems can probably be described fairly well [21].

Benthic Boundary Layer Transport

In many lakes and coastal seas, net sediment deposition can be very small. In these cases, capture of synthetic compounds may be driven primarily by the activities of the biota, which include physical disturbance by demersal fishes, irrigation of the sediments by benthic macrofauna [22] and sorption to surface layers followed by subduction and mixing of contaminated layers by crustacea and worm populations. In addition, physical stirring via sediment "bursting" [23,24] can facilitate exchange across the benthic interface. In general, transport across the benthic boundary layer can be described via an advection/dispersion equation [25], in which the advection term is used to account for groundwater flow, and the exchange events are described via a dispersion term.

The effect of benthic organisms on particle-size fractionation is of particular concern in exposure modeling. Under typical conditions, lacustrine oligochaete populations can be active to a depth of about 10–15 cm in the sediment deposits [26]. These organisms selectively ingest silts and fine particles and deposit a fecal layer on the surface of the sediment [27]. Most sediment organic matter, and thus most synthetic organic contaminants, are associated with these particle sizes. These facts imply that burial of contaminated sediment fractions may be retarded by the activities of burrowing organisms under slowly depositing regimes. Vertical size fractionation, although an acknowledged phenomenon [25], has not yet been incorporated explicitly into sediment transport or exposure models.

Estuarine systems are especially complex, because of their spatial mixture of fluvial, depositing, turbulent eroding and stable sediment regimes. In the James estuary, net sedimentation rates vary from < 1 to > 19 cm/yr; Kepone contamination accordingly varies from near-surface to burial depths greater than 50 cm [28]. Estuarine systems may afford significant research opportunities for testing alternative descriptions of exchange mechanisms across the benthic boundary layer in a wide variety of depositional regimes. The addition of sorption/desorption kinetics to sedimentation and benthic exchange models is a substantial research problem that thus far has received little attention.

Volatilization

The Whitman [29] two-film model for volatilization appears to work reasonably well for synthetic organic compounds. There are, however, at least six fairly distinct methods of computing transport resistance in the aqueous and gaseous phases of the air/water interface. In one of the initial applications of the Whitman model to environmental problems, Liss and Slater [30] suggested that simple kinetic theory could be used to index chemical transport resistances to piston velocities of water vapor and the normal dissolved gases. This idea leads to use of the square root of the molecular weight of the volatilization equations. A second method follows from strict application of the formal logic of the Whitman model; in this method, the diffusivity of the compound enters the equations as an index factor. In a third approach, derivations based on renewal theory or penetration theory result in the use of the square root of diffusivities. Dobbins [31] proposed a fourth approach, which is a hybrid of film and renewal theory, in which the root of diffusivity used in the equations varies between 0.5 and 1.0 as a function of hydrodynamic turbulence. More recently, Mackay et al. [32] have proposed direct use of compound-specific Schmidt numbers (incorporating viscosity and diffusivity) to compute transport resistances. Wolff and van der Heijde [33] have proposed a method for estimating boundary-layer thicknesses, thereby arriving at transport resistances via molecular diffusion constants.

In some ways, then, we have a superabundance of similar volatilization models; the time is certainly ripe for comparative field tests of these approaches. The required chemical data include the Henry's law constant, molecular weight and diffusivity of the compound in air and water. Some of these can be computed from other properties of the compound, e.g., computation of Henry's law constant via vapor pressure and solubility [34]. Partition coefficients and ionization constants are also needed when effects of sorption and ionization are significant. Environmental data needed for these models include reaeration rates, wind speeds, temperature, and, in some cases, wind fetch and water-current velocities.

The Whitman model has performed well in laboratory investigations. Because the flow fields in wind tunnels are much less turbulent than natural situations [35], the most immediate need for field testing of volatilization models is perhaps for application of the models to compounds whose total transport resistance is controlled by the gas phase.

The two-film approach has been regarded, in practice, as a method of computing average volatilization. Under these conditions, the Henry's law constant of a compound can be used to estimate the percentage of the total transport resistance lying in each phase of the air/water interface [36]. In the context of specific field validation sites, however, it should be remembered that strong diel shifts in transport resistance and its distribution between phases are possible. For example, volatilization of a compound that is nominally dominated by liquid-phase resistance could well be strongly retarded under very still atmospheric conditions (as can occur, for one case, in a small sheltered pond during nighttime hours), particularly if a shallow layer of stagnant contaminated air builds up over the surface of the water body.

CHEMICAL PROCESSES

Direct Photolysis

Direct photolysis, in which direct absorption of sunlight by a synthetic chemical induces transformation of the compound, has been studied intensively and incorporated into both process-specific [37] and more general [13] exposure models. The rate of sunlight absorption by the compound can be computed from the compound's absorption spectrum in pure water and sunlight intensity in a water body. The efficiency of conversion of the compound to daughter products can be evaluated in the laboratory [38] and coupled to the compound's rate of light absorption in the environment to yield the process transformation rate.

Computation of rates of direct photolysis requires a knowledge of spectral solar intensity at depth in the water body. These data can be computed from the solar spectrum at the surface of the water column and the absorption spectra of chlorophylls, natural dissolved organics, suspended organic detritus and sediments, and water itself. Thus, the chemical and environmental data needed to compute rates of direct photolysis in the environment are relatively easy to obtain.

Several research problems remain unresolved, however. The full effects of sorption on photolysis rates are imperfectly known. Surface sorption appears to accelerate immediate photolysis in some cases; this phenomenon may result from transfer of the chemical into a hydrocarbon-like microenvironment [39]. In addition, the effects of multistage sorption/desorption are not completely clear, although some experimental results can be rationalized via a two-stage desorption model [40]. Although it has been hypothesized that inorganic species present at high concentrations in marine water (e.g., chloride) may affect photolysis of organics, the possibility of such photonucleophilic reactions in marine waters has not been investigated systematically [41].

Finally, some nonlinear behaviors have been observed in direct photolysis studies, but have yet to be installed in operational models. As a fairly trivial example, the contribution of a pollutant to the overall absorption spectrum of a water body has generally been neglected; this phenomenon is probably important only for undiluted waste streams or heavily polluted systems. A less trivial example is the "quenching" of excited states of the chemical by dissolved oxygen. In practice, this problem has been overcome by measuring reaction quantum yields in air-saturated water [42], but the rates of direct photolysis in oxygen-deficient sunlit waters (a presumably unusual condition) may be somewhat understated. Finally, some instances of photosensitization by a photoproduct have been described [43]; the generality and environmental significance of this kind of nonlinearity has not been addressed quantitatively by process chemists or modelers.

Indirect Photolysis

Direct photolysis is not the sole pollutant transformation process driven by the solar flux in aquatic systems. The simultaneous occurrence of plant decomposition products (humic materials), dissolved oxygen and sunlight often results in an acceleration of the rate of transformation of organic pollutants. Zepp et al. [44], for example, found that methoxychlor, with a direct photolysis half-life of more than 300 hr, had a half-life of as little as 2.2 hr under irradiation in natural waters containing dissolved humic materials. Further, Ross and Crosby [45] found that a solution of aldrin in water from a taro paddy can be converted photochemically to dieldrin, despite the fact that aldrin does not absorb sunlight.

These kinds of reactions are usually termed "indirect" or "sensitized" photolysis. Indirect photolysis can be subdivided into two general classes of reactions. First, "sensitized photolysis" per se involves sunlight absorption and electronic excitation of a sensitizer (humic) molecule, followed by direct chemical interaction between the excited state of the sensitizer and a pollutant molecule. Possible chemical reactions include a direct energy transfer to the pollutant molecule, hydrogen atom transfer from pollutant to sensitizer to give free radicals, and union of sensitizer and pollutant yielding an excited-state complex or "exciplex" [41]. The resulting free radicals or exciplexes can then react with dissolved molecular oxygen, a process termed "type I sensitized photooxidation" by Zepp and Baughman [41].

The second class of indirect photolysis involves formation of chemical oxidants in natural waters, primarily via the interaction of sunlight, humic materials and dissolved oxygen (type II sensitized photooxidation of Zepp and Baughman [41]). The primary oxidants known to occur in natural waters are hydroxyl and peroxy radicals [46] and singlet oxygen [47]. Alkoxy radicals and diradicals may also contribute to environmental oxidation of some compounds, but their presence in natural waters has not been demonstrated conclusively [48]. Process mechanics and routes of transformation for one indirect pathway, photooxidation mediated by humic materials and singlet oxygen, have been described for air-saturated waters [49–51]. Much remains to be done, however, before a fully general treatment of indirect photochemical processes can be incorporated into routine exposure assessment models.

Hydrolysis

Hydrolysis of synthetic organic compounds also occurs via a variety of pathways. Specific-acid and -base-catalyzed processes can be easily described in terms of the pH and temperature of water bodies [52]. Neutral hydrolysis (i.e., pH-independent mechanisms) can proceed either via direct reactions involving the water molecule in the rate-limiting step, or via indirect molecular transformation involving water as a reactant. (Both are apparently first-order reactions in pure water solution.) The required chemical information can be encapsulated in a pH-rate profile (Figure 1) [53,54]; hydrolysis kinetics are readily described via second-order equations incorporating the effects of temperature and, where required, ionization on chemical reactivity.

In general, sorption to sediment phases would be expected to retard the rate of chemical transformations requiring the presence of water. Sorption has been observed to retard reactions involving hydroxide or hydronium ion in the rate-limiting step (specific-acid and -base hydrolysis). Neutral reactions, however, appear to be unaffected by sorption, at least in sediment suspensions of up to 20 g/l, for a variety of neutral pathways [55]. This behavior may not hold in fully benthic subsystems where the sediment/water ratio can be much greater than 20 g/l, but appropriate experimental investigations have not yet been accomplished. In addition, metal ion catalysis of hydrolytic pathways is known to occur

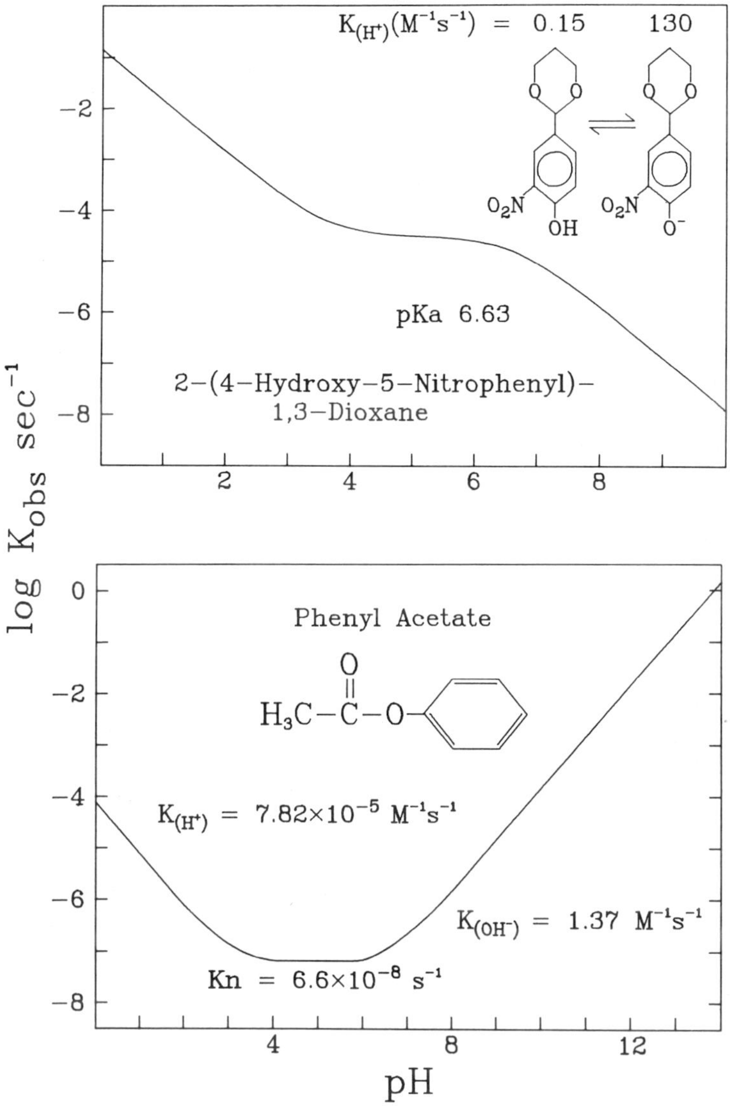

Figure 1. Hydrolysis pH-rate profiles of a substituted 2-phenyl-1, 3-dioxane (HND) (top); and phenyl acetate (bottom). HND profile constructed from data of Bender and Silver [53]; phenyl acetate profile constructed from data of Mabey and Mill [54]. HND hydrolyzes via an acid-catalyzed pathway; the phenolate anion is some 867 times more reactive than its conjugate acid. Phenyl acetate reacts via specific-acid-, neutral- and base-catalyzed transformation pathways.

for some compounds, but the environmental significance of this pathway is still imperfectly understood.

General acid/base catalysis is perhaps of less significance in natural freshwaters, although probably of some importance in special situations. This phenomenon can be described fairly well via the Bronsted law (relating rate constants to pK of general acids and bases) and probably can be incorporated into exposure models in the near future. Maximum rates of general acid/base catalysis can be deduced from a compound's specific-acid/base hydrolysis behavior, and actual rates can be determined from relatively simple laboratory experiments [56].

A number of cases are known of apparently sediment-catalyzed transformations. These reactions are inhibited by heat sterilization of the system, but are unaffected by mercuric chloride or azide bacteriostasis. One possible explanation of this phenomenon involves the presence of exoenzymes activated by attachment to sediment particles. This pathway is in need of considerable additional research if it is to be described in a system-independent way and incorporated into exposure assessment models.

Reduction

Reductive dechlorination of organic compounds (e.g., Toxaphene and DDT) in anaerobic soils and sediments has been observed by many investigators (e.g., the summary in Tinsley [57]). Although the redox state of these systems can be described via redox potential and pE, the suitability of this thermodynamic quantity as an accurate predictor of environmental kinetics is very uncertain. In addition, some reductive pathways are catalyzed by ferrous iron. The mechanics, pathways and kinetics of anaerobic reductions of synthetic organic compounds currently require substantial additional process research. At the moment, it is not even clear whether these pathways are intrinsically a biological phenomenon or a chemical phenomenon occurring under reaction conditions produced by the activities of the biota.

BIOLOGICAL PROCESSES

Microbial Transformations

Microbial communities are ubiquitous in almost all aquatic ecosystems. Microbiota play a central role in the remineralization of plant and animal debris; they have evolved the capacity to transform and harvest energy from an immense array of naturally occurring organic compounds. This ability does not extend to all synthetic compounds, however, and it is now "axiomatic that microorganisms are fallible and that many synthetic organic compounds are recalcitrant . . . and accumulating in some environments" [58]. In consequence, qualitative "biodegradability" tests [59] are now routine in the detergent industry, and "sludge" tests have been suggested [60] as a routine procedure for evaluating industrial synthetic organic compounds. The latter tests can also provide some assurance that the compound will not destroy the microbial communities that serve man in sewage treatment plants, as well as providing an evaluation of the biodegradability of the compound [61].

The discovery of microbial fallibility has led to a more critical and quantitative view of microbial metabolism of synethtic organics; it no longer suffices to know that a compound is "biodegradable." Instead, the rates, products, microbial populations and environmental conditions surrounding an observed biodegradation are of increasing interest to the microbiological and ecological community. Given that the natural microbial substrates can massively accumulate in special circumstances (e.g., peat bogs and swamps), a degree of skepticism is obviously warranted toward any claim of universal, unconditional biodegradability of a synthetic organic compound. The rates at which a compound is biodegraded must depend on the structure of the compound and the metabolic capacities of the microbial community resident in the ecosystem receiving the compound.

Microbial communities derive energy for metabolism and growth from the breakdown of organic compounds, and the kinetics of growth and substrate utilization have been of interest to microbiologists for many years. Population growth and substrate utilization rate have been described via Monod's analogy [62] with Michaelis-Menten enzyme kinetics [63]. In this approach, the growth of a microbial population in a nonlimiting environment is described by:

$$dN/dt = u\,N \tag{1}$$

where u = "specific growth rate"
N = microbial biomass or population size

The Monod equation modifies this formulation via the recognition that consumption of resources in a finite environment must at some point curtail the rate of increase (dN/dt) of the population. This modification is implemented via:

$$u = u(\max) \frac{\{S\}}{Ks + \{S\}} \tag{2}$$

where {S} = the concentration of the growth limiting substrate
u(max) = the "maximum specific growth rate" obtained when {S} is present in excess (i.e., nonlimiting)
Ks = the "saturation constant," which is that value of {S} allowing the population to grow at rate u(max)/2

An equation describing the behavior of the growth substrate {S} over time, and thus by implication the dynamics of a biodegradable synthetic compound, follows via a simple derivation [63]: assuming that a fixed amount of growth results from metabolism (and therefore loss from solution) of a unit quantity of S, then

$$dN = -Y\,dS \tag{3}$$

where Y equals the "yield coefficient" in cells or biomass produced per unit S metabolized. Taking:

$$\frac{dS}{dt} = \frac{dS}{dN}\frac{dN}{dt} \tag{4}$$

gives:

$$dS/dt = -(u/Y)\,N \tag{5}$$

or, substituting u(max){S}/(Ks + {S}) for u,

$$\frac{ds}{dt} = -\frac{u(\max)}{Y}\frac{\{S\}}{Ks + \{S\}} N \tag{6}$$

The observed microbial growth yield Y is often calculated from the amount of microbial biomass produced during the course of transformation of a measured quantity of substrate. Because a microbial population must satisfy maintenance requirements before any growth can occur, the apparent yield Y can fall quite drastically at low {S}. Expressions with separated growth and maintenance yields have been developed [64], and it is possible to compute separately the magnitudes of growth and maintenance yields from microbial growth experiments [63].

The Monod formulation has been applied successfully to biotransformation of synthetic chemicals in a laboratory setting. These studies have demonstrated that a Monod analysis need not be restricted to single-species populations; that is, the Monod equations can serve as adequate descriptors of substrate transformation by "mixed" or "heterogeneous" (i.e., multispecies) microbial populations. In one example, Paris et al. [65] isolated a four-species consortium of organisms able to use malathion as its sole source of carbon. Analysis of the growth response of the population and of concurrent malathion transformation rates then gave the Monod parameters u(max) = 0.37/hr, Ks = 2.17 μM/l (0.716 mg/l) and Y = 4.1×10^{10} cell/μM (1.2×10^{11} cell/mg). In a similar study, Smith et al. [66] investigated the degradation of *p*-cresol by mixed microbial cultures able to use the compound as a sole carbon source. These authors expressed their results in terms of Monod parameters, finding u(max) = 0.62/hr, Y = 1.8×10^{9} cell/mg and Ks = 0.84 mg/l.

Unfortunately, these elegant applications of classical microbiological methods to biotransformation (or "biolysis") of synthetic organic compounds are difficult to apply in a broader ecological context. The first difficulty, which has received some attention in the microbiological literature, is primarily mechanical: the Monod formulation is nonlinear in its parameters and thus imposes a high cost in computation time when used in a computer program or mathematical model. For trace concentrations of pollutants (i.e., $\{S\} \ll$ Ks), however, the term Ks + $\{S\}$ in the denominator of the Monod equation can be approximated by Ks, by giving the linear approximation:

$$\frac{d\{S\}}{dt} = -\frac{u(\max)}{Y\,Ks} N\{S\} \tag{7}$$

This formulation is similar to the "second-order" equations used to describe the kinetics of chemical reactions, and the term u(max)/Y*Ks can be termed, by analogy, a second-order biolysis rate constant Kb2, with units /hr/(cell/l) when population sizes N are expressed in cell/l.

The propriety of this substitution can be readily evaluated. For malathion, u(max)/(Y*Ks) = $0.37/(4.1 \times 10^{11} * 2.17) = 4.16 \times 10^{-12}$/hr/(cell/l)]. Paris et al. [65] also computed values of Kb2 from microbial population sizes and malathion transformation rates in their experimental studies, finding a mean value for Kb2 of $2.6 \pm 0.7 \times 10^{-12}$/hr/(cell/l) (95% confidence interval) for a series of eight experimental determinations spanning a malathion concentration range of 0.0273–0.33 μM. The measured Kb2 differed from u(max)/(Y*Ks) by less than a factor of 2, suggesting that simplified kinetic experiments could be used to develop second-order rate constants, in lieu of a detailed elaboration of u(max), Y and Ks, when the region of concern is restricted to levels of compound $\ll$ Ks.

Furthermore, dissolved pesticide concentrations in surface waters are often very low indeed. For example, dieldrin, lindane and DDT have been found primarily at nanogram-per-liter levels both in the United States [67] and in Britain [68]. For industrial chemicals, however, the situation can be somewhat different. For example, the release of phenolic wastes into the St. Lawrence River results in riverine concentrations of 0.01–0.15 mg/l [69]. Although these high concentrations are restricted to a dispersion cone immediately downstream of the effluents, Ks for *p*-cresol (0.84 mg/l) is uncomfortably close to the highest measured concentrations.

The need for a thorough evaluation of the adequacy of a linear approximation to the Monod formula disappears, however, on examination of the conceptual difficulties standing in the way of the application of either equation to environmental situations. A natural microbial community derives its energy from a large variety of organic detrital materials. A microbial species restricted to a trace-level synthetic compound as its sole carbon source would be at a severe competitive disadvantage; there is no way of predicting a priori the population densities the degrader could attain in real systems. Even when a synthetic compound is sufficiently similar to natural substrates that it can be degraded indifferently by the microbial community as a whole, the presence of multiple energy-yielding substrates in the environment violates a fundamental assumption underlying the Monod approach; {S}, the synthetic compound, limits growth. In many instances, moreover, a compound may be transformed or degraded in an energy-requiring detoxification process, making the concept of cell yield (Y) of dubious utility. When the compound, for whatever reason, is degraded in the absence of a change in population size, the apparent zero yield forces abandonment of either version of the Monod equation. (This phenomenon is sometimes called "cometabolism" [70,71].)

Despite these difficulties, the rate of transformation of organic pollutants must depend on the structure of the compound and the metabolic capacity of microbial communities. The simplest expression of this duality is the second-order equation:

$$d\{S\}/dt = -Kb2\,B\,\{S\} \tag{8}$$

which asserts that the rate of biolysis (d{S}/dt) is first-order in compound concentration {S} and in microbial activity B, and in which the identification of Kb2 with u(max)/(Y*Ks) is discarded. This approach requires that the second-order rate constant be determined via laboratory studies of relatively undisturbed samples drawn from natural microbial communities. It also requires demonstration that d{S}/dt is linearly proportional to {S} for fixed levels of B, and that appropriate measures of B be found.

One conventional microbiological technique for characterizing microbial populations is simple population size measured, for example, by colony counts on pour-plate agars [72]. To the extent that population size is an adequate measure of community metabolic capacity, simple microbiological techniques could serve to characterize the rate of biotransformation of synthetic compounds in natural environments. This hypothesis has been explored for three compounds (butoxyethyl ester of 2,4-D (2,4-DBE), malathion and chlorpropham) by Paris and co-workers [73,74]. These authors investigated biotransformation of their study compounds in samples of natural waters drawn from 40 locations in the continental United States. Ambient water temperatures at the collection points ranged from 1 to 29 °C, and the laboratory studies were conducted at the observed ambient temperature of the sampled environments. Ambient bacterial populations, as measured by 48-hr incubation on tryptone glucose extract (TGE) agar at 22 °C, ranged from 4×10^2 to 9×10^5 cell/ml [74]. Biolysis of these chemicals was demonstrably first-order in compound concentration and in population size (Figure 2) and was relatively independent of temperature. The 95% confidence intervals (based on between-site variation) from Kb2 (/hr/(cell/l)) and number of sites for these compounds were $5.42 \pm 0.97 \times 10^{-10}$ (2,4-DBE, 31 sites), $4.54 \pm 0.74 \times 10^{-11}$ (malathion, 14 sites) and $2.63 \pm 0.72 \times 10^{-14}$ (chlorpropham, 11 sites).

Evaluation of biolysis in such "river die-away" studies is particularly difficult when a combination of small populations and slow biotransformation rates leads to inconveniently long biolysis half-lives. In such cases, population densities must be augmented via centrifugation or supplemental nutrient broth. The latter procedure can have the disadvantage of favoring opportunistic heterotrophs at the expense of organisms better equipped to degrade more recalcitrant compounds. For example, the second-order biolysis rate constant (Kb2) for chlorpropham is unaffected when population densities are augmented via nutrient broth supplementation, but the measured Kb2 for phenanthrene decreases in approximate proportion with the population increase [75].

Steen et al. [76] reported second-order rate constants for aerobic biotransformation of synthetic chemicals by microbial populations derived from aquatic sediments. The observed rate constants for 2,4-DBE and malathion were consistent with the rate of degradation by water-column populations, at Kb2 of 2.3×10^{-10} and 4.0×10^{-11}/hr/(cell/l), respectively. Aerobic biolysis of chlorpropham by the sediment-derived populations, however, was almost an order of magnitude faster

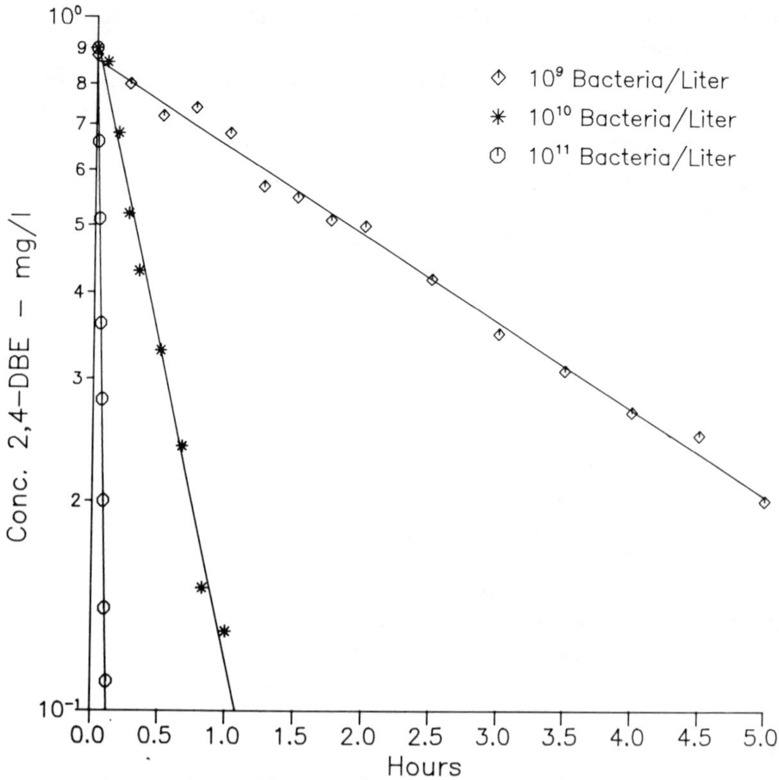

Figure 2. Microbial transformation kinetics of butoxyethyl ester or 2,4-dichlorophenoxyacetic acid (2,4-DBE). Bacterial populations from Overlook Lake, Georgia. Population densities measured via 48-hr incubations at 22°C on TGE agar [74].

[$Kb2 = 1.42 \times 10^{-13}$/hr/(cell/l)] than biolysis by the water-column populations [$Kb2 = 2.4 \times 10^{-14}$/hr/(cell/l)]. By varying the sediment/water ratio, these authors also adduced evidence suggesting that chlorpropham, methoxychlor and several phthalate esters, when sorbed to suspended sediments, became relatively unavailable to biolytic organisms.

Plate-count estimates of bacterial population densities are extremely selective and may in some circumstances underestimate total population sizes by three or more orders of magnitude [6,77,78]. Furthermore, although increases in pollutant flux (i.e., mg/l-time or mass reacted/time) with increasing chemical concentration (consonant with pseudo-first-order kinetics) have been observed in nature [79], it is also true that, at elevated pollutant concentrations, both a reduction in the apparent first-order rate consonant [57], and zero-order kinetics [69] have been observed. The latter phenomena are consonant (mathematically) with a Monod formulation. It is not clear, however, whether these kinetics are a reflection of the maximum growth rate of a specialized subpopulation of degraders or result from toxic effects of elevated concentrations on the microbial community.

In any case, it is clear that the studies of second-order kinetics cited above are a useful beginning in the detailed quantitative study of rates of biotransformation of pollutant chemicals. Little more extrapolative power can be developed until such time as the mechanisms of biotransformations, and the environmental coupling variables that govern the metabolic capacity of natural and laboratory microbial populations, have rigorously been determined and measured in both kinds of systems. Presumably metabolic capacity is related to the nutrient status of the cells, which in turn is related to the nutrient history of the water body. Functional descriptions of these relationships are an important subject for additional research. Furthermore, some compounds are degraded only by specific subpopulations of the natural microbial community. Although the second-order rate constant describing the biolysis capacity of specific transforming species is insensitive to the growth phase of the population (Figure 3) [80], it is not currently possible to predict the population density of functional subpopulations in natural communities. Finally, the in situ biolysis capacities of sediment microbial populations have not been characterized fully: fungi, aerobic bacteria and anaerobes may play important roles in benthic subsystems.

Algal Transformations

Little can be said about algal metabolism of synthetic organics beyond the simple observation that the phenomenon is known to occur [81]. The biolysis capacity of algae, its relationship to photosynthetic rates or light intensities, nutrient status, and other presumptive influences on this pathway, are all open areas for process-level research. Although perhaps of relatively slight significance to the overall fate of most compounds, algal degradation may in effect limit the degree of food-chain transmission of pollutants in some cases.

Biosorption and Bioaccumulation

Although perhaps somewhat beyond the scope of this discussion, transfer of synthetic organics to living organisms deserves at least passing mention. In most cases, the biota constitute a sufficiently small fraction

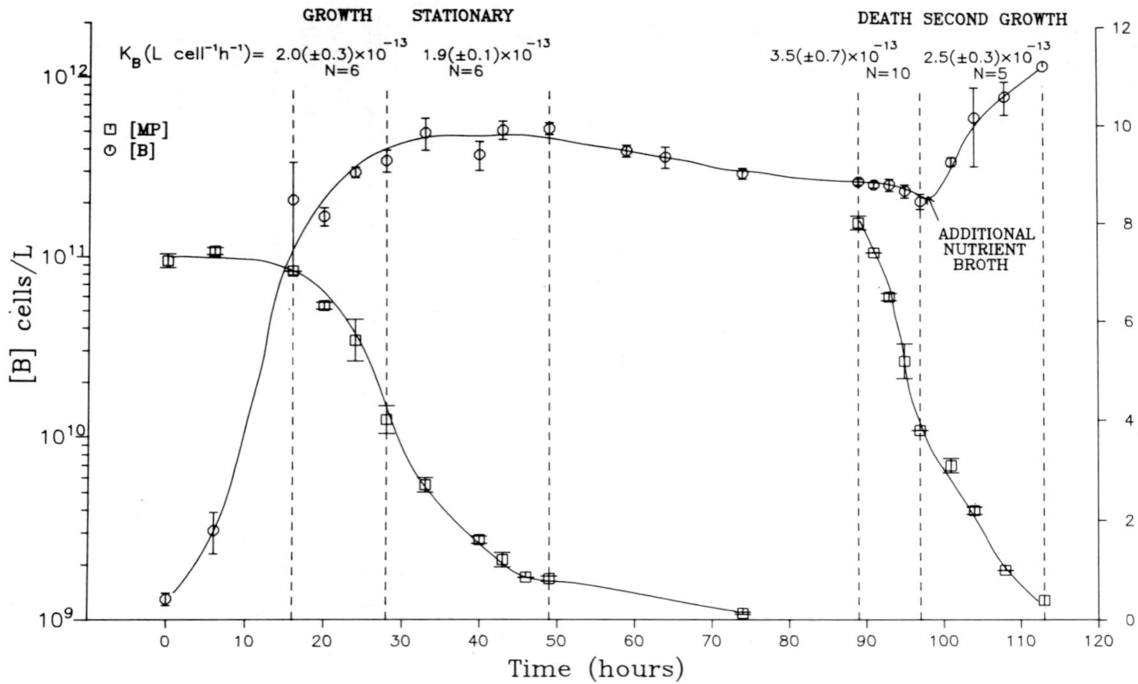

Figure 3. Time course of transformation of methyl parathion by *Flavobacterium aquatile* in nutrient broth during a variety of growth phases of the experimental population. Note relative stability of the second-order biolysis rate constant (courtesy D. L. Lewis [80]).

of the total volume of the system, so that the total mass of chemical associated with it is negligible. In a total mass balance sense, then, transport vectors (e.g., fish migration and diel vertical migration of zooplankton) usually have little effect on the overall dynamics of a synthetic organic pollutant. Still, transfer of pollutants up the food chain, and direct pollutant transfer to otherwise uncontaminated populations (e.g., human consumption of contaminated fish), suggest that this topic should not be neglected altogether in discussions of aquatic fate models. In the case of microorganisms, with their very high surface-to-volume ratio, capture of synthetic organics can be described via a simple partition coefficient [82]. Simple partitioning probably can be used to describe equilibrium uptake of synthetic organics by the vast majority of organisms occupying the base of aquatic food chains. The uptake of pollutant chemicals by higher organisms is complicated by the presence of multiple dietary and direct routes of exposure, including transport across gill membranes and, at least for marine fishes, direct water intake. Clearly the significance of partitioning phenomena, as compared with dietary uptake, must differ with the physiology and life history of the individual type of organism affected. For example, blue crabs (*Callinectes*) appear to take up little or no Kepone via direct absorption from seawater, but readily accumulate this compound via dietary intake [83]. Although simple biosorption or bioconcentration factors are adequate for many purposes, detailed exposure models probably should include a capability for representing dietary exposures, biosorption or membrane transfers, and depuration kinetics in at least a generalized sense. Some effort has been devoted to a "bioenergetic" dietary model [84]. The logic of this approach is similar to that used for human exposures; it holds considerable promise for inclusion in exposure models.

REFERENCES

1. Baughman, G. L., and L. A. Burns. "Transport and Transformation of Chemicals: A Perspective," in *The Handbook of Environmental Chemistry, Vol. 2, Part A,* O. Hutzinger, Ed. (New York: Springer-Verlag, 1980).
2. Shen, H. W., Ed. *Modeling of Rivers* (New York: John Wiley & Sons, Inc., 1979).
3. Nihoul, J. C. J., Ed. *Hydrodynamics of Estuaries and Fjords: Proceedings of the 9th International Liege Colloquium on Ocean Hydrodynamics* (Amsterdam: Elsevier Scientific Publishing Co., 1978).
4. Gibbs, R. J., Ed. *Transport Processes in Lakes and Oceans* (New York: Plenum Publishing Corporation, 1977).
5. Li, Y.-H. "Vertical Eddy Diffusion Coefficient in Lake Zurich," *Schweiz. Zeit. Hydrol.* 35:1–7 (1973).

6. Wetzel, R. G. *Limnology* (Philadelphia: W. B. Saunders Company, 1975).
7. Snodgrass, W. J. "Relationship of Vertical Transport Across the Thermocline to Oxygen and Phosphorus Regimes: Lake Ontario as a Prototype," in *Transport Processes in Lakes and Oceans,* R. J. Gibbs, Ed. (New York: Plenum Publishing Corporation, 1977), pp. 179–202.
8. Karickhoff, S. W., D. S. Brown and T. A. Scott. "Sorption of Hydrophobic Pollutants on Natural Sediments," *Water Res.* 13:241–248 (1979).
9. O'Connor, D. J., and J. P. Connolly. "The Effect of Concentration of Adsorbing Solids on the Partition Coefficient," *Water Res.* (in press).
10. Karickhoff, S. W. "Sorption Kinetics of Hydrophobic Pollutants in Natural Sediments," in *Contaminants and Sediments, Vol. 2, Analysis, Chemistry, Biology,* R. A. Baker, Ed. (Ann Arbor, MI: Ann Arbor Science Publishers, Inc., 1980), pp. 193–205.
11. Brown, D. S. "A Model for Predicting Sorption of Organic Cations in Soils and Sediments" (in preparation).
12. Stumm, W., and J. J. Morgan. *Aquatic Chemistry—An Introduction Emphasizing Chemical Equilibria in Natural Waters* (New York: Wiley-Interscience, 1970).
13. Burns, L. A., D. M. Cline and R. R. Lassiter. "Exposure Analysis Modeling System (EXAMS): User Manual and System Documentation," U.S. EPA Environmental Research Laboratory, Athens, GA (1981).
14. Wershaw, R. L., P. J. Burcar and M. C. Goldberg. "Interaction of Pesticides with Natural Organic Material," *Environ. Sci. Technol.* 3:271–273 (1969).
15. Perdue, E. M., and N. L. Wolfe. "Modification of Pollutant Hydrolysis Kinetics in the Presence of Humic Substances," *Environ. Sci. Technol.* (submitted).
16. Simons, D. B., and F. Senturk. *Sediment Transport Technology* (Ft. Collins, CO: Water Resources Publications, 1977).
17. Garde, R. J., and K. G. R. Raju. *Mechanics of Sediment Transportation and Alluvial Stream Problems* (New York: John Wiley & Sons, Inc., 1977).
18. Onishi, Y., D. L. Schreiber and R. B. Codell. "Mathematical Simulation of Sediment and Radionuclide Transport in the Clinch River, Tennessee," in *Contaminants and Sediments, Vol. 1, Fate and Transport, Case Studies, Modeling, Toxicity,* R. A. Baker, Ed. (Ann Arbor, MI: Ann Arbor Science Publishers, Inc., 1980), pp. 393–406.
19. Uchrin, C. G., and W. J. Weber, Jr. "Modeling of Transport Processes for Suspended Solids and Associated Pollutants in River-Harbor-Lake Systems," in *Contaminants and Sediments, Vol. 1, Fate and Transport, Case Studies, Modeling, Toxicity,* R. A. Baker, Ed. (Ann Arbor, MI: Ann Arbor Science Publishers, Inc., 1980), pp. 407–425.
20. Borland, W. M. "Reservoir Sedimentation," in *River Mechanics,* H. W. Shen, Ed. (Ft. Collins, CO: H. W. Shen, 1971), pp. 29–38.
21. Schnoor, J. L. "Fate and Transport of Dieldrin in Coralville Reservoir: Residues in Fish and Water Following a Pesticide Ban," *Science* 211:840–842 (1981).
22. Aller, R. C. "The Influence of Macrobenthos on Chemical Diagenesis of Marine Sediments," PhD Thesis, Yale University, New Haven, CT (1977).
23. Heathershaw, A. D. "'Bursting' Phenomena in the Sea," *Nature* 248:394–395 (1974).

24. Heathershaw, A. D. "Measurements of Turbulence in the Irish Sea Benthic Boundary Layer," in *The Benthic Boundary Layer,* I. N. McCave, Ed. (New York: Plenum Publishing Corporation, 1976), pp. 11-31.
25. Berner, R. A. *Early Diagenesis—A Theoretical Approach* (Princeton, NJ: Princeton University Press, 1980).
26. Jones, B. F. and C. J. Bowser. "The Mineralogy and Related Chemistry of Lake Sediments," in *Lakes—Chemistry, Geology, and Physics,* A. Lerman, Ed. (New York: Springer-Verlag, 1978), pp. 179-235.
27. Fisher, J. B. "Effects of Tubificid Oligochaetes on Sediment Movement and the Movement of Materials across the Sediment-Water Interface," PhD Thesis, Case Western Reserve University, Cleveland, OH (1979).
28. Custhall, N. H., I. L. Larsen and M. N. Nichols. "Man-made Radionuclides Confirm Rapid Burial of Kepone in James River Sediments," *Science* 213:440-442 (1981).
29. Whitman, W. G. "A Preliminary Experimental Confirmation of the Two-Film Theory of Gas Absorption," *Chem. Metal. Eng.* 29:146-148 (1923).
30. Liss, P. S., and P. G. Slater. "Flux of Gases across the Air-Sea Interface," *Nature* 247:181-184 (1974).
31. Dobbins, W. E. "Mechanism of Gas Absorption by Turbulent Liquids," in *Advances in Water Pollution Research, Proceedings of the International Conference, Vol. 2,* W. W. Eckenfelder, Ed. (New York: The MacMillan Co., 1964), pp. 61-76.
32. Mackay, D., W. Y. Shiu, A. Bobra, J. Billington, E. Chau, A. Yeun, C. Ng and F. Szeto. "Volatilization of Organic Pollutants from Water," U.S. EPA Project Report, Contract No. R805150010 (1980).
33. Wolff, C. J. M., and H. B. van der Heijde. "A Model to Assess the Rate of Evaporation of Chemical Compounds from Surface Waters," *Chemosphere* (submitted).
34. Mackay, D., and P. J. Leinonen. "Rate of Evaporation of Low-Solubility Contaminants from Water Bodies to Atmosphere," *Environ. Sci. Technol.* 9:1178-1180 (1975).
35. Thibodeaux, L. J. *Chemodynamics—Environmental Movement of Chemicals in Air, Water, and Soil* (New York: John Wiley & Sons, Inc., 1979).
36. Mackay, D. "Volatilization of Pollutants from Water," in *Aquatic Pollutants: Transformation and Biological Effects,* O. Hutzinger, I. H. van Lelyveld and B. C. J. Zoeteman, Eds. (Oxford: Pergamon Press, 1978), pp. 175-185.
37. Zepp, R. G., and D. M. Cline. "Rates of Direct Photolysis in Aquatic Environment," *Environ. Sci. Technol.* 11:359-366 (1977).
38. Zepp, R. G. "Quantum Yields for Reaction of Pollutants in Dilute Aqueous Solution," *Environ. Sci. Technol.* 12:327-329 (1978).
39. Miller, G. C., and R. G. Zepp. "Photoreactivity of Aquatic Pollutants Sorbed on Suspended Sediments," *Environ. Sci. Technol.* 13:860-863 (1979).
40. Zepp, R. G., and P. F. Schlotzhauer. "Effects of Equilibration Time on Photoreactivity of the Pollutant DDE Sorbed on Natural Sediments," *Chemosphere* 10:453-460 (1981).
41. Zepp, R. G., and G. L. Baughman. "Prediction of Photochemical Transformation of Pollutants in the Aquatic Environment," in *Aquatic Pollutants: Transformation and Biological Effects,* O. Hutzinger, I. H. van

Lelyveld and B. C. J. Zoeteman, Eds. (Oxford: Pergamon Press, 1978), pp. 237–263.

42. Zepp, R. G. "Assessing the Photochemistry of Organic Pollutants in Aquatic Environments," in *Dynamics, Exposure and Hazard Assessment of Toxic Chemicals,* R. Haque, Ed. (Ann Arbor, MI: Ann Arbor Science Publishers, Inc., 1980), pp. 69–110.
43. Wagner, P. J., I. E. Kochevar and A. E. Kemppainen. "Type II Photoprocesses of Phenyl Ketones: Procedures for Determining Meaningful Quantum Yields and Triplet Lifetimes," *J. Am. Chem. Soc.* 94:7489–7494 (1972).
44. Zepp, R. G., N. L. Wolfe, L. V. Azarraga, R. H. Cox and C. W. Pape. "Photochemical Transformation of the DDT and Methoxychlor Degradation Products, DDE and DMDE, by Sunlight," *Arch. Environ. Contam. Toxicol.* 6:305–314 (1977).
45. Ross, R. D., and D. G. Crosby. "The Photooxidation of Aldrin in Water," *Chemosphere* 4:277–282 (1975).
46. Mill, T., D. G. Hendry and H. Richardson. "Free-Radical Oxidants in Natural Waters," *Science* 207:886–887 (1980).
47. Zepp, R. G., N. L. Wolfe, G. L. Baughman and R. C. Hollis. "Singlet Oxygen in Natural Waters," *Nature* 267:421–423 (1977).
48. Mill, T. "Chemical and Photo Oxidation," in *The Handbook of Environmental Chemistry, Vol. 2, Part A,* O. Hutzinger, Ed. (Berlin: Springer Verlag, 1980), pp. 77–105.
49. Zepp, R. G., G. L. Baughman and P. F. Schlotzhauer. "Comparison of Photochemical Behavior of Various Humic Substances in Water: I. Sunlight Induced Reactions of Aquatic Pollutants Photosensitized by Humic Substances," *Chemosphere* 10:109–117 (1981).
50. Zepp, R. G., G. L. Baughman and P. F. Schlotzhauer. "Comparison of Photochemical Behavior of Various Humic Substances in Water: II. Photosensitized Oxygenations," *Chemosphere* 10:119–126 (1981).
51. Zepp, R. G., and P. F. Schlotzhauer. "Comparison of Photochemical Behavior of Various Humic Substances in Water: III. Spectroscopic Properties of Humic Substances," *Chemosphere* 10:479–486 (1981).
52. Wolfe, N. L. "Determining the Role of Hydrolysis in the Fate of Organics in Natural Waters," in *Dynamics, Exposure and Hazard Assessment of Toxic Chemicals,* R. Haque, Ed. (Ann Arbor, MI: Ann Arbor Science Publishers, Inc., 1980), pp. 163–178.
53. Bender, M. L., and M. S. Silver. "The Hydrolysis of Substituted 2-Phenyl-1,3-dioxanes," *J. Am. Chem. Soc.* 85:3006–3010 (1963).
54. Mabey, W., and T. Mill. "Critical Review of Hydrolysis of Organic Compounds in Water under Environmental Conditions," *J. Phys. Chem. Ref. Data* 7:383–345 (1978).
55. Wolfe, N. L. Personal communication.
56. Perdue, E. M., N. L. Wolfe and S. W. Karickhoff. "Prediction of Buffer Catalysis in Field and Laboratory Studies of Pollutant Hydrolysis Reactions," *Environ. Sci. Technol.* (submitted).
57. Tinsley, I. J. *Chemical Concepts in Pollutant Behavior* (New York: Wiley-Interscience, 1979).
58. Alexander, M. "Recalcitrant Molecules, Fallible Micro-organisms," in *Microbial Ecology: A Conceptual Approach,* J. M. Lynch and N. J. Poole, Eds. (Oxford: Blackwell Scientific Publishers, 1979), pp. 246–253.

59. Swisher, R. D. *Surfactant Biodegradation* (New York: Marcel Dekker, Inc., 1970).
60. Buzzell, J. C., Jr., C. H. Thompson and D. W. Ryckman. "Behavior of Organic Chemicals in the Aquatic Environment: III. Behavior in Aerobic Treatment Systems (Activated Sludge)," Manufacturing Chemists Association, Washington, DC (1969).
61. Baird, R. B., C. L. Kuo, J. S. Shapiro and W. A. Wanko. "The Fate of Phenolics in Wastewater—Determination by Direct-Injection GLC and Warburg Respirometry," *Arch. Environ. Contam. Toxicol.* 2:165–178 (1974).
62. Monod, J. *Recherches sur la Croissance des Cultures Bacteriennes* (Paris: Herman et Cie, 1942).
63. Slater, J. H. "Microbial Population and Community Dynamics," in *Microbial Ecology: A Conceptual Approach,* J. M. Lynch and N. J. Poole, Eds. (Oxford: Blackwell Scientific Publishers, 1979), pp. 45–63.
64. Pirt, S. J. *Principles of Microbe and Cell Cultivation* (Oxford: Blackwell Scientific Publishers, 1975).
65. Paris, D. F., D. L. Lewis and N. L. Wolfe. "Rates of Degradation of Malathion by Bacteria Isolated from Aquatic System," *Environ. Sci. Technol.* 9:135–138 (1975).
66. Smith, J. H., W. R. Mabey, N. Bohonos, B. R. Holt, S. S. Lee, T.-W. Chou, D. C. Bomberger and T. Mill. "Environmental Pathways of Selected Chemicals in Freshwater Systems, Part II, Laboratory Studies," U.S. EPA Report 600/7-78-074 (1978).
67. Matsumura, F. "Current Pesticide Situation in the United States," in *Environmental Toxicology of Pesticides,* F. Matsumura, G. M. Boush and T. Misato, Eds. (New York: Academic Press, Inc., 1972), pp. 33–60.
68. Brooks, G. T. "Pesticides in Britain," in *Environmental Toxicology of Pesticides,* F. Matsumura, G. M. Boush and T. Misato, Eds. (New York: Academic Press, Inc., 1972), pp. 61–114.
69. Visser, S. A., G. Lamontagne, V. Zoulalian and A. Tessier. "Bacteria Active in the Degradation of Phenols in Polluted Waters of the St. Lawrence River," *Arch. Environ. Contam. Toxicol.* 6:455–469 (1977).
70. Alexander, M. "Biodegradation of Toxic Chemicals in Water and Soil," in *Dynamics, Exposure and Hazard Assessment of Toxic Chemicals,* R. Haque, Ed. (Ann Arbor, MI: Ann Arbor Science Publishers, Inc., 1979), pp. 179–190.
71. Jacobson, H. W., N. L. O'Mara and M. Alexander. "Evidence for Cometabolism in Sewage," *Appl. Environ. Microbiol.* 40:917–921 (1980).
72. *Standard Methods for the Examination of Water and Wastewater,* 14th ed. (Washington, DC: American Public Health Association, 1976).
73. Baughman, G. L., D. F. Paris and W. C. Steen. "Quantitative Expression of Biotransformation Rate," in *Biotransformation and Fate of Chemicals in the Aquatic Environment,* A. W. Maki, K. L. Dickson and J. C. Cairns, Jr., Eds. (Washington, DC: American Society for Microbiology, 1980), pp. 105–111.
74. Paris, D. F., W. C. Steen, G. L. Baughman and J. T. Barnett, Jr. "Second-Order Model to Predict Microbial Degradation of Organic Compounds in Natural Waters," *Appl. Environ. Microbiol.* 41:603–609 (1981).
75. Paris, D. F. Personal communication.
76. Steen, W. C., D. F. Paris and G. L. Baughman. "Effects of Sediment

Sorption on Microbial Degradation of Toxic Substances,'' in *Contaminants and Sediments, Vol. 1, Fate and Transport, Case Studies, Modeling, Toxicity,* R. A. Baker, Ed. (Ann Arbor, MI: Ann Arbor Science Publishers, Inc., 1980), pp. 477–482.
77. Fletcher, M. ''The Aquatic Environment,'' in *Microbial Ecology: A Conceptual Approach,* J. M. Lynch and N. J. Poole, Eds. (Oxford: Blackwell Scientific Publishers, 1979), pp. 92–114.
78. Jannasch, H. W., and G. E. Jones. ''Bacterial Populations in Sea Water as Determined by Different Methods of Enumeration,'' *Limnol. Oceanog.* 4:128–139 (1959).
79. Sherrill, T. W., and G. S. Sayler. ''Phenanthrene Biodegradation in Freshwater Environments,'' *Appl. Environ. Microbiol.* 39:172–178 (1980).
80. Lewis, D. L. Unpublished observations.
81. O'Kelley, J. C., and T. R. Deason. ''Degradation of Pesticides by Algae,'' U.S. EPA Report 600/3-76-022 (1976).
82. Baughman, G. L., and D. F. Paris. ''Microbial Bioconcentration of Organic Pollutants from Aquatic Systems—A Critical Review,'' *CRC Crit. Rev. Microbiol.* 8:205–228 (1981).
83. Schimmel, S. C., J. M. Patrick Jr., L. F. Faas, J. L. Oglesby and A. J. Wilson, Jr. ''Kepone: Toxicity and Bioaccumulation in Blue Crabs,'' *Estuaries* 2:9–15 (1979).
84. Weininger, D. ''Accumulation of PCBs by Lake Trout in Lake Michigan,'' PhD Thesis, University of Wisconsin—Madison (1978).

CHAPTER 8

EVALUATIVE MODEL FOR ESTIMATING ENVIRONMENTAL FATE

W. B. Neely

Environmental Sciences Research
Dow Chemical Company
Midland, Michigan 48640

D. Mackay

Department of Chemical Engineering
University of Toronto
Toronto, Ontario M5S 1A4

Most industrial countries have recognized the need for some form of control of the use of hazardous chemicals. This need has caused the passage of legislation that gives governments authority over which chemicals will be introduced into the marketplace and, in extreme cases, to ban their use entirely. To handle this activity, industry and government require a set of tools to help establish the degree of hazard associated with the use of each chemical.

Since hazard is a function of exposure and toxicity, it is convenient to investigate these two elements for a new chemical in a stepwise manner. At each level, a comparison of results can be made to ascertain the degree of hazard. When it is determined that the risk is acceptable, further testing can be terminated. This is a tiered approach to testing, in which one proceeds from one level to the next; more sophisticated testing is required as the chemical is subjected to more detailed scrutiny. The early tests by their very nature should be easy to use, economical and understandable by all concerned, and the results should be comparable between laboratories.

There has been, and will continue to be, much activity in designing and building models to estimate environmental exposure. Only a few of the many investigations in this field will be cited [1–3]. We believe such studies should be encouraged; however, we also believe that the time has come to adopt a model for initial exposure assessment. Our purpose here is to describe and discuss an evaluative model that may be used in the initial assessment of exposure that fulfills the previously stated criteria.

THE MODEL

Most of the early effort with estimating environmental exposure was based on kinetic models. Although such models are valuable in predicting the time course of the chemical as it moves through the ecosystem, they can become rather complicated. Mackay has reported on an equilibrium model that is easy to conceptualize and requires only a minimum of mathematical manipulation [4]. This has been modified slightly and used by Neely [5] and is employed here as a starting point. The goal is to take a "slice of the earth," a *unit world,* so that the major compartments will be represented. The proposed model is shown in Figure 1. The first section of this chapter will deal with the dimensions and proper-

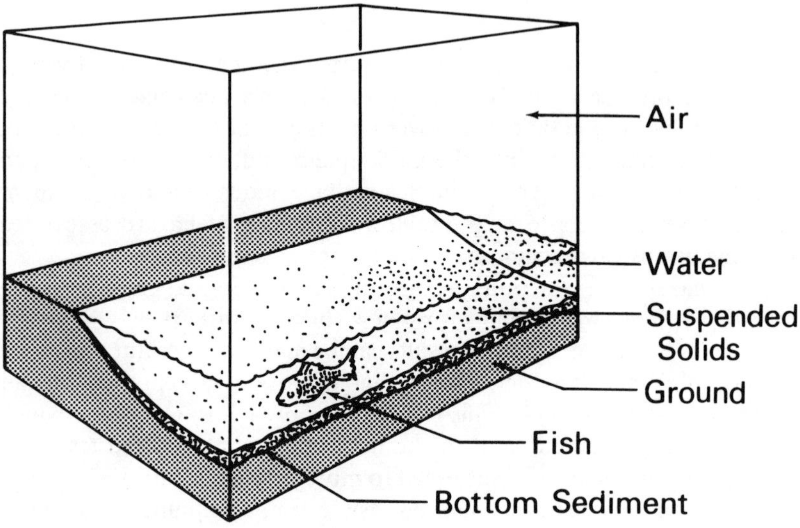

Figure 1. Schematic representation of the unit world.

ties of the unit world. The second will discuss the type of calculations that may be made.

The final section will illustrate with an example how the unit world can be used to assess environmental exposure. An important concept to emerge from this study is that the world should dictate the laboratory testing for a chemical. Although this may appear self-evident, it needs to be emphasized to prevent laboratory results that were generated for a different purpose from being forced into an environmental system where they are not appropriate.

Properties of the Unit World

Area

An arbitrary horizontal area of 1 km^2 is selected, largely on the basis that it is easily visualized.

Atmosphere

The earth's atmosphere is divided into regions primarily on the basis of temperature gradients. Typically, the temperature falls with increasing altitude in the troposphere, passes through a minimum at the tropopause, and then rises in the stratosphere. This temperature inversion provides a very stable barrier to mixing; therefore, the mixing depth for the troposphere is usually considered to be about 10 km.

When atmospheric concentrations are expressed on a mass/volume basis, the mixing depth of 10 km must be corrected for the decrease in density that is observed with increasing altitude. Neely [6] derived Equation 1 to express the mass of air as a function of height and may be used to make the correction:

$$\text{Mass} = \rho_0 A \int_0^h e^{-Kh}\, dh \qquad (1)$$

where ρ_0 = density of air at sea level (1220 g/m^3)
A = area (m^2)
K = constant (1.07×10^{-4}/m)
h = height (m)

With an area of 10^6 m^2, Equation 1 may be integrated between $h = 0$ and 10,000 to yield a value of 7.37×10^{12} g for the mass of air. Assuming a uniform density of 1220 g/m^3, 7.37×10^{12} g of air will require a column

6000 m (6 km) in altitude, which is the corrected atmospheric height producing a volume of 6×10^9 m^3. It is further assumed that the atmosphere is perfectly mixed and contains no suspended particulates.

Hydrosphere

The water column is assigned an average depth of 10 m. The area of 7×10^5 m^2 is based on the fact that water covers 70% of the earth's surface. The water volume is thus 7×10^6 m^3 and, like the air compartment, is perfectly mixed. Freshwater is assumed since this is of most interest.

Lithosphere

The bottom sediments in the unit world are represented by a completely mixed, active layer, which overlays the inactive deep sediments. The depth of this layer is controlled by many factors, such as wind-induced turbulence and the presence of benthic organisms. The actual thickness is deduced from the vertical distribution of suitable tracers [7]. For this model an average depth for the active layer of 3 cm will be assigned.

The terrestrial compartment has an area of 3×10^5 m^2. The mixing depth is based on observations that the furrow slice of a cultivated field is approximately 15 cm [8]. Since most chemicals will not penetrate below that level, a mixing depth of 15 cm is assumed, giving a total volume of 4.5×10^4 m^3.

The remaining lithospheric material is represented by the suspended matter in the water column at a concentration of 5 g/m^3 (5 ppm). This is an arbitrary number and represents an average figure of suspended matter in water.

Biosphere

Microorganisms and fish are the only biota included in the model. Fish are used to estimate the magnitude of bioconcentration, and the bacteria are essential to gain a perspective on the ability of the ecosystem to degrade the added chemical. The microbial population will be discussed later.

The concentration of fish can vary greatly, depending on the productivity of the water. For natural lakes of glacial origin, the areal density of fish is approximately 5.6–17 g/m^2 [9]. For the water volume of the "unit world," this is equivalent to an average fish-to-water volume ratio of 10^{-6}.

Media Densities

The following densities will be assumed:

- atmosphere: 1.22 kg/m^3
- hydrosphere: 1000 kg/m^3
- lithosphere: 1500 kg/m^3
- biosphere: 1000 kg/m^3

Media Composition and Conditions

Since most physicochemical data are reported at 25°C, this will be the temperature of the unit world. A pH of 7.0 will be assigned to the water compartment.

The organic carbon content of the soil is an important parameter for estimating the sorption of chemicals. Accordingly, the terrestrial soil will have an organic carbon content of 2%, while the bottom and suspended sediment in the water compartment will be 4%.

To calculate partition coefficients, the correlation Equation 2 of Veith et al. [10] is used for fish; Equation 3, established by Karickhoff et al. [11], is used for soils, sediment and suspended matter.

$$\log BF = 0.85 \log K_{ow} - 0.70 \tag{2}$$

$$K_{oc} = 0.6 K_{ow} \tag{3}$$

where BF = bioconcentration factor between water and whole fish
K_{ow} = octanol/water partition coefficient
K_{oc} = partition coefficient between water and soil organic carbon

Transport

Two transport mechanisms from the unit world will be considered. The first is exit from the troposphere by diffusion into the stratosphere, with a rate constant of 0.015/yr [12]. The second is the sedimentation rate constant, which measures the speed that material adsorbed to the particulates becomes buried in the bottom sediments. Once the chemical is covered to a depth greater than 3 cm, it will be assumed to have been removed from the dynamics of the unit world. A value of 0.08/yr (2×10^{-4}/day) will be assigned to this rate constant [7]. Both mechanisms operate on all chemicals added to the unit world. Since the water body is a pond, advection of water in or out is ignored.

Transformation

Three actions are considered: photodegradation, hydrolysis and biodegradation. Since the hydrosphere is assumed to be at pH 7.0, this becomes the condition under which hydrolysis should be studied.

One of the important atmospheric degradations is via a second-order reaction with hydroxyl radicals. The rate constant for the reaction shown in Equation 4 is usually reported in units of cm^3/mol-sec.

$$R + OH^{\bullet} \rightarrow R^{\bullet} + H_2O \tag{4}$$

To estimate a first-order rate constant, it is necessary to know the value of [$OH^{\bullet}$]. For the purpose of this model, a concentration of 5×10^5 molecules of hydroxyl radical/cm^3 will be assigned to the atmospheric compartment [6,13].

Microbial degradation will be treated as a pseudo-first-order reaction, which necessitates knowing the concentration of microorganisms in the soil and water compartments. The microbial concentrations adopted are those suggested by the National Academy of Sciences [14] of 10^6 cell/ml for sediment and 10^5 cell/ml for water, based on the studies of Baughman and Lassiter [15].

For aqueous photolytic calculations, it is suggested that the unit world be located at 40°N latitude at the time of the equinox. This facilitates application of the solar model of Zepp and Cline [16] to estimate a photodegradation rate in the water column. The absorption of light will be attenuated, depending on the type of water. Obviously, clear water will allow more light to pass through and thus increase photodegradation. Zepp and Cline [16] investigated a number of different water bodies and presented an average curve for ten rivers. Equation 5 summarizes their data and provides a convenient method for estimating the decrease in light absorption with depth in the water column of the proposed model:

$$\alpha = 0.45\, e^{-0.0045\lambda} \tag{5}$$

where α = attenuation coefficient at wavelength λ (/cm)

Transport Between Media

The simplest assumption is that all media are in equilibrium, i.e., a common fugacity or chemical potential prevails. If information on intermedia transport rates is available, it is possible to include them and allow for nonequilibrium situations. Unfortunately, there are considerable doubts from average values of soil/air and sediment/water transport rates. Air/water rates have been suggested by Liss and Slater [17]. It is probably a fair assumption that biotic and suspended solids are in equilibrium with the water.

Calculations

Before examining the model, it is useful to clarify the concepts of equilibrium and steady state as they apply to the calculations. This is illustrated by the tank analogy in Figure 2, in which each tank represents an environmental medium, and the volume of the water is analogous to the mass of chemical. In Figure 2A, the levels are equal and unchanging; there is no reaction, no inflow or outflow, and there is a fixed amount of water (solute). This equilibrium, steady-state, no-flow system is termed "Level I."

In Figure 2B, there is loss of water (solute) by leakage (reaction), but this is at a rate exactly balanced by an inflow (emission). The tanks (media) are in equilibrium, but their contents are changing, the levels adjusting to values such that there is no net change of amount in the system. It is possible to calculate an average residence time (persistence) of the water (solute) in the system. This equilibrium, steady-state flow system is termed "Level II."

In Figure 2C, there is a transfer resistance between the tanks, thus, there is a tendency for the tank into which the water is introduced to have a higher level, and flow is always from a high to low level at a rate dependent on the resistance of the valve. The loss rates again adjust to equal the inflow. This is a steady-state, nonequilibrium flow system termed "Level III."

Only Level I and II calculations are described here. Rarely are sufficient data available to permit Level III analyses, except in site-specific cases.

There are at least three ways of dealing with equilibrium processes, and they are all identical. The choice depends on personal preference. For chemists, the usual method is to work in concentrations. However, such units tell nothing about equilibrium, since there is a discontinuity in concentration when an interface is crossed. For example, at equilibrium the oxygen concentration in air is 8 g-mol/m^3, and in water it is 0.03 g-mol/m^3.

The second approach uses fugacity, which is related to the partial pres-

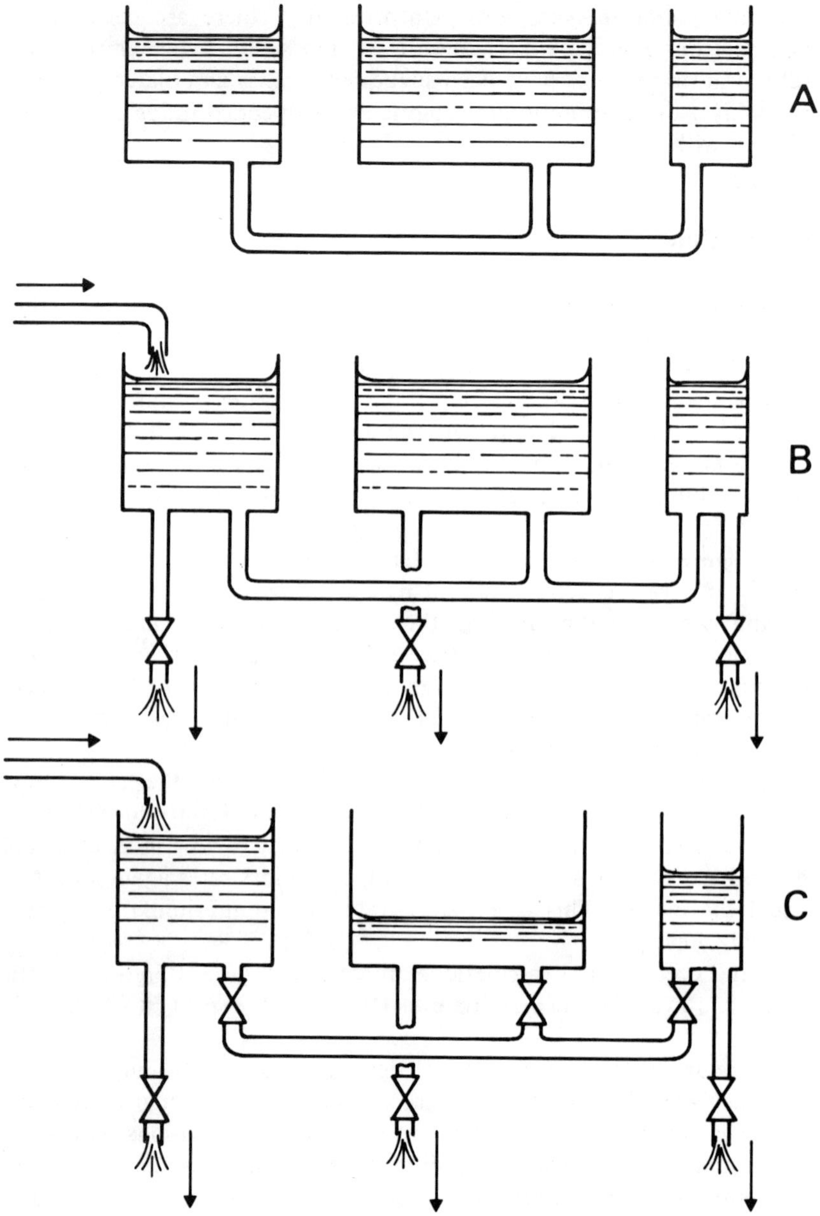

Figure 2. Tank analogy illustrating equilibrium (A) steady-state, no-flow system; (B) steady-state, flow system; and (C) nonequilibrium, steady-state, flow system.

sure of the chemical and is a measure of the escaping tendency of the molecule. At equilibrium, the fugacities are equal in all phases. In the example of oxygen cited above, the fugacity is 0.2 in water and air at equilibrium. The level of the tanks in Figure 2 may be regarded as the fugacity.

The third approach is similar to fugacity in that it deals with chemical potential. However, this has units of energy/mol which is conceptually very difficult to grasp.

The big advantage of using fugacity over concentration is in dealing with equilibrium situations where all phases have identical fugacities. This advantage has reduced significance for investigators who are not trained to think in terms of partial pressure. In the hope of gaining greater clarification, concentration units in this chapter will be in the more recognizable form of moles rather than partial pressure.

Level I Calculations

This level assumes a constant amount of material added to the unit world (Figure 1). There will be no further input and there will be no outflow and no reaction (Figure 2A). In essence, the Level I analysis asks the question, "How does the chemical partition itself among the various phases?"

Equilibrium may be described by a series a partition coefficients presented in Equations 6 through 8.

$$H = C_a/C_w \tag{6}$$

$$K_p = C_s/C_w \tag{7}$$

$$BF = C_f/C_w \tag{8}$$

where
H = Henry's constant
C_a = concentration in air
C_w = concentration in water
C_s = concentration in soil
C_f = concentration in fish
BF = bioconcentration factor (Mg/m^3)
K_p = soil sorption coefficient (Mg/m^3)

Henry's constant is estimated from Equation 9 from Dilling [18].

$$\frac{P \times M \times 16.04}{T \times S} \tag{9}$$

where P = vapor pressure (mm Hg)
M = molecular weight
T = temperature (°K)
S = solubility [g/m³ (ppm)]

The soil adsorption coefficient K_p is derived from K_{oc} (Equation 3) multiplied by the percent organic carbon in the soil, and BF is evaluated by Equation 2.

The total moles in the unit world may be expressed by Equation 10 where the summation is over all the compartments represented in the model of Figure 1:

$$M = \Sigma V_i C_i \tag{10}$$

Equation 10 may be expanded as illustrated in Equation 11. The volumes and densities for the world are summarized in Table I.

Table I. Volumes and Densities of Various Compartments in the Unit World

Compartment	Volume (m^3)	Density (Mg/m^3)
Air	6×10^9	
Water	7×10^6	
Bottom Sediment	2.1×10^4	
Suspended Sediment	23.3	
Terrestrial Ground	4.5×10^4	
Fish	7	1
Soil		1.5

$$M = C_w[V_\alpha H + V_w + (V_{bs}K_p + V_{ss}K_p + V_gK_p)\rho_s + V_fBF\rho_f] \tag{11}$$

where V = volume of the compartment
a = atmosphere
w = water
bs = bottom sediments
ss = suspended sediments
g = terrestrial ground
ρ_s, ρ_f = density of soil and fish

Thus, for Level I calculations, knowledge of the molecular weight, water solubility, vapor pressure and the octanol/water partition coefficient is required. Using this information, the partition coefficients in Equation 11 may be calculated, and the ratio of the moles in each compartment to the total moles may be evaluated. It should be pointed out that if the octanol/water partition coefficient is not known, it can be calculated by Equation 12 derived by Banerjee et al. [19]:

$$\log K_{ow} = 6.5 - 0.89(\log S) - 0.015(mp) \tag{12}$$

where mp = melting point (°C)
S = solubility (μmol/l)

Level II Calculations

At this level, a constant input of material is balanced by a constant outflow. The outflow is a summation of all possible reactions that may occur in the compartments, including advection. Once the model is in equilibrium, the amounts in the various compartments may be calculated, including a residence time. This latter item is of particular interest, since it is a measure of the persistence of the chemical. It should be recalled that no intermedia transport resistance applies at Level II. Accordingly, it is immaterial into which compartment the substance is introduced.

The removal rate from a compartment is estimated by:

$$\text{removal rate} = M_i \Sigma k_j \tag{13}$$

where M_i = number of moles in compartment i
Σk_j = summation of dissipating rate constants in compartment i

and the total removal rate (R_t) from the unit world is the double summation:

$$R_t = \Sigma_i [M_i \Sigma k_j] \tag{14}$$

The residence time is defined as the average length of time that a molecule remains in the unit world. It may be estimated from a knowledge of the total moles divided by the total rate of removal (i.e., Equation 11 divided by Equation 14). Furthermore, the water concentration (C_w) at steady state is given by:

$$C_w = \frac{\text{input (mol/day)}}{\Sigma_i (V_i K_i \rho_i \Sigma k_j)} \tag{15}$$

where K_i = appropriate partition coefficient
ρ_i = appropriate density

Once C_w is known, the remaining concentrations may be estimated by using the appropriate partition coefficient.

USING THE MODEL

Level I and II calculations will be illustrated with data on quinoline, for which an extensive data base has been collected by Smith et al. [20].

The physical properties for a Level I calculation are shown in Table II along with the derived values of the partition coefficients. Using Equation 11 and the volume data in Table I, the Level I results are shown in Table III.

Table II. Physical Properties of Quinoline

Structure	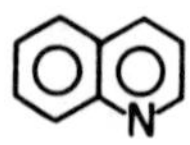
Molecular Weight	129.15
Melting Point (°C)	−15.6
Vapor Pressure[a] at 25°C (mm Hg)	9×10^{-3}
Water Solubility (g/m^3)	6100
K_{ow}	100.7[b]
H (from Equation 9)	2.3×10^{-6}
K_{oc} (from Equation 3)	225
BF (from Equation 2)	31

[a] There is considerable doubt as to the true value of the vapor pressure. It could be higher.
[b] This is an experimental value reported by Leo et al. [22].

Table III. Level I Calculation for Quinoline

Compartment	%
Air	0.85
Water	96.82
Bottom Sediments	1.20
Ground	1.12
Fish	0.0012
Suspended Sediments	0.0010

These results indicate the strong preference that quinoline has for water. When looking for degradation reactions, therefore, the water phase should be investigated first. It would be an inefficient use of resources to explore the atmosphere for possible dissipating mechanisms.

Level II Calculation

In the water compartment, three main mechanisms (hydrolysis, photodegradation and biodegradation) should be investigated as possible degradation mechanisms for quinoline. Quinoline contains no groups that are hydrolyzable; consequently, this mechanism need not be investigated as a degradation route [20]. Smith et al. [20] investigated the destruction of quinoline in water by photolysis and biodegradation. Their results will be summarized. The absorption spectrum of quinoline in the visible part of the spectrum is given in Table IV. Along with the absorption coefficient, the attenuation coefficient as calculated from Equation 5 is also provided. The quantum yield as reported by Smith et al. [20] is 3.3×10^{-4}. Using this information and the computer program of Zepp and Cline [17], the photodegradation rates shown in Table V were calculated.

Biodegradation can be regarded as a second-order reaction [21] with a

Table IV. Adsorption Spectrum of Quinoline in Water (pH 6.8)

Wavelength (nm)	Absorption (/M-cm)	Attenuation Coefficient (/cm)
297.5	2910	0.118
300	3050	0.117
302.5	2740	0.116
305	2480	0.114
307.5	2050	0.113
310	2440	0.112
312.5	2920	0.111
315	1680	0.109
317.5	622	0.108
320	269	0.107
323	119	0.106
330	26	0.102
340	9	0.090

Table V. Photodegradation of Quinoline as a Function of Season at Latitude 40° N

Season	Half-Life Integrated over Full Day	k[a] (/day)[a]
Spring	34	0.02
Summer	25	0.03
Fall	59	0.012
Winter	129	0.005

[a] $k = 0.69/t_{1/2}$.

rate constant in units of ml/cell-hr. Once the second-order constant is determined, a pseudo-first-order constant may be generated for the unit world by dividing the experimental constant by the concentration of microbial cells that are present in the particular compartment. These calculations are shown in Table VI.

In performing Level II calculations, the first number that should be estimated is the expected water concentration from Equation 15. The calculations may be conceptualized more easily by examining the data in Table VII. The summation of line 6 is the denominator of Equation 15. By dividing this into the assumed input of 1 mmol/day, a value of 1.56×10^{-8} mmol/m^3 is arrived at for C_w. By using C_w, and the data in lines 7 (total moles in the respective compartment) and 8 (removal rate in each compartment), the total rate of removal may be estimated. The residence time for quinoline in the unit world is simply line 7 divided by line 8, or 0.11 day.

From line 8 it is readily apparent that biodegradation in the water

Table VI. Biodegradation Rate Constant for Quinoline in the Unit World

Experimental Second-Order Constant[a] (ml/cell/hr)	3.1×10^{-6}
Concentration (cell/ml) of Cells in	
Water Column	10^5
Bottom Sediments	10^6
Terrestrial	10^6
First-Order Rate Constant (/day)	
Water Column	7.44
Bottom Sediments	74.4
Terrestrial	74.4

[a] From Smith et al. [20].

Table VII. Summary of Level II Calculations for Quinoline

	Compartment				
	Air	Water	Bottom Sediments	Ground	Total
Volume (m^3)	6×10^9	7×10^6	2.1×10^4	4.5×10^4	
K[a]	1×10^{-5}	1	2.57	1.28	
Density, ρ[b]	1	1	1.5	1.5	
VK_ρ	6×10^4	7×10^6	8.1×10^4	8.60×10^4	
K (per day)	4×10^{-5}	0.02[c] 7.44[d]	74.4	74.4	
$VK_\rho \times k$ (m^3/day)[e]	2.45	1.4×10^5[c] 5.2×10^7[d]	5.65×10^6	6.06×10^6	6.38×10^7
VK_ρ/C^f(mmol)	9.39×10^{-4}	1.09×10^{-1}	1.26×10^{-3}	1.35×10^{-3}	
$k/VK_\rho^w/C_w$)(mmol degr./day)	3.7×10^{-8}	8.13×10^{-1}	9.39×10^{-2}	1.00×10^{-1}	1.0

[a] Partition coefficient between compartment and water.
[b] Density for air is 1 since partition coefficient is dimensionless.
[c] Photolysis.
[d] Biodegradation.
[e] The sum of line 6 divided into the input gives water conc C_w.
[f] Total moles in each compartment. $C_w = 1$ mmol/$(6.38 \times 10^7) = 1.56 \times 10^{-8}$ mmol/m^3.

column is the most important route for degradation. This follows from the observation that Level I indicated the great preference of quinoline for water, and since there is a fast biodegradation rate, it follows that most of the quinoline (81%) will disappear from the unit world via this route. Photodegradation plays a minor role in this ecosystem. The importance of photodegradation would increase as the body of water becomes more oligotrophic.

SUMMARY

The unit world described in this report has the ability to assess the distribution of chemicals in a realistic manner. The Level I calculation based on solubility, vapor pressure and molecular weight allows a quick profile to be made of the chemical. Such a profile focuses attention on the environmental compartments where degradation mechanisms will have the greatest opportunity to play a role.

By accepting the characteristics of this unit world, it becomes possible for investigators to rank chemicals on a scale that will have the same meaning, regardless of where they are located. Furthermore, the calculation as presented may be performed on any programmable hand calculator.

REFERENCES

1. Blau, G. E., and W. B. Neely. "Mathematical Model Building with an Application to Determine the Distribution of DURSBAN® Insecticide Added to a Simulated Ecosystem," in *Advances in Ecological Research,* A. MacFayden, Ed. (New York: Academic Press, Inc., 1975), pp. 133–163.
2. Southworth, G. R. "Transport and Transformation of Anthracene in Natural Waters," in *Aquatic Toxicology,* L. L. Marking and R. A. Kimerle, Eds. (Philadelphia: American Society for Testing Materials, 1979), pp. 359–380.
3. Eschenroeder, A., E. Irvine, A. Lloyd, C. Tashima and K. Tran. "Computer Simulation Models for Assessment of Toxic Substances," in *Dynamics, Exposure and Hazard Assessment of Toxic Chemicals,* R. Haque, Ed. (Ann Arbor, MI: Ann Arbor Science Publishers, Inc., 1980), pp. 323–368.
4. Mackay, D. "Finding Fugacity Feasible," *Environ. Sci. Technol.* 13:1218–1223 (1979).
5. Neely, W. B. "Complex Problems—Simple Solutions," *Chemtech* 11:249–251 (1981).
6. Neely, W. B. *Chemicals in the Environment—Distribution, Transport, Fate, Analysis* (New York: Marcel Dekker, Inc. 1981).
7. DiToro, D. M., D. J. O'Connor, R. V. Thomann and J. P. St. John.

"Analysis of Fate of Chemicals in Receiving Waters—Phase I," Hydro-Qual, Inc., Mahwah, NJ (1981).
8. Laskowski, D., Agricultural Department, Dow Chemical Co., Midland, MI. Personal communication (1981).
9. Lennon, H., Department of Biology, Central Michigan University, Mt. Pleasant, MI. Personal communication (1979).
10. Veith, G. D., D. L. DeFoe and B. V. Bergstedt. "Measuring and Estimating the Bioconcentration Factor of Chemicals in Fish," *J. Fish. Res. Board Can.* 36:1040–1048 (1979).
11. Karickhoff, S. W., D. S. Brown and T. A. Scott. "Sorption of Hydrophobic Pollutants on Natural Sediments," *Water Res.* 13:241–248 (1979).
12. *Halocarbons: Effects on Stratospheric Ozone* (Washington, DC: National Academy of Sciences, 1976).
13. Crutzen, P. J., I. A. S. Isaksen and J. R. McAfee. "The Impact of the Chlorocarbon Industry on the Ozone Layer," *J. Geophys. Res.* 83:345–363 (1978).
14. *Polychlorinated Biphenyls* (Washington, DC: National Academy of Sciences, 1979), p. 153.
15. Baughman, G. L., and R. R. Lassiter. "Prediction of Environmental Pollutant Concentration," in *Estimating the Hazard of Chemical Substances to Aquatic Life,* ASTM Technical Publication 657, J. Cairns, Jr., K. L. Dickson and A. W. Maki, Eds. (Philadelphia: American Society for Testing and Materials, 1978), pp. 35–54.
16. Zepp, R. G., and D. M. Cline. "Rates of Direct Photolysis in Aquatic Environment," *Environ. Sci. Technol.* 11:359–366 (1977).
17. Liss, P. S., and P. G. Slater. "Flux of Gases across the Air-Sea Interface," *Nature* 247:181–184 (1974).
18. Dilling, W. L. "Interphase Transfer Processes. II. Evaporation Rates of Chloro Methanes, Ethanes, Ethylenes, Propanes, and Propylenes from Dilute Aqueous Solutions—Comparisons with Theoretical Predictions," *Environ. Sci. Technol.* 11:405–409 (1977).
19. Banerjee, S., S. H. Yalkowski and S. C. Valvani. "Water Solubility and Octanol/Water Partition Coefficients of Organics—Limitations of the Solubility-Partition Coefficient," *Environ. Sci. Technol.* 14:1227–1229 (1980).
20. Smith, J. H., W. R. Mabey, N. Bohonos, B. R. Holt, S. S. Lee, T.-W. Chou, D. C. Bomberger and T. Mill. "Environmental Pathways of Selected Chemicals in Freshwater Systems, Part II, Laboratory Studies," U.S. EPA Report 600/7-78-074 (1978), pp. 93–121.
21. Paris, D. F., W. C. Steen, G. L. Baughman and J. T. Barnett, Jr. "Second-Order Model to Predict Microbial Degradation of Organic Compounds in Natural Waters," *Appl. Environ. Microbiol.* 41:603–609 (1981).
22. Leo, A., C. Hansch and D. Elkins. "Calculation of Partition Coefficient," *Chem. Rev.* 71:525 (1971).

CHAPTER 9

FATE AND TRANSPORT MODELING FOR TOXIC SUBSTANCES

Jerald L. Schnoor, Narasinga B. Rao, Kathyrn J. Cartwright and Richard M. Noll

Civil and Environmental Engineering
University of Iowa
Iowa City, Iowa 52242

One of the most widely used herbicides in the United States is the soluble acetanilide alachlor (Lasso®). In Iowa, it is applied at 1.1 kg/ha as a preemergent for grass control in corn during the period from mid-April to mid-May (see Table I for structure). Concentrations in Iowa waters range 10–100 μg/l in headwater streams, with dilution and biodegradation accounting for lower concentrations ($\sim$1 μg/l) in large rivers directly after application. It is a very water-soluble herbicide (242 ppm) and with moderately low vapor pressure (2.2×10^{-5} mm Hg).

Other herbicides widely used in Iowa for broadleaf control are atrazine (Aatrex®) and cyanazine (Bladex®), triazine herbicides. These herbicides are also quite soluble in water (33 and 150 ppm, respectively) and with low vapor pressures. Atrazine is applied at 0.5–0.8 kg/ha (active ingredient) from mid-April to mid-May. Concentrations in the Iowa River at Iowa City are 1–2 μg/l during May–August. Concentrations of atrazine and cyanazine in the headwaters of the East Solidier River in Iowa were 54 and 130 μg/l, respectively, on June 18, 1979.

The purpose of this study was to test and validate a toxics fate and transport model. A field validation was undertaken, and microcosm and laboratory studies were used to supplement the field data collection under spring runoff (time-variable) conditions. One objective of the research was to determine if laboratory-derived rate constants could be

Table I. Kinetic Biodegradation Rate Constants

Sample	Alachlor Initial Concentration ($\mu g/1$)	Zero-Order k″ ($\mu g/1$-day)	Zero-Order r	First-Order k′ (day^{-1})	First-Order r	Second Order Rate ($\times 10^{-10}$) (liter/organism-day)[a]	Microbe Conc. ($\times 10^6$) (organism/1)[b]
After Rain							
No Nutrients	226.4	8.10	−0.99	0.071	−0.99	5.3	135 ± 104
Dextrose		7.83	−0.98	0.062	−0.97	15.4	40 ± 39
Dextrose, N&P		7.91	−0.99	0.048	−0.99	2.5	195 ± 121
Before Rain							
Dextrose	218.7	6.20	−0.97	0.023	−0.96	3.5	69 ± 70
Dextrose, N&P		6.12	−0.99	0.044	−0.99	6.7	66 ± 39
Riverwater	244.5	6.88	−0.99	0.030	−0.97	6.4	45 ± 40
Reservoir Water	213.5	3.68	−0.98	0.021	−0.97	20.7	10 ± 11
Sediment/Reservoir Water	247.1	8.01	−0.98	0.074	−0.99	8.1	91 ± 49
Activated Sludge/Riverwater	277.8	7.20	−0.97	0.046	−0.98	5.7	79 ± 41
Soil 2/Riverwater	264.4	7.94	−0.99	0.060	−0.98	11.3	53 ± 32
Soil 1/Riverwater	144.6	4.20	−0.99	0.062	−0.98	0.1	5000 ± 1908
1/8 Inoculum/Riverwater	128.7	5.26	−0.99	0.083	−0.99	11.3	73 ± 68
3/4 Inoculum/Riverwater	114.2	3.98	−0.97	0.099	−0.99	3.5	280 ± 176
Mean		6.41 ± 1.65		0.055 ± 0.023		7.7 ± 4.4	
Coefficient of Variation (× 100)		25.7		41.8		56.4	
Geometric Mean		6.11		0.051		6.07	

[a] $k = k'/X$.

[b] Concentration is averaged for linear portion of degradation over the time period for which first-order calculations were made.

used directly in a field validation effort. Soluble, biodegradable chemicals were investigated to complement previous studies with dieldrin.

Field data on alachlor were gathered on Rathbun Reservoir, a south-central Iowa reservoir on the Chariton River, by the University of Iowa Hygienic Laboratory [1]. The study was conducted during a high-flow runoff period. Under these conditions, Rathbun Reservoir had a mean depth of approximately 29 ft, a mean hydraulic detention time of 162 days, and a mean volume of 351,000 ac-ft. Rathbun's 535-mi^2 watershed is primarily in row-crop agriculture and pasture land.

MODEL DEVELOPMENT

The model uses a completely mixed flow-through system of one compartment. Kinetic interactions of the model are depicted by Figure 1. The model assumes a rapid sorptive equilibrium between dissolved and particulate pesticide. Sorbed pesticide undergoes sedimentation and scour. Dissolved pesticide is more available to volatilization, photolysis and hydrolysis. Equations 1 and 2 must be solved with a time-variable mass loading W(t) and detention time. The herbicides are approximately 99% in the dissolved phase, so that the last term of Equation 2 is of little consequence. The time-variable equation becomes a coupled set of ordinary differential equations, shown below, that were solved via a variable-step-size fourth-order Runge-Kutta numerical technique.

$$\frac{dV}{dt} = Q_{in} - Q_{out} \tag{1}$$

$$\frac{d(VC_T)}{dt} = W(t) - Q_{out}C_T - \Sigma kCV - k_sC_pV \tag{2}$$

where
- t = time (days)
- V = volume (liters)
- Q = flowrate (liter/day)
- C_T = total herbicide concentration (μg/l)
- W = herbicide loading in mass per time (μg/day)
- k = sum of the pseudo-first-order reaction rate constants in the dissolved phase (hydrolysis, oxidation, photolysis, volatilization and biotransformation) (liter/day)
- k_s = sedimentation rate constant (liter/day)
- C_p = particulate herbicide concentration = K_pCM (μg/l)
- C = dissolved herbicide concentration (μg/l)

Equation 2 must be differentiated by the chain rule:

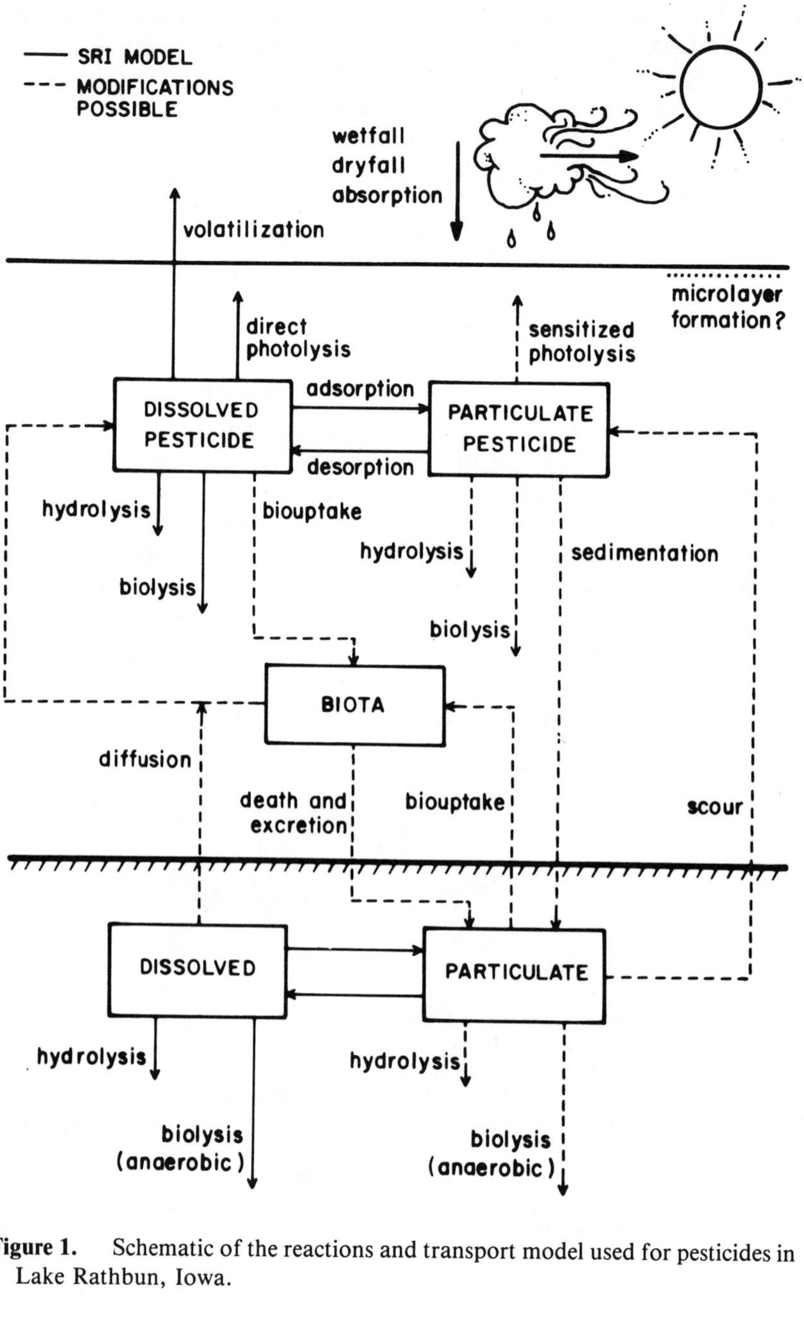

Figure 1. Schematic of the reactions and transport model used for pesticides in Lake Rathbun, Iowa.

$$\frac{d(VC_T)}{dt} = V\frac{dC_T}{dt} + C_T\frac{dV}{dt} \tag{3}$$

The final equation is:

$$\frac{dC_T}{dt} = \left[Q_{in}C_{in} - Q_{out}C_T - \Sigma KCV(t) - K_sC_pV(t) - C_T\frac{dV}{dt}\right]\cdot\frac{1}{V(t)} \tag{4}$$

Relationships between the total, particulate and dissolved pesticide concentrations can be calculated assuming local equilibrium:

$$C = \frac{C_T}{1 + K_pM} \tag{5}$$

$$C_p = \frac{k_pMC_T}{1 + K_pM} \tag{6}$$

where M = suspended solids concentration (kg/l)
K_p = solids/water partition coefficient [(μg/kg)/(μg/l)]

GASP-IV Simulation Language

Ordinary differential equations can be coded easily into the simulation language, GASP-IV. GASP-IV is a simulation language developed for handling discrete, continuous and continuous/discrete time systems. The integration method provided by GASP-IV is a fourth-order, variable-step-size Runge-Kutta routine for integrating systems of first-order, ordinary differential equations with initial values.

Equations 1 and 4 can be coded directly into GASP-IV as:

```
DD(1) = QIN - QOUT

DD(2) = (QIN*CIN - QOUT*SS(2) - SS(2)*DD(1)/SS(1)

        - (KP*M/(1 + KP*M))*SS(2)*KS - (1/1 + KP*M))*SS(2)*K
```

where SS(1) ≡ V
DD(1) ≡ dV/dt
SS(2) ≡ C_T
DD(2) ≡ dC_T/dt

These equations are coded in subroutine STATE of GASP-IV. The initial conditions and the input values are set in subroutine INTLC. The values QIN, QOUT and CIN are obtained from a user-written FORTRAN function that looks up the values from a table with independent variable TNOW (simulation time). Simulation for several pesticides can be performed simultaneously and their concentrations will be represented by SS(3), SS(4), etc. Outputs are obtained in the form of a table and/or plot (with user-specified scale).

As an example of model validation, consider the herbicides atrazine and alachlor in Rathbun Reservoir, Iowa. To assess the herbicide dynamics, a three-step effort was undertaken, including (1) biotransformation tests in natural waters, (2) microcosm testing, and (3) field data collection.

MATERIALS AND METHODS

Biotransformation Tests

Biodegradation tests were performed in unfiltered Iowa River water with spiked additions of alachlor to 100–300 μg/l under the following conditions: with and without bacterial seed from the Iowa River; with and without bacterial seed from treated soil; with and without nutrient additions; and during and after rainfall events in the Iowa River. Controls were used to determine the extent of gas stripping, chemical hydrolysis and sorption to glassware. Samples were aerated in the dark at ~23°C.

A study of the microbial community was also performed during the degradation tests. Samples were spread-plated, incubated at 20°C for 7 days and counted at 10× magnification using a Quebec counter; colony morphologies were recorded. Colonies were transferred to nutrient broth for 3 days at 20°C, observed and gram-stained, and all morphologies were noted. In addition, samples were streaked on BBL-brand MFC, M-Enterococcus and M-Endo agars to test for fecal coliforms, fecal streptococci and total coliform.

Microcosm Test

Fish used in this study were sunfish (*Lepomis* sp.) collected by shocking from the Coralville Reservoir on the Iowa River in early November 1979. Algal growth was mostly *Microspora* sp. indigenous to the Iowa River. The algae became a test parameter by default after they choked out the macrophyte *potamogeton* originally intended for use. Clams used were *Amblema* sp. collected by hand from a bed south of Fort Madison, Iowa, in Pool 19, Mississippi River, during November 1979. Sediment was collected from the same clam bed and used to cover the bottoms of the aquaria to a depth of about 4 in. Water for the study was withdrawn from Iowa River influent to the University of Iowa Water Plant. Water quality varied greatly over the course of the two experimental phases. All equipment and organisms were arranged in a manner to simulate productive areas of the Mississippi River. Into each aquarium were placed 30 fish and 30 clams. Organisms were allowed to acclimate to the aquaria for a period of 2 weeks prior to beginning the herbicide experiment. Fish were fed a combination of live mealworms and commercial pellets as a supplement to natural food organisms present in the microcosm.

For the microcosm studies, Iowa River water was drawn into a 100-gal cylindrical Nalgene tank, and chemicals were mixed in a batch process. Delivery to the aquaria was accomplished by a Masterflex peristaltic pump at 132 ml/min (~50 gal/day). Aquaria were 55-gal all-glass rectangular tanks ($18 \times 48 \times 12$ in.) fitted with glass tubing at one end 1.5 in. from the top to allow positive overflow into a drainage system. Total water volume of each prepared aquarium was approximately 5 ft^3 or 37–40 gal (140–150 liters) with a depth of 16 in. Stable water temperature was maintained by two 200-W aquarium heaters per tank. During the colder months, a temperature gradient resulting from the flow of cold influent into the warmer tank water was evident. Mixing was accomplished by placement of an airstone near the center of the tank. The photoperiod was automatically controlled with a 16-hr light and 8-hr dark cycle. Illumination was provided by two 40-W fluorescent bulbs per aquarium suspended 6 cm (2.25 in.) above the water surface. Alachlor was added as Lasso herbicide, the emulsifiable liquid, diluted in a carrier of acetone. Inflow concentrations averaged 8–10 μg/l, but varied because of mixing conditions in the feed tank.

Field Tests

Samples were prepared, collected and preserved according to *Standard Methods* [2] and National Pollutant Discharge Elimination System (NPDES) criteria. Grab samples were collected from four locations representing the two principal inflows from the Chariton River and South Chariton River, a depth-composite sample from Rathbun Reservoir, and a downstream sample. Weekly samples of water, fish and sediment were performed for the herbicides alachlor, atrazine and the insecticide dieldrin. The sampling period consisted of a series of runoff events from May to July 1978. Flow was from gauge records of the U.S. Geological

Survey (USGS), and samples were analyzed by the University of Iowa Hygienic Laboratory.

RESULTS AND DISCUSSION

Biotransformation Test

Loss of alachlor by gas stripping proved to be a relatively small but detectable loss mechanism. The degradation rate was surprisingly independent of the source of inoculum, substrate or nutrient addition, and water quality. Zero-order and first-order kinetics fit the laboratory data best with rate constants of 6.4 μg/l-day and 0.055/day, respectively. Second-order kinetics did not hold very well, since 100-fold increases in viable cells (as measured by the spread-plate technique in petri plates filled with BBL-brand nutrient agar) did not result in increased degradation rates. Paris et al. [3–5] have noted second-order kinetics for biodegration, first-order in total cell count and first-order in pesticide concentration.

Results are shown in Table I from all of the biotransformation studies. The first-order rate constants are in good agreement with those calculated from Beestman and Deming [6] and those of Rao [7] for alachlor degradation in soil. An average half-life of 12.6 days can be computed. The second-order rate constant is quite variable and is an order of magnitude larger than that determined in laboratory protocol measurements by Paris and Baughman [8]. However, the procedures differ considerably and their measurement was a much shorter-duration test (3–5 days).

A typical degradation curve (a semilog plot of concentration vs time) is shown in Figure 2. Usually a lag time of approximately 5–8 days was evident, in which the degradation rate was low, followed by a more rapid and linear phase. Zero-order kinetics also fit the experimental data quite well, as reported by Pitter [9].

Results of the microbial community study indicated that spiked additions of alachlor, dextrose or nutrients did promote significant changes in community structure, and that succession caused large changes in the taxa of controls during the course of the degradation experiments. This technique also demonstrated a source of bacterial contamination coming through the filtered air supply [10].

Microcosm Test

Results from the microcosm experiment are shown in Figures 3 and 4 [11]. Alachlor and atrazine were not detected in fish, sediments, clams or algae, probably as a result of their low octanol/water partitioning and

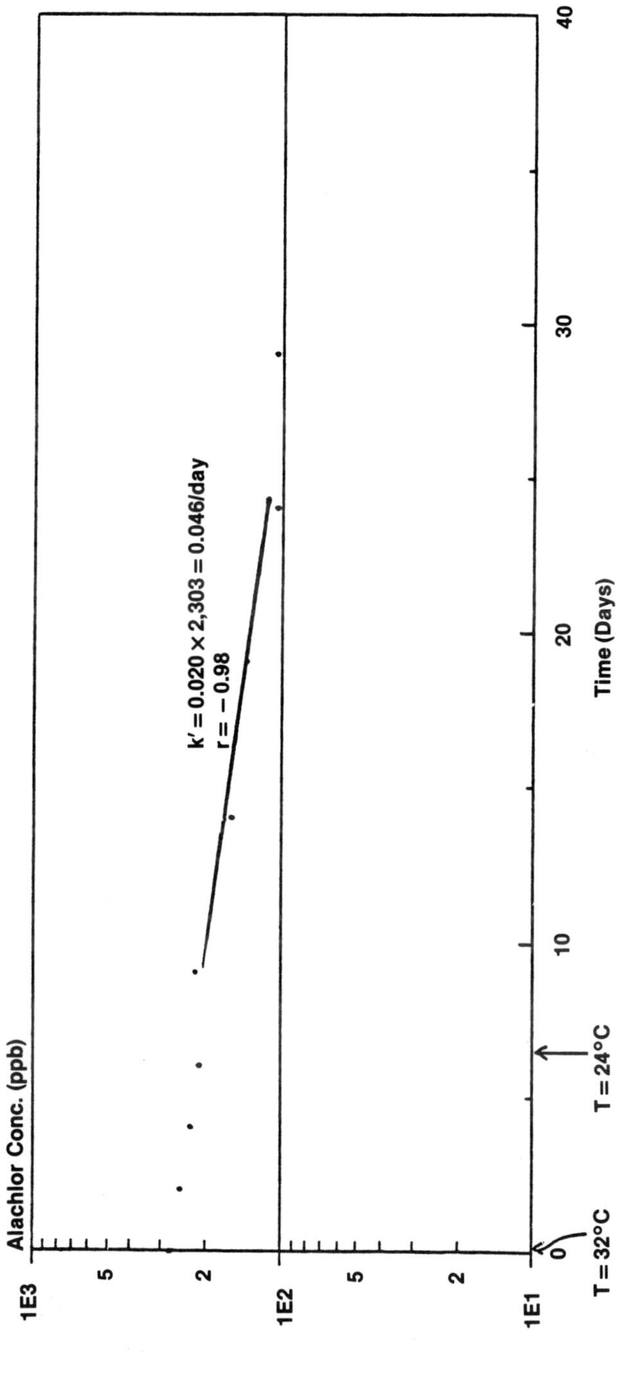

Figure 2. Alachlor degradation in Iowa riverwater with an activated sludge inoculum.

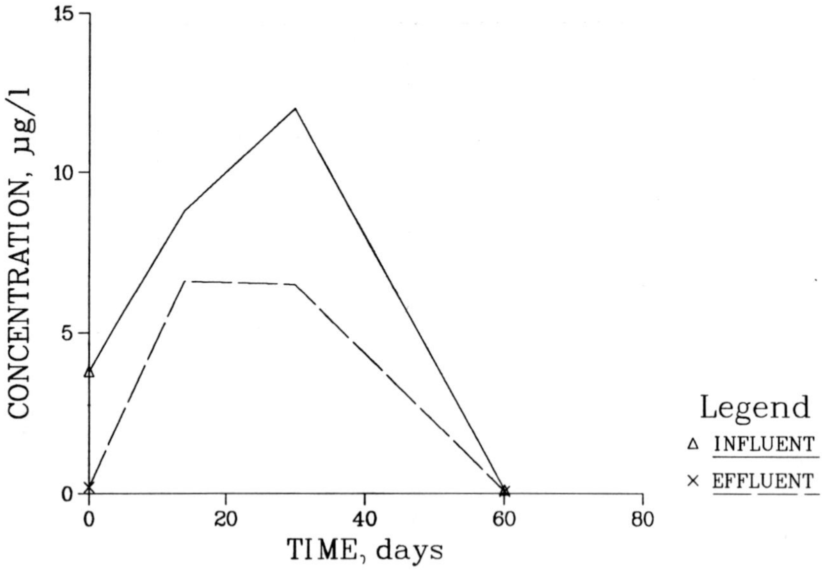

Figure 3. Alachlor microcosm concentrations in filtered water samples.

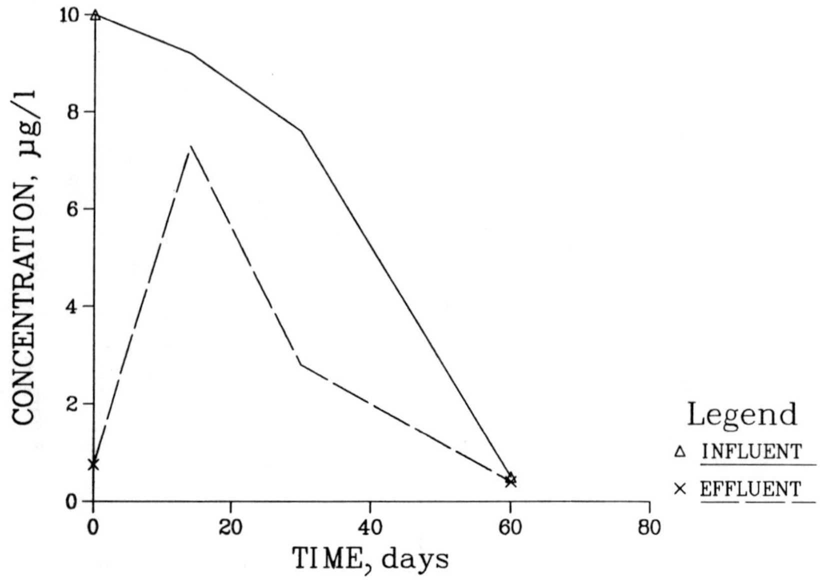

Figure 4. Atrazine microcosm concentrations in filtered water samples.

rapid metabolic transformations. After 30 days, the feed solution was changed to settled riverwater, and washout of herbicides was observed from days 30 to 60. The percent of alachlor removed during the experiment was approximately 35% (Figure 3). At near steady state (days 15–30), the removal was also ~35%. Therefore, it is possible to estimate the degradation rate for a system with a 0.8-day detention time to be 0.67/day. This is much greater than the estimated 0.055/day from the biotransformation study. This points up the need for similarity in laboratory microcosms. Microcosms should have the same biomass-to-volume and production-to-biomass ratios to mimic the ecosystem. In this case, the large mass of algae that developed and the large biomass were probably responsible for the rapid loss of herbicides from sorption and biotransformations. Algae are known to biodegrade many organics and herbicides. Atrazine was ~40% removed in the microcosm study, also indicating sorption and possible biotransformation (Figure 4).

Field Test

Modeling efforts on Rathbun Reservoir in Iowa for the herbicide alachlor and atrazine are much different than for hydrophobic pollutants. Being quite soluble, these herbicides are shown to undergo negligible sedimentation, but to degrade biologically and to be transported out of the reservoirs via the outflow. Figures 5 to 8 are the results for alachlor in Rathbun Reservoir. Time-variable loadings and flow were required to accurately calibrate the model to in situ and outflow concentration data. The rate of degradation in Figure 8 was significant, and the pseudo-first-order biodegradation rate constant was 0.04/day in relative agreement with laboratory biotransformation measurements. Equations 1 and 4 were solved simultaneously to plot the results shown in Figure 8. At least for the case of alachlor in Rathbun Reservoir, the use of a laboratory-derived biotransformation rate would have been appropriate for modeling purposes.

The results for atrazine in Rathbun Reservoir are less satisfying. Literature values for atrazine degradation in soils and waters range from 0.01 to 0.05/day. Using the lowest of these literature values (0.01/day) and the measured input concentrations of Figure 9, the calculated concentration is less than the 0.5 to 1.2-μg/l measured lake concentration (Figure 10). The model results of Figure 10 use double-the-measured inputs to fit the field data. The reason why model results do not reflect the field data cannot immediately be attributed to channeling or incomplete mixing, because the alachlor mass balance worked nicely (Figure 8).

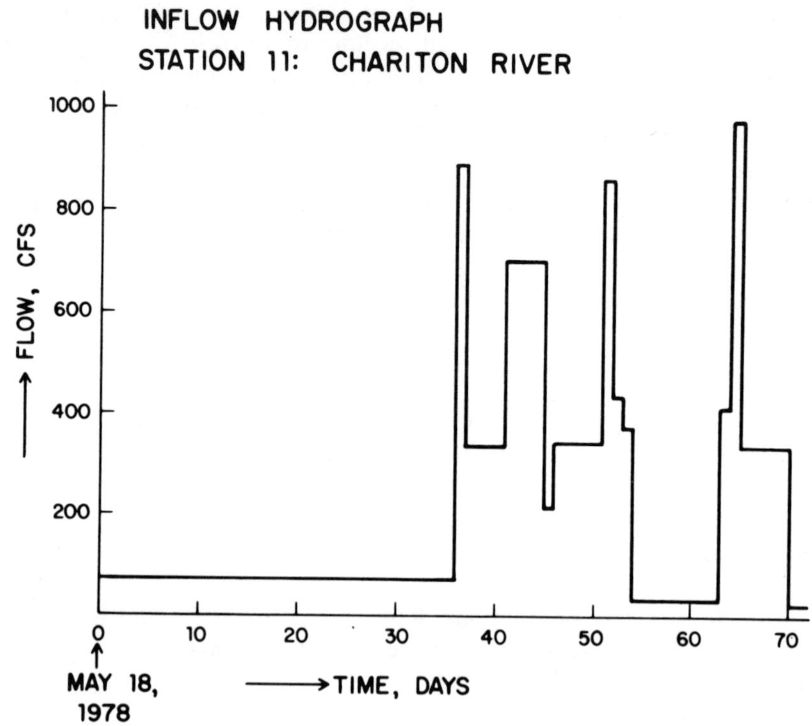

Figure 5. Flow measured at USGS gauging station and used as model inflow to Lake Rathbun.

Other possible reasons include: (1) an unmeasured source of atrazine, such as overland runoff or bed sediment, and (2) analytical chemistry problems at these very low concentrations near the detection limit of 0.2 μg/l.

For comparison, a final simulation was performed for dieldrin in Rathbun Reservoir. Measured inflow concentrations were used as model input (Figure 11) and the results are plotted in Figure 12. Once again, model results are lower than measured values. However, these concentrations range from 0.4 to 1.0 ng/l and are extremely low.

As an observation, one should not search for biological effects of pesticides in large rivers and reservoirs. It is the headwaters and small tributaries that have 50- to 100-fold higher exposure concentrations. It is also the tributaries that often have the most habitat diversity and, hence,

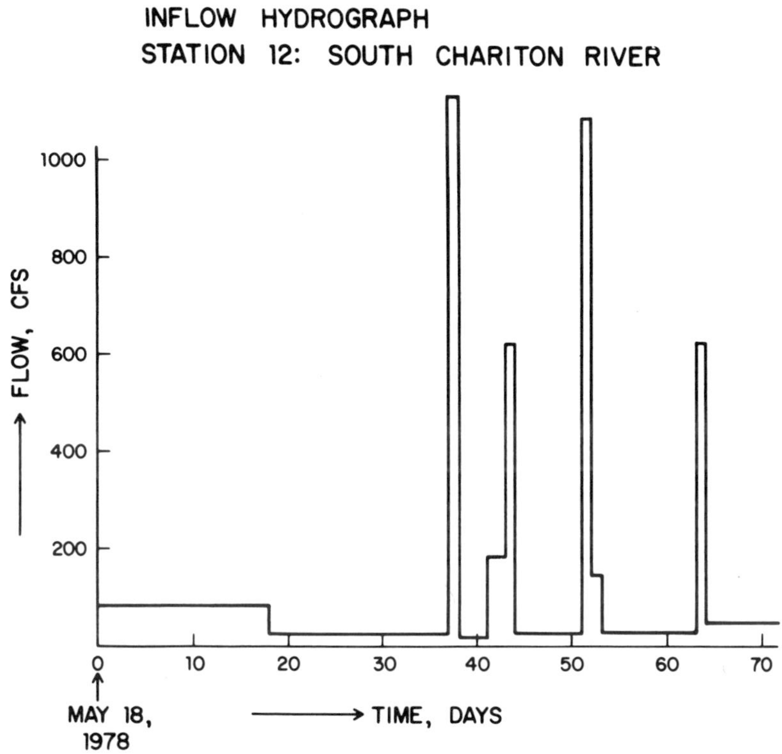

Figure 6. Flow measured at USGC gauging station and used as model inflow to Lake Rathbun.

species diversity, as well as spawning areas of fish. Nevertheless, the pesticide concentrations found in Iowa during the tenure of this research are well below all acute toxic thresholds and LC-50 values.

This field validation demonstrates the use of laboratory measurements in a modeling framework. A further improvement would be to track the daughter products of herbicide biodegradation, as well as the parent compound. It is worth noting that approximately 75–90% of the transport occurs over less than 20% of the time. Because of the transient nature of runoff events, it is necessary to perform time-variable simulations to represent accurately the mass fluxes, as well as concentrations. Annual average or steady-state representations are not accurate in these cases.

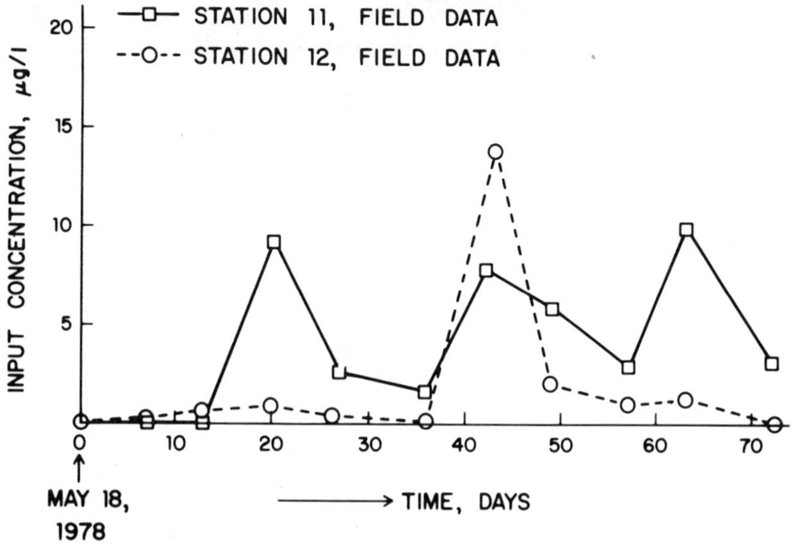

Figure 7. Field measurements and model input concentrations at Lake Rathbun: Lasso.

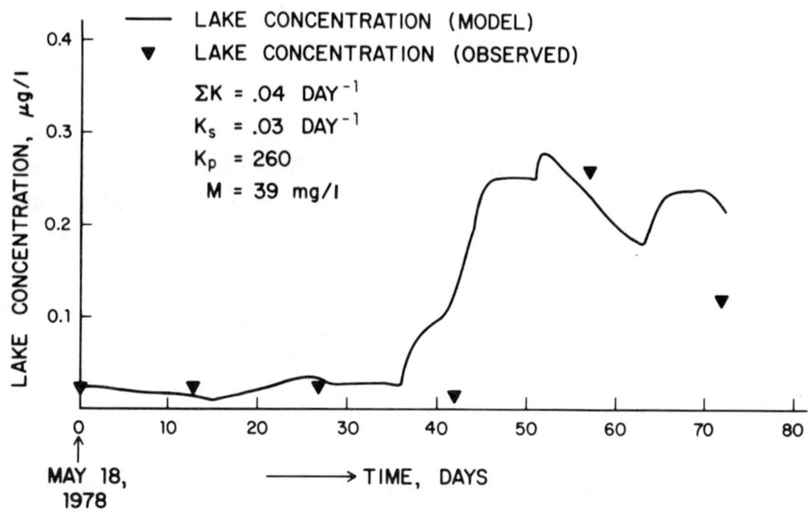

Figure 8. Field measurements and model results for Lake Rathbun: Lasso.

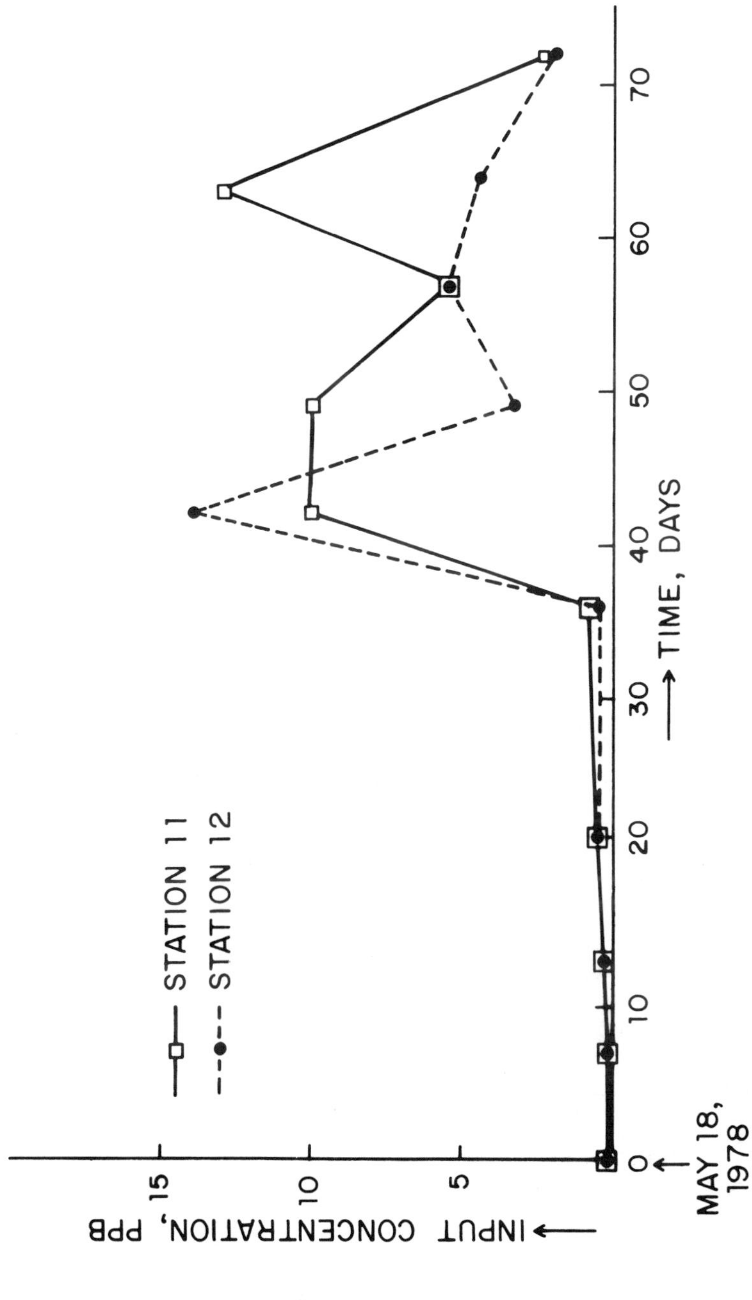

Figure 9. Field measurements and model input concentrations to Lake Rathbun: atrazine.

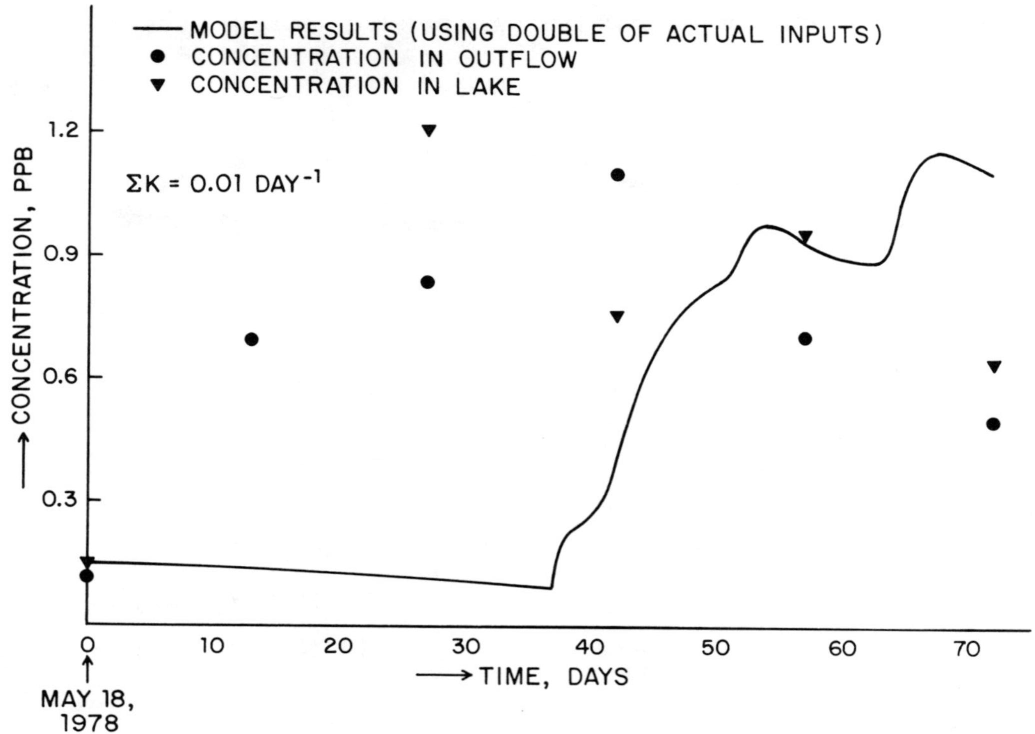

Figure 10. Field measurements and model results for Lake Rathbun: atrazine.

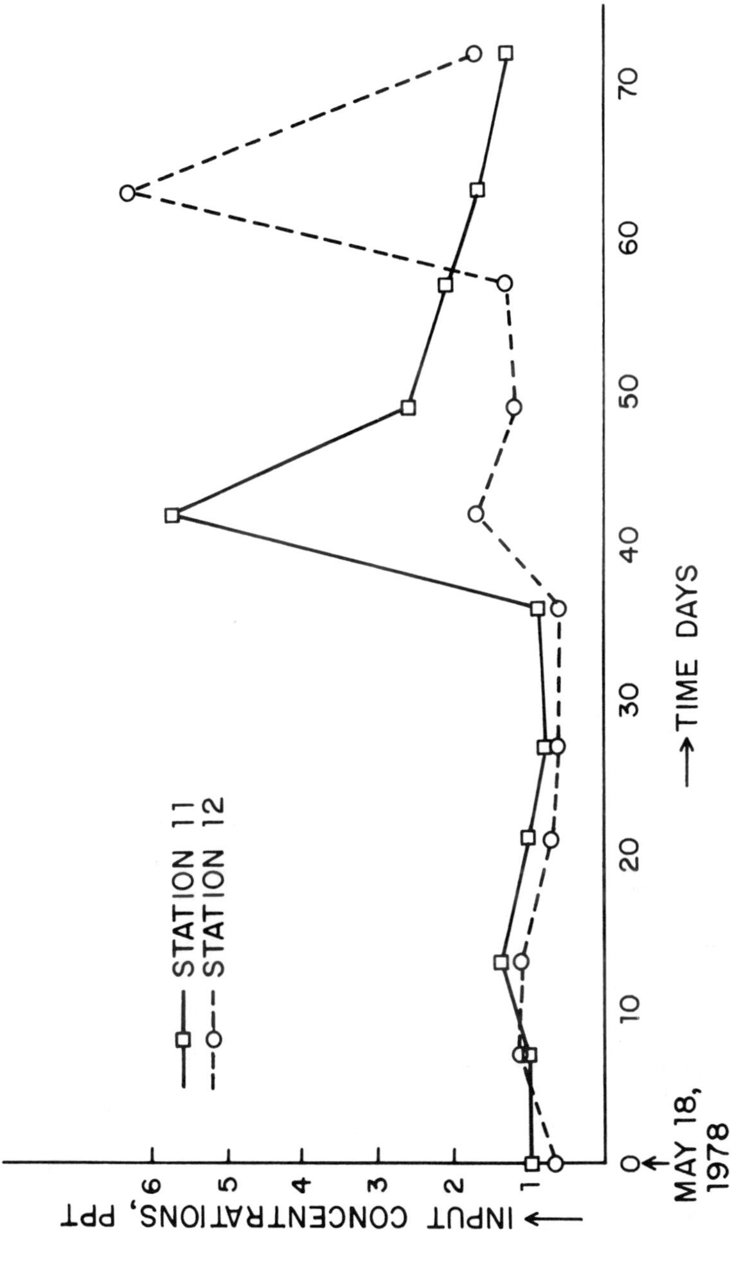

Figure 11. Field measurements and model input concentration to Lake Rathbun: dieldrin.

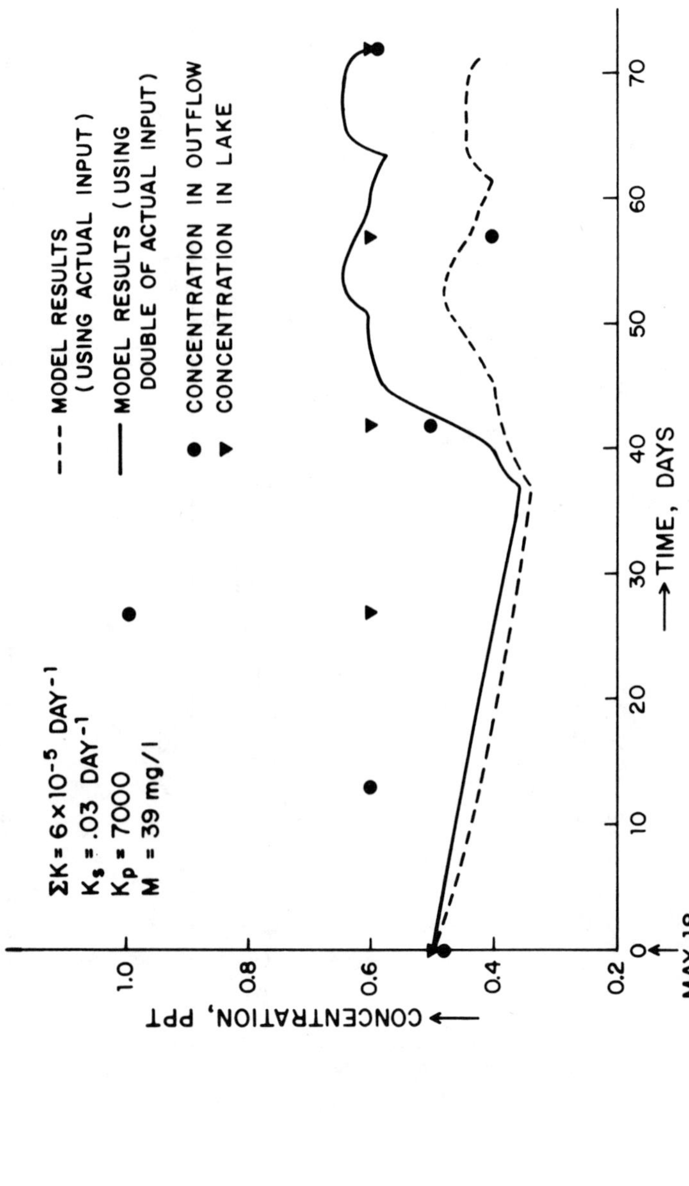

Figure 12. Field measurements and model results for Lake Rathbun: dieldrin.

SUMMARY

A fate and transport model has been calibrated and field-validated with pesticide data from Rathbun Reservoir, Iowa. Alachlor, atrazine and dieldrin were simulated. Laboratory biodegradation tests were accurate within a factor of five in a field application of the model. Microcosm studies were successful in reproducing bioconcentration effects, but microcosms should not be relied on for fate studies because of the large growth of biomass that inevitably results.

REFERENCES

1. Kennedy, J. O. "Lake Rathbun 1978 Pesticide Study," Final Report to the U.S. Army Corps of Engineers, Kansas City District, Contract No. DACW-41-78-M-1097 (1978).
2. *Standard Methods for the Examination of Water and Wastewater,* 14th ed. (Washington, DC: American Public Health Association, 1976).
3. Paris, D. F., A. L. Lewis and N. L. Wolfe. "Rates of Degradation of Malathion by Bacteria Isolated from Aquatic Systems," *Environ. Sci. Technol.* 9(2):135–138 (1975).
4. Paris, D. F., D. O. Lewis, J. T. Barnett and G. L. Baughman. "Microbial Degradation and Accumulation of Pesticides in Aquatic Systems," U.S. EPA Report 660/3-75-007 (1975).
5. Paris, D. F., W. C. Steen and G. L. Baughman. "Prediction of Microbial Transformation of Pesticides in Natural Waters," paper presented before the Division of Pesticide Chemistry, American Chemical Society, Anaheim, CA (1978).
6. Beestman, G. B., and J. M. Deming. "Dissipation of Acetanilide Herbicides from Soils," *Agron. J.* 66(2):308–311 (1974).
7. Roa, P. S. C., and J. M. Davidson. "Adsorption and Movement of Selected Pesticides at High Concentrations in Soils," *Water Res.* 13:375–380 (1979).
8. Paris, D. F. and G. L. Baughman, U. S. EPA Environmental Resource Laboratory, Athens, GA. Personal communication (1980).
9. Pitter, P. "Determination of Biological Degradability of Organic Substances," *Water Res.* 10:231–235 (1976).
10. Cartwright, K. J. "Microbial Degradation of Alachlor Using River Die-Away Studies," MS Thesis, University of Iowa, Iowa City, IA (1980).
11. Noll, R. M. "Pesticides and Heavy Metals: Fate and Effects in a Laboratory Microcosm," MS Thesis, University of Iowa, Iowa City, IA (1980).

CHAPTER 10

SIMPLIFIED MODEL OF THE FATE OF PARTITIONING CHEMICALS IN LAKES AND STREAMS

Dominic M. Di Toro, Donald J. O'Connor and Robert V. Thomann

HydroQual, Inc.
Mahwah, New Jersey 07430
and Environmental Engineering and Science Program
Manhattan College
Bronx, NY 10471

John P. St. John

HydroQual, Inc.
Mahwah, New Jersey 07430

This chapter addresses the problem of analyzing the fate of chemicals discharged into receiving waters, specifically lakes and streams. The focus is on contemporary industrial chemicals and the mechanisms that affect their distribution in the aquatic environment. These chemicals are associated, to a greater or lesser degree, with suspended and sedimented particles that are present in all receiving waters. Thus, particle transport mechanisms markedly affect the chemical's fate. The analysis that has been developed includes these mechanisms as an integral part of the formulation, as well as the other kinetic, transfer and transport processes of importance. It is similar in many ways to previously developed models [1–6]; however, the focus of this effort is to analyze exhaustively the solutions, so that an intuitive grasp of the problem can be developed.

The methodology is based on the principle of conservation of mass. It is expressed in mathematical form for application to lakes, impoundments and flowing streams. The solutions have been analyzed in detail and cast into forms that are suitable for desktop calculations. The results are comprehensible without elaborate numerical computer-based calculations. Thus, they provide direct insight into the relative importance of the various mechanisms that determine the fate of a particular chemical in a specific body of water.

DEFINITIONS, MECHANISMS AND MASS–BALANCE EQUATIONS

The receiving water models for lakes and streams include the mechanisms illustrated in Figure 1 and defined in Table I. The loading rate of chemical W_T is the primary input to the water column. The relevant reactions that remove or transform the chemical are: hydrolysis, oxidation, biodegradation, photolysis and volatilization. These are summed to yield the total degradation rates in the water column K_1 and sediment K_2. The

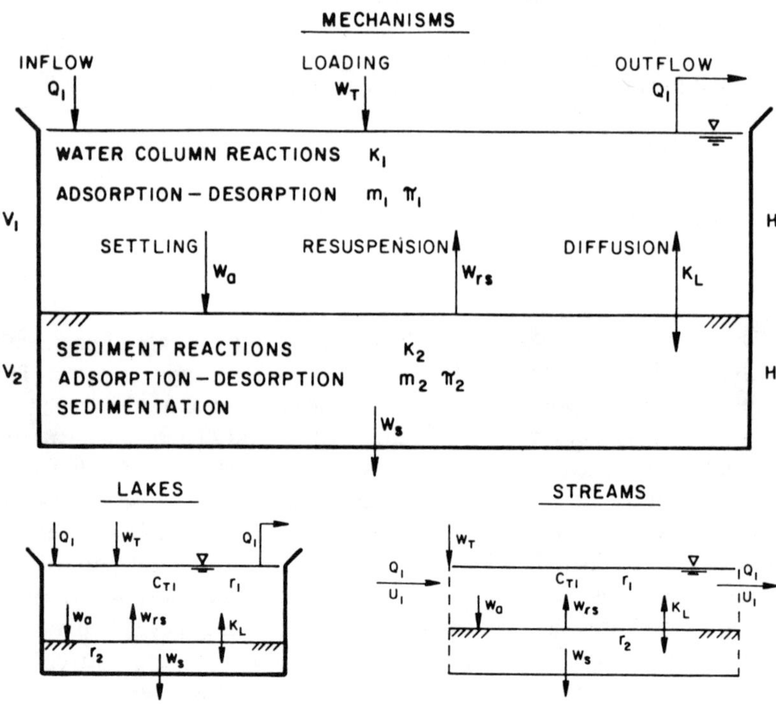

Figure 1. Receiving water volume segment: mechanisms and application to lakes and streams.

Table I. Definitions

Parameter	Water Column	Sediment
Chemical/Biological		
Loading rate (kg/day)	W_T	
Sum of hydrolysis, oxidation biodegradation, photolysis and volatilization rates (/day)	K_1	K_2
Partition coefficients (liter/kg)	π_1	π_2
Physical		
Solids concentration (mg/l)	m_1	m_2
Depths (m)	H_1	H_2
Volumes (m^3)	V_1	V_2
Flowrate (m^3/day)	Q_1	
Velocity (m/day)	$U_1 = Q_1 H_1 / V_1$	
Detention time (days)	$t_{01} = V_1/Q_1$	
Settling velocity (m/day)	w_a	
Resuspension velocity (mm/yr)		w_{rs}
Diffusion exchange coefficient (cm/day)		K_L
Sedimentation		
Velocity (mm/yr)		w_s
Rate coefficient (/day)		$K_s = w_s/H_2$
Concentrations		
Total dissolved + particulate (μg/l)	c_{T1}	c_{T2}
Particulate (μg chemical/g solids)	r_1	r_2
Fractions[a]		
Particulate $f_p = m\pi/(1 + m\pi)$	f_{p1}	f_{p2}
Dissolved $f_d = 1/(1 + m\pi)$	f_{d1}	f_{d2}

[a]For π in units of liter/kg and m in units of mg/l, a conversion factor of 10^{-6} kg/mg is necessary, so that these fractions are dimensionless, i.e., m (mg/l) $\times$ π(liter/kg) $\times$ 10^{-6} (kg/mg).

adsorption/desorption reaction is assumed to be at equilibrium, and the fraction of the chemical mass that is either dissolved or adsorbed to particulates is determined by the mass of adsorbing solids m and the partition coefficients π in the water column (m_1 and π_1) and the sediment (m_2 and π_2), respectively.

The geometry of the receiving water segment is specified by the depths of the water column H_1, and the active sediment layer H_2, together with the volumes of these segments V_1 and V_2, respectively. The aqueous transport is specified by the flowrate of the water column Q_1 and the velocity U_1, in streams, or the hydraulic detention time t_{01}, in lakes.

The sediment/water column transport of particulates is specified by the settling velocity of adsorbing particles to the sediment layer w_a and their resuspension velocity from the sediment w_{rs}. Dissolved-phase transport is specified by the diffusive exchange coefficient K_L, which is a function of the interstitial and overlying water diffusion coefficients.

The mass-balance equation for total chemical concentration in the water column (c_{T1}) and sediment (c_{T2}) segments are constructed from the sum of all the rates of change produced by each of these mechanisms. The mass-balance equation for the water column is shown by:

$$\frac{dc_{T1}}{dt} = \left.\frac{dc_{T1}}{dt}\right|_{\text{kinetic}} + \left.\frac{dc_{T1}}{dt}\right|_{\text{outflow}} + \left.\frac{dc_{T1}}{dt}\right|_{\text{diffusion}} + \left.\frac{dc_{T1}}{dt}\right|_{\text{particle}} + \left.\frac{dc_{T1}}{dt}\right|_{\text{discharge}} \tag{1}$$

The mass-balance equation for the sediment is:

$$\frac{dc_{T2}}{dt} = \left.\frac{dc_{T2}}{dt}\right|_{\text{kinetic}} + \left.\frac{dc_{T2}}{dt}\right|_{\text{diffusion}} + \left.\frac{dc_{T2}}{dt}\right|_{\text{particle}} \tag{2}$$

The water column kinetic derivative is expressed in terms of the sum of all water column decay rates K_1, which is computed as the sum of the dissolved hydrolysis, oxidation, biodegradation, photolysis and volatilization rates, multiplied by the fraction of total chemical that is dissolved f_{d1}; plus the analogous particulate rates multiplied by the particulate fraction f_{p1}. The sediment decay rate K_2 is computed in similar fashion using f_{d2} and f_{p2}. The reactions are all assumed to be first-order in the chemical concentration; they may also be functions of pH (hydrolysis) or microbial biomass (biodegradation), but that does not affect the mass-balance equations.

The outflow term is expressed in terms of the volumetric outflow rate in the water column (Q_1). The diffusive exchange of dissolved chemical between the overlying water column and the interstitial water of the sediment is expressed in terms of a mass transfer coefficient K_L (cm/day), which can be shown to be the ratio of interstitial water diffusion coefficient and the characteristic length of the gradient of dissolved chemical [7]. It multiplies the gradient of dissolved chemical concentration to produce the diffusion flux. The particle-exchange term in the water column is the difference between the loss via settling (with velocity w_a) of particulate chemical $f_{p1} c_{T1}$ and the gain via particle resuspension of sediment particulate chemical, $f_{p2} c_{T2}$ (with velocity w_{rs}). The analogous expressions in the sediment equations are the negative of the water column expressions, since a loss from the water column is a gain for the sediment layer and vice versa. The division by the depths of each layer properly accounts for the differing volumes of these two layers. Finally, the loss of sediment particulate chemical via burial is expressed in terms of a sedimentation velocity (w_s), which accounts for particles leaving the active sediment layer as sediment depth increases via sedimentation. Substituting the appropriate expressions into Equations 1 and 2 yields:

$$\frac{dc_{T1}}{dt} = -K_1 c_{T1} - \frac{Q_1}{V_1} c_{T1} - \frac{K_L}{H_1}(f_{d1} c_{T1} - f_{d2} c_{T2}) - \frac{w_a}{H_1} f_{p1} c_{T1} + \frac{w_{rs}}{H_1} f_{p2} c_{T2} + \frac{W_T}{V_1} \tag{3}$$

$$\frac{dc_{T2}}{dt} = -K_2 c_{T2} + \frac{K_L}{H_2}(f_{d1} c_{T1} - f_{d2} c_{T2}) + \frac{w_a}{H_2} f_{p1} c_{T1} - \frac{w_{rs}}{H_2} f_{p2} c_{T2} - \frac{w_s}{H_2} f_{p2} c_{T2} \tag{4}$$

These two differential equations specify the mass-balance relationships that are implied by the mechanisms of degradation, outflow, interstitial and water column diffusion, particle transport, and chemical mass discharge to the receiving water segment. It remains to solve these equations and interpret the results.

STEADY-STATE SOLUTION

The differential equations for c_{T1} and c_{T2} specify the response of the water column and sediment segment concentrations to a constant chemical discharge of magnitude W_T. The time-variable behavior has been investigated elsewhere [7]. In this chapter, solutions for the steady-state concentrations are presented and discussed.

For steady-state conditions, $dc_{T1}/dt = 0$ and $dc_{T2}/dt = 0$, and what remains are two simultaneous linear algebraic equations that can be solved directly. The results are given in Table II. Although these expressions do indeed give the solution to the steady-state behavior of the receiving-water segment, they provide no insight at all into the behavior of the solution as a function of the physical, transport and chemical parameters that describe the receiving-water segment. This is a very uncomfortable situation for the analyst, since he is essentially in the dark concerning the influence of each parameter on the predicted behavior of

Table II. Receiving Water Segment Equations and Solutions

Simultaneous Equations

$$\left(\frac{Q_1}{V_1} + K_1 + \frac{w_a}{H_1} f_{p1} + \frac{K_L}{H_1} f_{d1}\right) c_{T1} - \left(\frac{w_{rs}}{H_1} f_{p2} + \frac{K_L}{H_1} f_{d2}\right) c_{T2} = \frac{W_T}{V_1}$$

$$\left(\frac{w_a}{H_2} f_{p1} + \frac{K_L}{H_2} f_{d1}\right) c_{T1} - \left(K_2 + \frac{w_{rs}}{H_2} f_{p2} + \frac{K_L}{H_2} f_{d2} + \frac{w_s}{H_2} f_{p2}\right) c_{T2} = \emptyset$$

Solutions:

$$c_{T1} = \frac{\dfrac{W_T}{V_1}\left(\dfrac{w_{rs}}{H_2} f_{p2} + \dfrac{w_s}{H_2} f_{p2} + \dfrac{K_L}{H_2} f_{d2} + K_2\right)}{\left(\dfrac{Q_1}{V_1} + K_1 + \dfrac{w_a}{H_1} f_{d1} + \dfrac{K_L}{H_1} f_{d1}\right)\left(\dfrac{w_{rs}}{H_2} f_{p2} + \dfrac{w_s}{H_2} f_{p2} + \dfrac{K_L}{H_2} f_{d2} + K_2\right) - \left(\dfrac{w_{rs}}{H_1} f_{p2} + \dfrac{K_L}{H_1} f_{d2}\right)\left(\dfrac{w_a}{H_2} f_{d1} + \dfrac{K_L}{H_2} f_{d1}\right)}$$

$$c_{T2} = \frac{c_{T1}\left(\dfrac{w_a}{H_2} f_{p1} + \dfrac{K_L}{H_2} f_{d1}\right)}{K_2 + \dfrac{w_{rs}}{H_2} f_{p2} + \dfrac{K_L}{H_2} f_{d2} + \dfrac{w_s}{H_2} f_{p2}}$$

and

$c_{d1} = f_{d1}c_{T1}$; $\quad c_{p1} = f_{p1}c_{T1}$

$c_{d2} = f_{d2}c_{T2}$; $\quad c_{p2} = f_{p2}c_{T2}$

the fate of the chemical. Of course it is possible to vary each parameter systematically and to observe the variation of the predicted solution, but there is no substitute for the insight that can be gained from a properly simplified solution.

It is possible to cast the solution into a form that conveys an intuitive understanding of its properties. The algebraic details are given elsewhere [7]. To understand the motivation that leads to the final form of the results, consider the following simple situation. A discharge of magnitude W_T enters the water column segment of volume V_1 with outflow Q_1, and therefore hydraulic detention time $t_{01} = V_1/Q_1$. The chemical undergoes no interaction with the suspended particles present and does not interact in any way with the sediment. It is only subject to water column decay, with first-order rate constant K_1. The solution for this case is well known to be:

$$c_{T1} = \frac{W_T/Q_1}{1 + K_1 t_{01}} \tag{5}$$

and its behavior is evident by inspection.

The surprising result is that the solution for the general case, given in Table II can be cast into exactly this form:

$$c_{T1} = \frac{W_T/Q_1}{1 + K_T t_{01}} \tag{6}$$

where the apparent total removal rate K_T is given by the equation:

$$K_T = K_1 + \beta \frac{r_2}{r_1} (K_2 + K_s) \tag{7}$$

The apparent total removal rate K_T is seen to be the sum of two expressions: K_1 (the total water column decay rate) and $\beta r_2/r_1$ $(K_2 + K_s)$. This latter expression has three terms, each of which have intuitive meanings. $K_2 + K_2$ is the sum of the total sediment decay rate (K_2) and the decay rate due to loss by sedimentation:

$$K_s = \frac{w_s}{H_2} f_{p2} \tag{8}$$

These are the only mechanisms that actually remove the chemical from the sediment. Therefore, it is not surprising that they play a major role in determining the apparent total removal rate. However, their effect is modified by the remaining terms.

The parameter β:

$$\beta = \frac{m_2 H_2 f_{p1}}{m_1 H_1 f_{p2}} \tag{9}$$

is essentially the ratio of the total mass of solids in the sediment segment $m_2 AH_2 = m_2 V_2$ to that in the water column $m_1 AH_1 = m_1 V_1$, where A is the interfacial surface area. It is modified by the ratio of particulate chemical in the water column f_{p1} to that in the sediment segment f_{p2}. The parameter β accounts for the relative importance of the quantity of solids and chemical in the sediment mass relative to the water column. For a small, active sediment mass relative to the water column mass of suspended solids mass, $m_2 V_2 \ll m_1 V_1$ and β (which is termed the sediment capacity factor) is small, so that the sediment has a negligible effect on the water column concentration, unless the sediment decay rates $K_2 + K_s$ are large. A more detailed analysis of the behavior of the sediment capacity factor is given below.

The ratio r_2/r_1 is the ratio of particulate chemical concentration in the sediment r_2 to that in the overlying water r_1 in units of mass of chemical per unit mass of solids. This ratio is itself determined by the solution of the steady-state equations since, by definition:

$$r_2 = f_{p2} c_{T2}/m_2 \tag{10}$$

$$r_1 = f_{p1} c_{T1}/m_1 \tag{11}$$

As shown elsewhere [7], r_2/r_1 is given by:

$$\frac{r_2}{r_1} = \frac{(w_{rs} + w_s) f_{p2} + K_L (\pi_2/\pi_1) f_{d2}}{(w_{rs} + w_s) f_{p2} + K_L f_{d2} + K_2 H_2} \tag{12}$$

Although this expression is somewhat formidable, it has some very interesting properties that are immediately apparent. Remembering that this equation specifies the ratio of particulate adsorbed chemical concentrations in the sediment and overlying water, it is surprising which parameters are *not* part of the expression: the discharge rate of chemical W_T, the aqueous decay reaction rate K_1 and the hydraulic detention time t_{01}. Hence, r_2/r_1 is not dependent on these water column parameters. It is determined by the particulate transport parameters: (1) resuspension velocity w_{rs}, and sedimentation velocity w_s, each of which is modified by the particulate fraction in the sediment layer f_{p2}; (2) the diffusive exchange coefficient K_L, modified by the fraction dissolved in the sediment layer

f_{d2}; (3) and the total sediment decay rate–sediment depth product K_2H_2, which expresses this process as an equivalent loss velocity. These are all sediment-related parameters. (Note that the subscripts in Equation 12 all relate to the sediment segment.) The only water column parameter involved is π_1, which appears as a ratio π_2/π_1. Therefore r_2/r_1 is determined entirely by the relative magnitudes of the particulate and diffusive mass transfer coefficients, the sediment decay rate, and the partition coefficient ratio.

The reason that r_2/r_1 appears as part of the apparent total removal rate K_T (Equation 7) is that it expresses the degree of chemical contamination of the sediment particles relative to the water column particles. Since K_T is the apparent removal rate for the water column concentration c_{T1}, the ratio r_2/r_1 makes the connection between the degree of contamination in the sediment relative to that in the water column. If sediment particles are highly contaminated relative to water column particles, removal mechanisms for these particles (e.g., sedimentation K_s) have an enhanced effect on the apparent total removal rate K_T. Conversely, if r_2/r_1 is small, then removing these particles has a lesser effect on the overlying water concentration.

In fact, an alternative expression for the apparent total removal rate K_T, which clarifies this relationship, can be obtained directly from its definition (Equation 7) and those of β (Equation 9) and r_2/r_1 (Equations 10 to 12). The result is:

$$K_T = K_1 + \frac{H_2 c_{T2}}{H_1 c_{T1}} (K_2 + K_s) \tag{13}$$

or, since volumes and depths are related via the interfacial cross-sectional area A ($V_1 = AH_1$ and AH_2), K_T can be expressed as:

$$K_T(V_1 c_{T1}) = K_1(V_1 c_{T1}) + (K_2 + K_s)(V_2 c_{T2}) \tag{14}$$

The expression $K_T(V_1 C_{T1})$ has units mass of chemical removed/unit time. Hence the apparent total removal rate of chemical mass in the water column is the sum of the water column and sediment mass removal rates. The effectiveness of each segment's removal mechanism is in proportion to the total mass of chemical in that segment, an intuitively reasonable result.

The key to understanding the behavior of the steady-state solution is an appreciation of the mechanisms that control the sediment capacity factor (β) and the particulate concentration ratio (r_2/r_1). This is discussed in more detail below.

SEDIMENT CAPACITY RATIO

The effectiveness of sediment removal mechanisms (decay and sedimentation) as chemical sinks is determined to a large extent by the magnitude of the sediment capacity factor, β. Therefore, its variation as a function of the relevant physical and chemical parameters is an important component in the understanding of chemical fate. The capacity factor is given by the expression:

$$\beta = \frac{m_2 H_2 f_{p1}}{m_1 H_1 f_{p2}} \tag{15}$$

where
m_1, m_2 = particle concentrations in the water column and sediment, respectively
H_1, H_2 = water column and sediment depths, respectively
f_{p1}, f_{p2} = particulate fractions of total chemical concentration in the water column and sediment, respectively

These latter fractions are given by:

$$f_{p1} = \frac{m_1 \pi_1}{1 + m_1 \pi_1} \tag{16}$$

$$f_{p2} = \frac{m_2 \pi_2}{1 + m_2 \pi_2} \tag{17}$$

Hence, the sediment capacity factor is a linear function of the depth ratio H_2/H_1 and a more complicated function of the water column and sediment solids concentrations m_1 and m_2, and partition coefficients π_1 and π_2.

A dimensionless plot of β vs $m_1\pi_1$ and $m_2\pi_1$ is presented in Figure 2. The depth ratio chosen for this plot is $H_1/H_2 = 1000$, corresponding for example, to a water column depth $H_1 = 10$ m and an active sediment depth $H_2 = 1$ cm. The solid contours are for $\pi_2 = \pi_1$ and the dotted contours are for $\pi_2 = 0.01\ \pi_1$, which illustrates the effect of differing partition coefficients in the water column and sediment layers that might result from sediment concentration–dependent partitioning [8]. As can be seen, this latter effect is only important for small β. The principal determinants of β are: (1) the depth ratio H_1/H_2, which is inversely proportional to β; (2) the product $m_2\pi_1$, which increases β as it increases; and (3) $m_1\pi_1$ if this product exceeds 0.1 corresponding to a dissolved column fraction f_{d1} becoming appreciably less than one. The magnitude of β can vary from 0.01 to 100 as illustrated; therefore, its value should

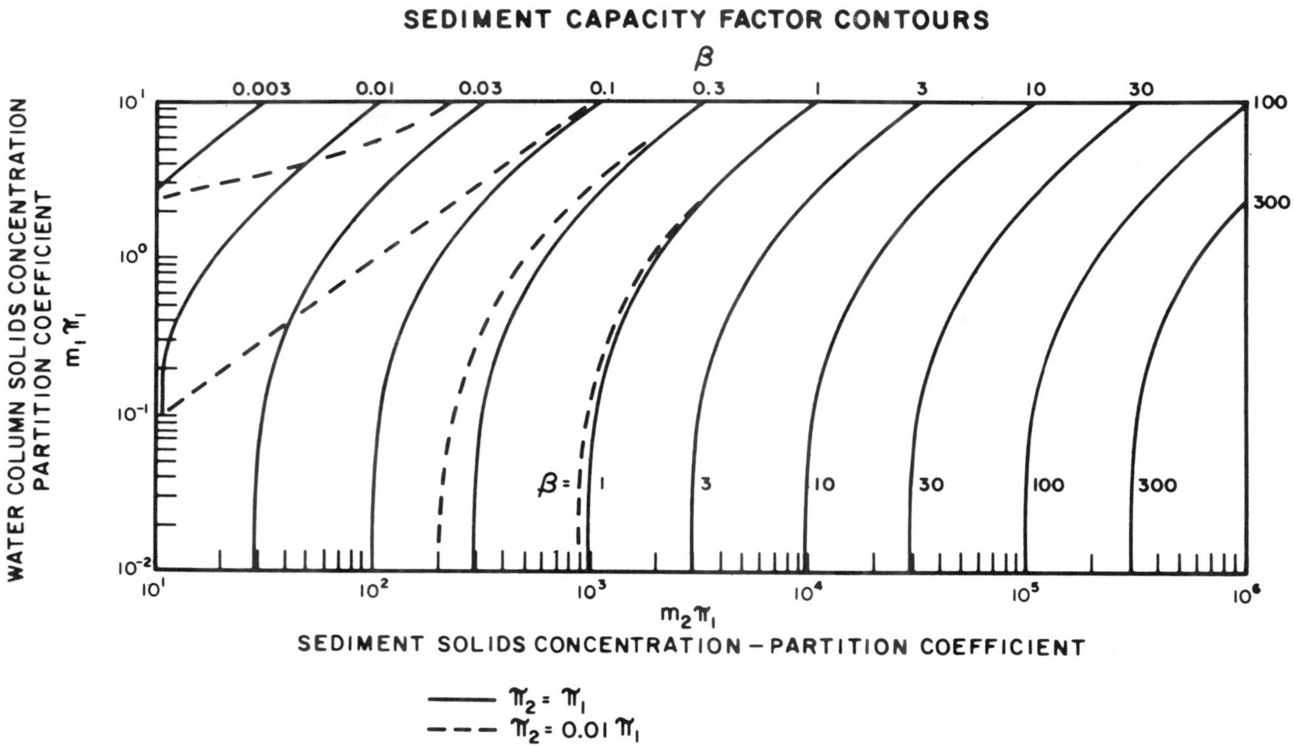

Figure 2. Contours of sediment capacity factor β vs dimensionless solids concentrations for $H_1/H_2 = 1000$.

be calculated for each situation of interest. Small values indicate that sediment-related removal mechanisms are insignificant relative to the water column concentration; large values magnify sediment effects.

PARTICULATE CONCENTRATION RATIO: THE CONDITIONS FOR EQUALITY

Certain properties of the particulate concentration ratio r_2/r_1 are apparent from inspection of its formula (Equation 12): if the resuspension and sedimentation velocity are zero, then the settling velocity is also zero; and if, in addition, the interstitial water/overlying water diffusion $K_L = 0$, then $r_2 = 0$ as expected, since these are the only mechanisms (settling and diffusion) that transport chemical to the sediment.

A more interesting question is: what conditions are necessary for $r_2 = r_1$? Conversely, what mechanisms cause the particulate concentrations to differ? The conditions for equality are:

1. Solid mass fluxes in equilibrium, that is:

$$m_1 w_a = m_2(w_{rs} + w_s) \tag{18}$$

The net flux of solids delivered to the sediment segment $m_1 w_a - m_2 w_{rs}$ is equal to the sedimentation rate $m_2 w_s$. Thus, this condition requires that there be no solids buildup in the sediment segment, which, in turn, means that there is no net storage of chemical occurring in the sediment segment because of an imbalance of solids fluxes. This condition is essentially forced by the requirement of steady state applied to the solids themselves. Hence, it should be viewed not as a restriction, but as a necessary condition implied by solids mass balance. The solids flux equality (Equation 18) is not a mathematically necessary condition. It is possible to compute r_2/r_1 without this restriction. The result is that $w_a m_1/m_2$ replaces $w_{rs} + w_s$ in the numerator of Equation 12. The effect of solids flux equality is to equate the numerator expression $w_a m_1/m_2$ to the expression $w_{rs} + w_s$.

2. The partition coefficient for sediment solids (π_2) equals the overlying water partition coefficient (π_1), or

$$\pi_2 = \pi_1 \tag{19}$$

3. There is no sediment decay:

$$K_2 = 0 \tag{20}$$

If these three conditions (solids flux equilibrium, equal partition coefficients and no sediment decay) are met, then despite the presence of settling, resuspension, sedimentation and diffusion, the particulate chemical concentrations in the water column and the sediment are equal.

This surprising result requires some explanation. The particulate concentrations are related to the dissolved concentrations via the partition coefficients: $r_2 = \pi_2 c_{d2}$. Since it is assumed that $\pi_1 = \pi_2$, the dissolved concentrations are also equal: $c_{d1} = c_{d2}$. Hence, it is not surprising that diffusion plays no role, since there is no gradient of dissolved chemical concentration between overlying water and interstitial water.

Consider the influence of particle mixing. The effect of increasing the resuspension velocity is that more particles are brought from the sediment to the water column. By the solids flux equality, the same number of particles are brought to the sediment from the water column by the increased settling velocity required by solids flux balance. These particles are transferred between segments with identical dissolved concentrations, so that they experience no change in environment and their chemical content is unchanged. Thus, there is no gradient of particulate concentration and, therefore, no driving force. Hence the increased particle mixing caused by increased resuspension and settling has no effect.

For a particulate concentration ratio $r_2/r_1 = 1$, the total apparent decay rate becomes:

$$K_T = K_1 + \beta K_s \tag{20}$$

What is remarkable about this expression is, again, which parameters are *not* involved in determining the steady-state concentrations: settling velocity, resuspension and diffusion play no role in the ultimate fate of the chemical. It is determined only by the degradation mechanisms in the water column K_1 and the loss by sedimentation and burial K_s. The effectiveness of the latter loss mechanism is determined by the magnitude of the sediment capacity factor β.

UNEQUAL PARTICULATE CONCENTRATION RATIO

If any of the conditions specified above are not met, r_2 differs from r_1. Consider the case for which the solids flux equilibrium is met. As pointed out above, this is required by the steady-state assumption. Therefore, the only mechanisms to be considered are sediment decay and unequal partitioning.

Sediment Decay

For this case, sediment decay K_2 serves to lower r_2 relative to r_1, which is reasonable, since that decay occurs only in the sediment. Consider first the case that $\pi_2 = \pi_1$. Then:

$$\frac{r_2}{r_1} = \frac{1}{1 + \dfrac{K_2 H_2}{(w_{rs} + w_s) f_{p2} + K_L f_{d2}}} \qquad (21)$$

The amount by which r_2 is less than r_1 is determined by the ratio of the total sediment decay velocity $K_2 H_2$ to the sum of resuspension and sedimentation velocities, modified by the fraction particulate f_{p2} and the diffusion velocity K_L modified by the fraction dissolved f_{d2}. These velocities represent the losses of chemical from the sediment segment and suggest a definition of a total loss velocity w_{T2} as the sum of the individual velocities:

$$w_{T2} = (w_{rs} + w_s) f_{p2} + K_L f_{d2} \qquad (22)$$

and an equivalent sediment detention time t_{02} where:

$$t_{02} = \frac{H_2}{w_{T2}} \qquad (23)$$

which is analogous to the hydraulic detention time in the water column. Then the particulate ratio can be expressed as:

$$\frac{r_2}{r_1} = \frac{1}{1 + K_2 t_{02}} \qquad (24)$$

which is precisely the form of the solution to be expected in a situation with a decay rate K_2 and a detention time t_{02}. The water column particulate concentration r_1 is analogous to the influent concentration. The resulting sediment particulate concentration is reduced relative to the influent concentration by an amount that depends on the decay rate-detention time product: $K_2 t_{02}$. Hence, a sediment decay mechanism causes $r_2/r_1 < 1$.

Unequal Partitioning

Alternatively, consider the case for which no sediment decay occurs, $K_2 = 0$, but for which the sediment partition coefficient is less than the water column partition coefficient, or $\pi_2 < \pi_1$. This is normally the case because of the large increase of solids concentration in the sediment relative to the overlying water ($m_2 \gg m_1$) and the empirical fact that this can lower partition coefficients [8]. Then the particulate concentration ratio becomes:

$$\frac{r_2}{r_1} = \frac{1 + \gamma \pi_2/\pi_1}{1 + \gamma} \tag{25}$$

where

$$\gamma = \frac{K_L f_{d2}}{(w_{rs} + w_s) f_{p2}} \tag{26}$$

which is the ratio of diffusive to particle exchange velocities. Since $\pi_2/\pi_1 < 1$, and γ is always positive, it follows that $r_2 < r_1$. The reason is that as particles enter the sediment segment, they enter an environment where the partition coefficient is lower and they desorb a portion of their chemical burden. The result is that the dissolved concentration in the sediment segment increases relative to the overlying water:

$$\frac{c_{d2}}{c_{d1}} = \frac{\pi_1}{\pi_2}\frac{r_2}{r_1} = \frac{\dfrac{\pi_1}{\pi_2} + \gamma}{1 + \gamma} > 1 \tag{27}$$

If there is very little diffusion relative to particle mixing, then $\gamma \ll 1$ and:

$$\frac{c_{d2}}{c_{d1}} \simeq \frac{\pi_1}{\pi_2} \qquad \gamma \ll 1 \tag{28}$$

and

$$\frac{r_2}{r_1} \simeq 1 \qquad \gamma \ll 1 \tag{29}$$

The small amount of aqueous-phase mixing is insufficient to reduce the dissolved concentration gradient, whereas the particle mixing is sufficient to equilibrate the particle concentrations. Conversely, if there is very little particle mixing relative to diffusion mixing, then $\gamma \gg 1$:

$$\frac{r_2}{r_1} \simeq \frac{\pi_2}{\pi_1} \qquad \gamma \gg 1 \tag{30}$$

and

$$\frac{c_{d2}}{c_{d1}} \simeq 1 \qquad \gamma \gg 1 \tag{31}$$

In this case, the particles retain their gradient, but the dissolved mixing is sufficient to reduce the aqueous gradient.

Regardless of the magnitude of γ, it is true that

$$\frac{r_2}{r_1} < 1 \tag{32}$$

and this relationship is also true for $K_2 > 0$. Hence, it is expected that the sediment particulate concentration be less than the overlying water particulate concentration.

The only mechanism that can cause $r_2 > r_1$ is the unlikely situation that $\pi_2 > \pi_1$, or that the solids fluxes are not in balance, that is:

$$m_1 w_a > m_2 (w_{rs} + w_s) \tag{33}$$

which means that solids are being delivered to the sediment segment more rapidly than the sediment is being removed via sedimentation. This implies a nonsteady-state behavior that may be realistic at short time scales but cannot persist indefinitely. Therefore, one can conclude that, for the general case considered in this receiving water segment, $r_2 < r_1$, and the sediment solids should be less contaminated than overlying water solids at steady state. Note that this result applies only for the situation where chemical is being discharged to the water column.

SEDIMENT DECAY AND UNEQUAL PARTITIONING

Physical Parameter Effects

The behavior of r_2/r_1 when both mechanisms, mass-dependent partitioning ($\pi_2 < \pi_1$) and sediment decay ($K_2 > 0$), are present is of interest. Consider first the behavior of r_2/r_1 as a function of the physical parameters: K_L and $(w_{rs} + w_s)$. Contours of r_2/r_1 are shown in Figure 3a for π_1 and π_2 and in Figure 3b for $\pi_1 = 10\pi_2$. The behavior of r_2/r_1 as $w_{rs} + w_s$

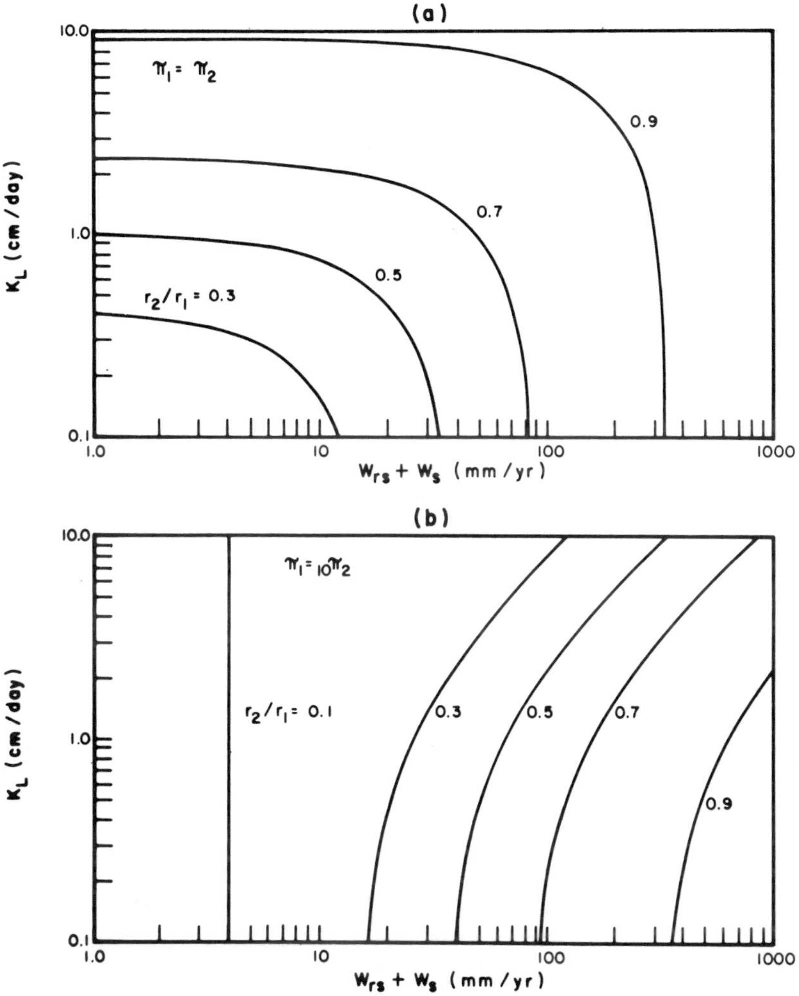

Figure 3. Contours of particulate concentration ration r_2/r_1 for $m_2 = 100{,}000$ mg/l, $H_2 = 1.0$ cm, $\pi_1 = 1000$ liter/kg, $K_2 = 0.01$/day.

increases is the same for both cases, i.e., r_2/r_1 increases toward unity as $w_{rs} + w_s$ increases, since increasing the particle mixing tends to equalize the particulate concentrations.

However, the behavior of r_2/r_1 as K_L increases depends on π_2/π_1. For $\pi_2 = \pi_1$ (Figure 3a), the overlying and interstitial water-dissolved concentrations are in the same ratio as r_2/r_1, and increasing the diffusion tends to increase r_2/r_1 toward unity since the gradients of dissolved and particulate chemical are in the same direction. With $\pi_1/\pi_2 = 10$ (Figure 3b), the interstitial water to overlying water concentration ratio is:

$$\frac{c_{d2}}{c_{d1}} = \frac{\pi_1}{\pi_2} = 10\,\frac{r_2}{r_1} \tag{34}$$

Thus although r_2 is less than r_1, c_{d2} is greater than c_{d1} (for $r_2/r_1 > 0.1$), and increasing the diffusive mixing tends to remove chemical mass from the sediment, since relatively uncontaminated overlying water is now exchanging with more contaminated interstitial water. The effect is to transfer chemical mass from the sediment to the overlying water, thereby decreasing r_2 relative to r_1. This effect occurs so long as $r_2/r_1 > 0.1$, so that c_{d1} is smaller than c_{d2}. However at $r_2/r_1 = 0.1$, $c_{d1} = c_{d2}$, since it is assumed in this example that $\pi_1/\pi_2 = 10$ and varying K_L has no effect on r_2/r_1, since no dissolved chemical concentration gradient exists. This corresponds to the vertical contour in Figure 3b. For $r_2/r_1 < 0.1$, the contours would resemble Figure 3a, and increasing K_L would tend to increase r_2/r_1.

Chemical Parameter Effects

The behavior of r_2/r_1 as a function of the chemical parameters K_2 and π_2 is shown in Figure 4a for $\pi_1 = \pi_2$ and in Figure 4b for $\pi_1 = 10\pi_2$. Increasing K_2 decreases r_2/r_1 in both cases, as would be expected However, the effect of increasing π_2 depends on whether $\pi_1 = \pi_2$, in which case increasing π_2 reduces dissolved mixing and increases the residence time in the sediment so that decay can be more effective (Figure 4a). If $\pi_1 = 10\pi_2$ (Figure 4b), increasing π_2 inhibits the transfer of chemical mass from the sediment to the overlying water, thus increasing r_2/r_1.

Based on these results, it appears that the probable range for r_2/r_1 is 0.1–1.0 for $K_2 < 0.1$/day. Combining this information with the probable range of β of 0.01–100 suggests that the range of $\beta(r_2/r_1)$, which is the parameter group that directly determines the importance of the sediment removal mechanisms, $K_2 + K_s$ is in the range $\beta(r_2/r_1) = 0.001$–100.

OVERALL REACTION RATE

The probable range of K_T is determined by sediment decay rate (K_2) and the equivalent sedimentation removal rate (K_s). From reservoir sedimen-

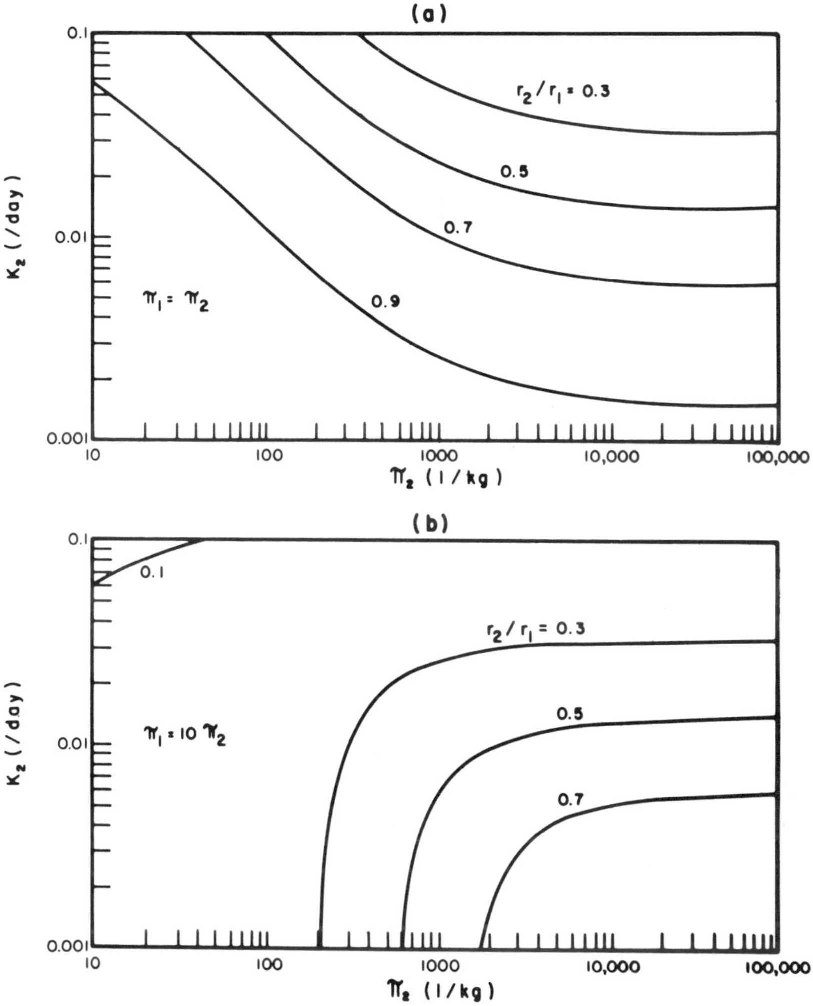

Figure 4. Contours of particulate concentration ratio r_2/r_1 for $w_{rs} + w_s = 50$ mm/yr, $K_L = 1.0$ cm/day, $m_2 = 100{,}000$ mg/l, $H_2 = 1.0$ cm.

tation data, has been estimated to be in the range: $\beta K_s = 0.025\text{–}0.5/\text{day}$ [4]. For r_2/r_1 in the range 0.1–1.0, the sedimentation sink is estimated to be in the range $\beta(r_2/r_1)K_s = 0.0025\text{–}0.5/\text{day}$.

Not much is known concerning the range of K_2 for the chemicals of concern; but if the probable range is $K_2 = 0\text{–}0.1/\text{day}$, then the range for $\beta(r_2/r_1)K_2 = 0\text{–}10/\text{day}$, the overall removal rate is:

$$K_T = K_1 + \beta \frac{r_2}{r_1} K_2 + \beta \frac{r_2}{r_1} K_s \tag{35}$$

$$K_T \simeq (0 - 0.1) + (0 - 10) + (0.0025 - 0.5)\,(/\text{day})$$

Hence even conservative chemicals (e.g., heavy metals, for which K_1 and K_2 are zero) can exhibit nonconservative behavior because of loss via sedimentation. Adding sediment decay can significantly increase the overall removal rate as can direct water column decay.

APPLICATION TO STREAMS AND RIVERS

The analysis of the fate of chemicals in streams and rivers is similar in many ways to the situation in lakes with what appears to be at first glance a significant difference. Whereas lakes can be represented, in many cases, by two completely mixed volumes, i.e., the water column and active sediment segments, the concern in a stream is with the longitudinal distribution of chemical downstream from the point of discharge. The key to the solution for this situation is shown in Figure 5. Consider the stream as being represented by a sequence of completely mixed volume segments. Let superscripts denote the segment being considered. Thus $c_{T1}^{(1)}$ is the water column chemical concentration in segment 1; $Q_1^{(1)}$ is the outflow rate; $t_{01}^{(1)}$ is the detention time, and so on for all parameters.

The first stream segment receives a loading rate of chemical $W_{T1}^{(1)}$. Its outflow is $Q_1^{(1)}$, the flowrate of the stream. If longitudinal dispersion is neglected, then the first stream segment is exactly analogous to the situation analyzed previously. The equation for $c_{T1}^{(1)}$ is, as before:

$$c_{T1}^{(1)} = \frac{W_T^{(1)}/Q_1^{(1)}}{1 + K_T^{(1)} t_{01}^{(1)}} \tag{36}$$

where $c_{T1}^{(1)}$ = water column concentration
$t_{01}^{(1)}$ = detention time of the first stream segment
$K_T^{(1)}$ = overall apparent removal rate evaluated using the parameters characteristic of the first segment

Consider the second completely mixed stream segment. The principal simplifying assumption to be employed in this analysis is that there is no horizontal bed motion. Although vertical sediment and interstitial water mixing is still being considered, horizontal bed motion is assumed not to be occurring. The reason for this assumption is that, for the fixed-bed case, the second stream segment receives only a water column input of chemical as a result of the stream flow from the upstream segment.

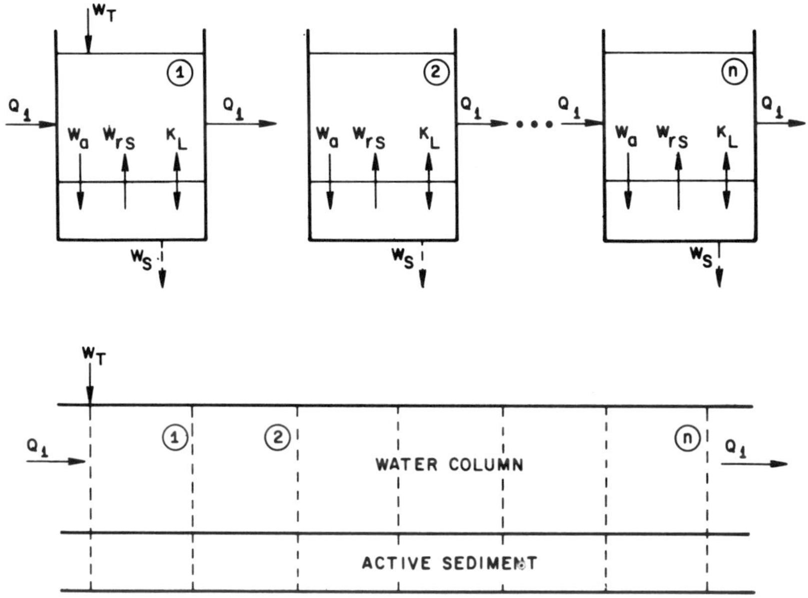

Figure 5. Representation of flowing streams as a sequence of receiving water segments.

$$W_{T1}^{(2)} = Q_1^{(1)} c_{T1}^{(1)} \tag{37}$$

The key observation is that the second segment is also exactly analogous to the lake situation, and therefore its water column concentration is given by:

$$c_{T1}^{(2)} = \frac{W_{T1}^{(2)}/Q_1^{(2)}}{1 + K_T^{(2)} t_{01}^{(2)}} \tag{38}$$

where the superscript denotes the fact that all parameters are evaluated using the appropriate values for the second stream volume element. Note that it is not necessary that $Q_1^{(2)} = Q_1^{(1)}$ so that lateral inflow augmenting the streamflow is permitted in the analysis.

A similar argument applies to each stream volume element so that for the ith element:

$$c_{T1}^{(i)} = \frac{W_{T1}^{(i)}/Q_1^{(i)}}{1 + K_T^{(i)} t_{01}^{(i)}} \tag{39}$$

and

$$W_{T1}^{(i)} = Q_1^{(i-1)} c_{T1}^{(i-1)} \tag{40}$$

For each volume element, $K_T^{(i)}$ is computed using the parameters characteristic of the ith segment.

Equations 39 and 40 represent a complete solution of the fixed-bed stream case and are well suited to numerical computation. The stream volume elements are chosen to be small enough so that the longitudinal change in all parameters is small between adjacent segments and the longitudinal gradient of concentration is well represented. A simple criterion that ensures this to be the case is that $K_T^{(i)} t_{0i}^{(i)} < 0.1$. However, as with most general solutions, these equations provide little insight into the solution behavior.

Consider an idealized situation for which all physical, transport and reaction parameters are constant with respect to distance. This may well be an adequate representation of a section of a stream that has fairly uniform characteristics. In particular it is assumed that the water column and active sediment solids concentrations and depths are constant, together with reaction rates and vertical transport coefficients. For this case, the water column concentrations become, from Equation 36:

$$c_{T1}^{(1)} = \frac{W_{T1}/Q_1}{1 + K_T t_{01}} \tag{41}$$

and, from Equations 37 and 38:

$$c_{T1}^{(2)} = \frac{Q_1 c_{T1}^{(1)}/Q_1}{1 + K_T t_{01}} = \frac{W_{T1}}{Q_1} \frac{1}{(1 + K_T t_{01})^2} \tag{42}$$

so that the nth volume element concentration becomes:

$$c_{T1}^{(n)} = \frac{W_{T1}}{Q_1} \frac{1}{(1 + K_T t_{01})^n} \tag{43}$$

The arbitrary parameter in this formulation is the detention time (t_{01}) since it depends on the volume (V_1) of each stream segment. Note that no other parameter involves V_1.

The method of removing V_1 from this solution uses the following device. Let the volume size shrink to zero and at the same time let the number of volumes increase to infinity. The mathematical form of the argument is to define V_1 as the product of the stream water column cross-sectional area A_l and a length Δx. That is, $V_l = \Delta x A_l$, and the detention time becomes:

$$t_{0l} = \frac{V_l}{Q_l} = \Delta x \frac{A_l}{Q_l} = \frac{\Delta x}{U_l} \tag{44}$$

where U_l = water column longitudinal velocity

Consider a location at a distance x from the location of the discharge. If Δx is the length of each stream volume segment, then the number of stream volume segments between the discharge location, which is taken to be $x=0$, and location x is:

$$n = \frac{x}{\Delta x} \tag{45}$$

so that the water column concentration $c_{Tl}^{(n)} = c_{Tl}(x)$ becomes, from Equation 43:

$$c_{Tl}(x) = \frac{W_{Tl}}{Q_l}\left(1 + \frac{K_T \Delta x}{U_l}\right)^{-x/\Delta x} \tag{46}$$

Now letting $\Delta x \rightarrow 0$ and using the limit:

$$\lim_{y \rightarrow 0}\left(\frac{1}{1 + ay}\right)^{1/y} = e^{-a} \tag{47}$$

yields:

$$c_{Tl}(x) = \frac{W_{Tl}}{Q_l} \exp\left(\frac{-K_T x}{U_l}\right) \tag{48}$$

This remarkably simple result is exactly analogous to the solution for the longitudinal distribution of a nonconservative substance, which decays following a first-order reaction with rate coefficient K_T in the water column only. Hence, K_T plays exactly the same role in the stream model for partitioning chemicals as it does in the lake model. It is the apparent overall decay rate that accounts for both water column decay and sediment water interaction and decay. The expression for K_T is the same as before, as is r_2/r_1; and since all of the parameters are assumed to be constant in the reach of stream being considered, r_2/r_1 is seen to be independent of location. This is, at first glance, quite unexpected, since the longitudinal distribution of water column concentration is decreasing exponentially, and it should follow that the sediment concentration is decreasing as well. That is indeed the case since:

$$c_{T2}(x) = \beta \frac{r_2 H_1}{r_1 H_2} c_{T1}(x) \tag{49}$$

which follows directly from the definitions of r_2 and r_1. Thus c_{T2} is also decreasing exponentially, but r_2/r_1 is constant.

The reason that r_2/r_1 is independent of water column concentration and stream flow, just as it was independent of loading rate and detention time in the lake setting, is as follows: at any location x, particles being mixed vertically by resuspension and settling, and interstitial water mixing with overlying water react locally with whatever the overlying water concentration is at that location. At steady-state equilibrium, r_2/r_1 is independent of water column conditions, as is the ratio of interstitial water to overlying water $[c_{d2}/c_{d1} = (\pi_1/\pi_2)(r_2/r_1)]$ for the same reason that they are independent of water column conditions in the lake setting: all reactions are assumed to be first-order, and transport and chemical partitioning are linear with respect to concentration. Hence, considering ratios eliminates the absolute dependence on concentrations.

Therefore, analysis of the distribution of chemicals in flowing streams follows the same principles outlined in the analysis of lakes, and all the investigations of the behavior of the sediment capacity factor and r_2/r_1 as a function of parameter values are directly applicable. It is for this reason that the behavior of these fundamental parameters is crucial to the understanding of fate of chemicals in general. It can be shown that this reasoning applies to estuaries as well, as long as horizontal bed transport is not significant.

POSTSCRIPT

The above analysis is a good example of the insight that can be gained from a properly formulated analytical solution. The relative importance and interactions of the 13 parameters W_T/Q_1, t_{01}, m_1, H_1, m_2, H_2, K_1, K_2, π_1, π_2, K_L, w_{rs} and w_s in determining the resulting receiving water and sediment concentrations can be pursued rationally. The importance of the dimensionless expressions for sediment capacity factor β and the central role played by the ratio of particulate concentrations r_2/r_1 becomes clear. The symmetry between lake and stream models is also apparent. An intuitive understanding of the situation can be gained, which, together with a feel for the relative magnitudes involved, can provide a degree of insight into a specific problem setting that is not otherwise attainable.

Contrast this situation to that which one faces when the problem is apparently so complex that only computer-based solutions are possible. Even when this is the case and a numerical computation is required, without the insight gained from the analysis of simplified situations, it is not possible to react intelligently to the numerical results. Apparently counterintuitive results are difficult to explain and even more difficult to check. Imagine one's response, for example, to computational results that indicated that increasing or decreasing sediment-water transport had no effect on the resulting water column and sediment concentrations. For the models analyzed in this paper, the reason for such behavior is clear: it occurs whenever $r_2 = r_1$. Whether this would be evident after an investigation of purely numerical results is uncertain, but our experience has been that insight is most rapidly and surely gained by an investigation of the structure of a properly posed and analyzed idealization.

ACKNOWLEDGMENTS

The results described in this chapter are part of a project report "Analysis of Fate of Chemicals in Receiving Waters" (CMA Project ENV-7-W) prepared by HydroQual for the Chemical Manufacturers Association. The authors gratefully acknowledge the assistance of the Chemical Manufacturers Association's Aquatic Research Task Group past and present members: R. E. Bailey, W. Bishop, J. L. Hamelink, R. A. Kimerle, E. C. Ladd, A. H. Lasday, G. Loewengart, R. J. Moolenaar, C. F. Muska, W. Pagano, C. P. Priesing, R. J. Samelson and W. F. Tully; and Chemical Manufacturers Association personnel: G. V. Cox and R. R. Romano for their interest, technical guidance and support throughout this project. Members of the HydroQual technical staff who have contributed to this project are James J. Fitzpatrick, Michael T. Kontaxis, Paul R. Paquin and Thomas W. Gallagher.

The project report and a computer program that implements the calculations discussed in this paper are available for a nominal fee from the Chemical Manufacturers Association.

REFERENCES

1. Smith, J. H., W. R. Mabey, N. Bohonos, B. R. Hold, S. S. Lei, T. W. Chow, D. C. Bomberger and T. Mill. "Environmental Pathways of Selected Chemicals in Freshwater Systems," U.S. EPA Report 600/7-77-113 (1977).

2. Baughman, G. L., and R. R. Lassiter. "Prediction of Environmental Pollutant Concentration," in *Estimating the Hazard of Chemical Substances to Aquatic Life,* ASTM Technical Publication 657, J. Cairns, K. L. Dickson and A. W. Maki, Eds. (Philadelphia: American Society for Testing and Materials, 1978), pp. 35–70.
3. Burns, L. A., C. M. Cline and R. R. Lassiter. "Exposure Analysis Modeling System (EXAMS): User Manual and System Documentation," U.S. EPA Environmental Research Laboratory, Athens, GA (1981).
4. O'Connor, D. J., and J. L. Schnoor. "Distribution of Organic Chemicals and Heavy Metals in Lakes and Reservoirs," Manhattan College Technical Report (1980).
5. Thomann, R. V., and D. M. DiToro. "Preliminary Model of Recovery of the Great Lakes Following Toxic Substances Pollution Abatement," paper presented at the Workshop on Scientific Basis for Dealing with Chemical Toxic Substances in the Great Lakes, Great Lakes Basin Commission, Ann Arbor, MI (1979).
6. Neely, W. B. *Chemicals in the Environment—Distribution, Transport, Fate, Analysis* (New York: Marcel Dekker, Inc., 1981).
7. DiToro, D. M., D. J. O'Connor, R. V. Thomann and J. P. St. John. "Analysis of Fate of Chemicals in Receiving Waters—Phase I," HydroQual, Inc., Mahwah, NJ (1981).
8. O'Connor, D. J., and J. P. Connolly. "The Effect of Concentration of Adsorbing Solids on the Partition Coefficient," *Water Res.* 14:1517 (1980).

CHAPTER 11

SYNOPSIS OF DISCUSSION SESSION: MODELING FATE PROCESSES

R. E. Bailey
Session Chairman

L. A. Burns, J. L. Schnoor, W. B. Neely and A. F. Werner

Successful modeling (prediction) of chemical fate requires an accurate equation for each significant process, and valid chemical, biological and physical data. Simple equilibrium partitioning models such as that of Neely and Mackay [1] require only a few input parameters: the relatively reliable physicochemical properties. This simple partitioning model predicts the most significant compartments to examine for transformation processes. The different processes used in chemical fate models are listed in Table I. Our present perception of confidence in the process equations, need for further development and ease of input data acquisition are included.

TRANSPORT

Advective transport refers to the movement of chemicals entrained in the current. Current velocity is the easiest of all transport parameters to measure. Mean current velocities (or flow divided by cross-sectional areas) yield reasonable approximations of advective transport for many uses. Dispersive transport consists of mixing caused by shear forces in the water caused by wind, bottom friction, density gradients, stream sinuousity and irregular geometries. Our level of knowledge regarding longitudinal dispersion in streams, rivers and estuaries is good. However,

Table I. Physical, Chemical and Biological Fate Processes

Process	Level of Confidence in Process Equations	Further Development Work Necessary	Facile Data Gathering to Use Process Equations
Transport[a]			
Advective	High	No	Yes
Dispersive	Medium	Yes	No
Water			
Sediment/Water	Low	Yes	No
Sediment			
Cohesive	Low	Yes	No
Non-cohesive	High	No	Yes
Partitioning			
Sorption Equilibrium			
Neutral	High	No	Yes
Ionic	Low	Yes	Moderate
Bioaccumulation	High	Yes[b]	Yes
Henry's Constant	High	No	Yes
Kinetics			
Sorption Kinetics	Medium	Yes[c]	Moderate
Volatilization	Medium	Yes[d]	Yes
Biotransformation, microbial			
Water Column	Medium	Yes	Moderate[e]
Sediment	Low	Yes	No
Photobiological, Algae	NA[f]	Yes	No
Higher Organisms	Low	Yes	Moderate
Specific Acid/Base and Neutral Photolysis	High	No	Yes
Direct Photolysis	High	No[g]	Yes
Indirect Photolysis	Medium	Yes[g]	No
Reductive Processes	Low	Yes	No
Acid/base Catalysis	Medium	Yes	No
Metal Catalysis	NA[f]	Yes	No
Sediment "Catalysis"	NA[f]	Yes	No

[a] Transport process designations are for surface waters; groundwater transport equations excluded.
[b] Invertebrate bioconcentrate requries further evaluation.
[c] Particularly mechanistic concerns.
[d] Validation of models.
[e] Too many protocols in use; no clearly defined test.
[f] NA = process equation not available.
[g] Role of oxygen quenching needs more work.

compartmentalization of models results in an artificial or numerical dispersion of which the user should be aware [2]. More development is needed for lateral and vertical mixing. Vertical dispersion (diffusion) between water and sediments is dependent on bioturbation, bed characteristics (including sediment character and roughness) and water-velocity fluctuations. It presently can be estimated using tracers (e.g., temperature, dissolved oxygen or nutrient releases). A first-principle equation for the process is desirable, but presently not feasible. Theories of bed load transport of fluvial, noncohesive sediments (sand) are well developed for rivers, but cohesive sediment (clays and silts) transport is not.

PARTITIONING

Partition coefficients between sediment and water have been reported in the literature for a large number of compounds. Although there is some range in reported values for a given substance, most of this is because of differences in the nature of the sediments and laboratory measurement techniques. For neutral hydrophobic pollutants, the octanol/water partition coefficient K_{ow} and the organic carbon content of the sediment are quite reliable estimators of the sediment/water partition coefficient. Ionic chemicals are not easily predicted because of ion exchange and competitive elution phenomena on natural sediments.

Bioaccumulation refers to both uptake from dissolved pollutants in water and uptake from food items and sediment ingestion. Bioconcentration is known to be proportional to the exposure concentration and the lipid content of the organism according to both laboratory and field measurements. Residues in food items of long food chains for hydrophobic substances may need to be considered, such as PCB in lake trout. Otherwise, bioaccumulation from the ingestion of food items can be neglected for most purposes. Many soluble chemicals are rapidly metabolized by fish and do not bioconcentrate; the role of metabolism needs further exploration. Measured bioconcentration factors (BCF) for these substances are lower than one would predict from K_{ow}–BCF correlations. More work is needed on bioaccumulation by invertebrates, particularly for heavy metals.

Henry's law constant K_H represents an equilibrium partitioning between vapor and solute. It is a function of pressure, temperature and solvent characteristics, but for the pertinent range of environmental characteristics, it can be assumed constant. Ranges of reported laboratory values can be wide, especially for complex mixtures of isomers such as polychlorinated biphenyls. The use of vapor pressure and solubility information is a good way to estimate Henry's law constant.

KINETICS

Whitman's "two-film" theory [3] appears adequate for environmental modeling of volatilization. Several versions of the theory are available. These should be compared under field and time-varying conditions. Significant net transfers between water bodies and the atmosphere can occur in either direction; the models account for the possibility that the atmosphere is a source of aquatic contamination.

Homogeneous phase-specific acid/base and neutral hydrolysis reactions are well understood. Sorption appears to arrest the reactions, but has little effect on neutral pathways up to 0.02 kg sediment/l. Neutral hydrolysis in benthic sediments requires further elucidation.

General acid/base catalysis can be estimated via the Bronsted equation and simple laboratory tests. Its environmental significance is unclear, but probably unimportant in the water column. Its importance in sediments depends on the effect of sorption on chemical reactivity.

Metal catalysis is conceptually simple, but its significance is unknown, as with general acid/base catalysis.

Reductive processes are probably important in sediments of anaerobic and hypolimnetic waters. The exact nature of the process (chemical or biological) is unknown.

Natural sediments accelerate or "catalyze" some thermal reactions. The mechanism may be biologically mediated, but purely chemical pathways may also contribute to this phenomenon.

An accurate process description of direct photolysis is available. Low dissolved oxygen may accelerate this process, but the situation is atypical.

Indirect photolysis is a class of reactions, rather than a single reaction. The singlet oxygen pathway has been described in a model-commensurate fashion; other pathways require further development.

Sorption kinetics require a multiphasic treatment. Current formulations, although of considerable predictive value, may receive significant revision.

Photobiological transformations of synthetic organics have been observed in the laboratory. What, where, when, why, how and how much are open questions.

Higher-organism metabolism, although of major significance to the organism, is usually of little moment to the overall fate of a compound. In a food-chain vector, however, these pathways may be a major effect pathway.

Microbial biotransformations are a class of reactions, rather than a single phenomenon. Although population numbers play a role in kinetics, and metabolic energy yield controls some cases, current formulations lack the secure theoretical bases available for chemical processes. In particular, the degradation rate in sediments has been shown [4] to be the dominant biodegradation route of linear alkylbenzene sulfonate, a moderately sorbed chemical. Most of the present biotransformation and mineralization rates are determined in the water column. Models suggest that a surprisingly large number of chemicals reside in benthic sediments. Therefore, the lack of biotransformation rates in sediments limits the application of models to chemicals strongly sorbed.

APPLICATIONS

One of the most useful applications of models is in the identification of critical processes for testing. Prediction of environmental biotransformation rates is still not readily translatable to all environments. The second-order rate formulation of Paris et al. [5], using standard plate counts to enumerate bacterial population, has been applied successfully in some cases. Other workers have found that alternative factors for estimation of microbial activity seem to work as well. Determination of first-order biotransformation rates in the environment of interest is currently the most useful approach to site-specific projects. The generalization of rates to different aquatic environments, sediments and soil awaits further research. Biotransformation in groundwater is presently believed to be small, but the exact role needs to be defined by further research. Development of more general biotransformation estimation techniques is presently under development in many laboratories.

The fate and significance of daughter products from the transformation of chemicals is conceptually straightforward to handle in models. The product from one step is simply fed to the next. Potential environmental problems could arise from a persistent minor product that is overlooked in the analysis because it appears to be of little importance in the fate of the parent compound.

In general, most of the models have been applied for organic compounds, particularly neutral organics. Relatively little work has been performed for high-molecular-weight organic polymers.

INORGANICS

Also notably absent from the process equation table, with the exception of some abiotic kinetics, is inorganic compounds, including nutrients (such as ammonia, phosphorus, sulfur and nitrogen) and the heavy metals (such as lead, nickel, cadmium and zinc). Because nutrients and heavy metals play such an integral part in biological systems, they are readily accumulated and incorporated into all aquatic and terrestrial trophic levels. Thus, studies of the fate of these elements are hampered by difficulties in decoupling the abiotic and biotic processes in meaningful ways. We are of the opinion that heavy metals will require specific modeling that may not easily be generalized to include other elements. However, because there is a finite number of elements, we feel that this is a reasonable endeavor that is already well underway.

The rare earth elements, because of their relative infrequency, have not been identified as important environmentally. However, because of the radioactive compounds within this family, special attention has been directed to this area. Extensive monitoring and modeling programs are already underway within the nuclear industry. Because of the facile nature of analytical detection, radionuclides provide the best opportunity for validation of models. It appears that hydrological models have the most to gain from this area of expertise in the immediate future.

REFERENCES

1. Neely, W. B., and D. Mackay. "An Evaluative Model for Estimating Environmental Fate," Chapter 8, this volume.
2. Schnoor, J. L., N. B. Rao, K. J. Cartwright and R. M. Noll. "Fate and Transport Modeling for Toxic Substances," Chapter 9, this volume.
3. Whitman, W. G. "A Preliminary Experimental Confirmation of the Two-Film Theory of Gas Absorption," *Chem. Metal. Eng.* 29:146–148 (1923).
4. Games, L. M. "Field Validation of Exposure Analysis Modeling Systems in a Flowing Stream," Chapter 18, this volume.
5. Paris, D. F., W. C. Steen, G. L. Baughman and J. T. Barnett. "Second-Order Model to Predict Microbial Degradation of Organic Compounds in Natural Waters," *Appl. Environ. Microbiol.* 41:603–609 (1981).

SECTION 3

STATE-OF-THE-ART AND CASE STUDY EXAMPLES

CHAPTER 12

HYDROLOGIC TRANSPORT MODELING

Norman H. Crawford
Hydrocomp International, Inc.
Mt. View, California 94040

Compounds become pollutants to the environment when they appear where they are not wanted. Most of these compounds are highly desirable in some places in the environment and highly undesirable in other places. Compounds move from place to place in two major ways: transport in air or water. Compounds that are transported in the air may evaporate from the land surface or may be released by a smokestack or an auto exhaust. They may be carried in the air for some distance before depositing on the land surface or in lakes and streams.

Transportation of compounds in water can be in surface runoff, or subsurface or groundwater flows. Rivers and streams have been used for decades to transport unwanted waste products from factories and cities. Water flow also causes inadvertent transport of compounds that become undesirable pollutants in new locations. Nutrients moved from agricultural lands by surface and subsurface flows cause algae problems in lakes and reservoirs. Topsoil eroded by surface runoff from farmlands deposits in river channels and is harmful to navigation.

Modeling the hydrologic cycle and hydrologic transport of pollutants was discussed and attempted early in this century, but lack of technology for intensive computation and massive data management prevented progress. Shortly after the introduction of digital stored-program computers, successful hydrologic models were developed [1]. The use of continuous hydrologic models to follow the transport of pollutants was a logical next step. One of the first studies of transport modeling [2] developed a digital model for the movement of radionuclides. Huff [2] was also the first person to use the term "hydrologic transport model."

Modeling has now become a standard tool in environmental analysis. Hydrologic transport modeling has been expanded to include almost all of the commonly considered environmental pollutants. Continuous models are now available that handle biological variables, sediment, conservative ions, biochemical oxygen demand (BOD), dissolved oxygen (DO), pesticides, nutrients and water temperature.

This chapter is a summary of the development of hydrologic transport modeling and the current state-of-the-art, and of the prospects for large-scale use of models in environmental regulation. Historic developments and new trends in model algorithms are discussed in the first section. The regulatory use of models is described in the second section. The last section proposes future directions for model development and use.

DEVELOPMENT OF MODELING TOOLS

The primary goal in modeling is to create algorithms that represent the physical processes that occur in nature. These algorithms are combinations of physical laws and empirical relationships. Models are tested against field data and are refined until the algorithms reproduce the observed phenomena as closely as possible. Hundreds, and perhaps thousands, of environmental models have been developed in the last 20 years. In hydrologic modeling, the first comprehensive hydrologic model that viewed continuous hydrologic processes on a detailed conceptual level was the Stanford Watershed Model. This model was easily adapted to hydrologic transport modeling because the physical processes of surface runoff, interflow and groundwater were all represented, and the Stanford models for hydrologic transport of sediments, water quality and the biological cycle were all accomplished within 10 years of the first publication of this model. Many variations of the Stanford model were developed and published [3–8] and many other hydrologic transport models have been developed [9–16].

Continuous and Steady-State Models

Hydrologic and water-quality processes are shown in Figure 1. The physical processes that drive hydrologic transport are continuous. Rainfall, overland flow, temperature and streamflow all vary continuously in time, and measured pollutant concentrations also vary continuously in time (Figure 2).

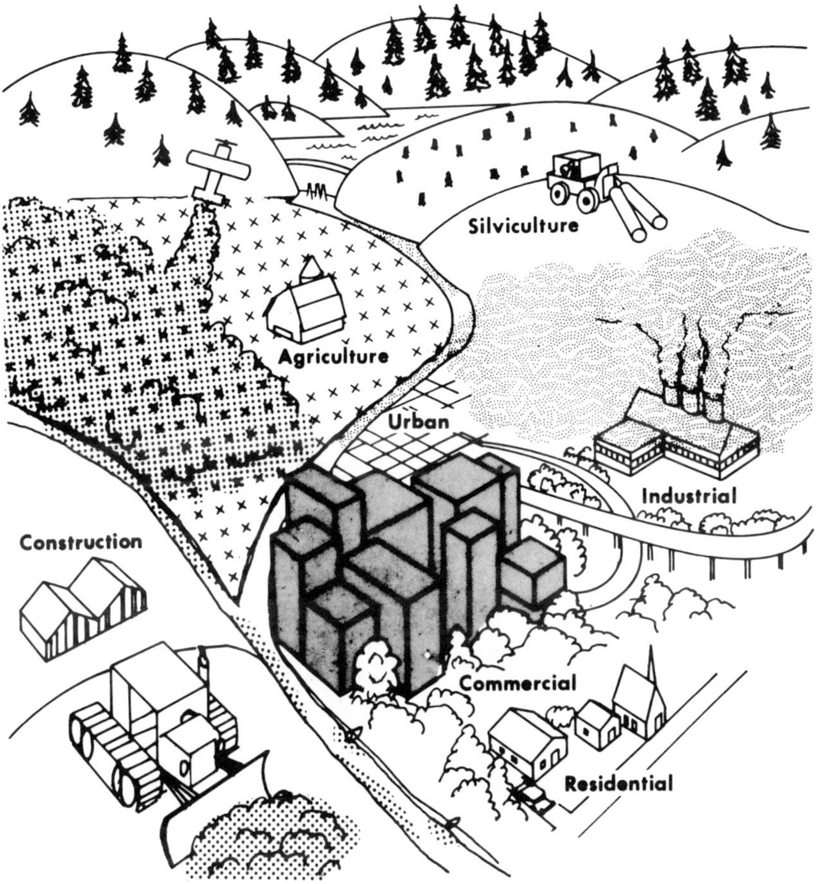

Figure 1. Hydrologic and water-quality processes.

Continuous models dynamically reproduce physical processes like infiltration, sediment transport and degradation of compounds. The Hydrological Simulation Program—FORTRAN (HSPF) modeling system uses "land segments" and "stream reaches." Representative process flowcharts for HSPF are given in Figures 3 and 4.

Continuous inflows to and outflows from land segments and stream reaches are calculated, and these physical elements can be interconnected to form networks of any complexity (Figure 5). In Figure 5, each of the arrows represents a time series of flow and pollutant transport. Reservoirs in HSPF are considered to be a special stream reach.

Many modeling techniques reduce the number of time-dependent variables by assuming that one or more of the driving physical processes are

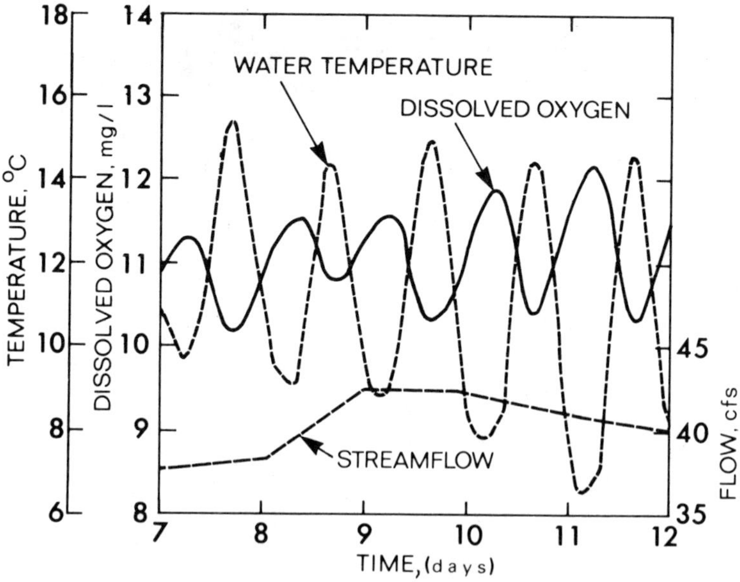

Figure 2. An example of diurnal fluctuations in quality factors. Big Soos Creak near Auburn, Washington (data courtesy CH2M/Hill).

not time-dependent. "Worst-case" design assumptions, like concurrent low stream flow/high stream temperature, have been used to study oxygen depletion below outfalls. Eliminating time dependency of certain variables can expedite analysis, but information on the true frequency of events is usually lost.

Benefits and Cost Analysis

The advantage of continuous simulation compared to a worst-case design study is that the frequency of damaging conditions can be studied and the cost/benefit of alternatives can be found using methods common to other water-resource investigations. When continouous data on a pollutant are calculated for different alternatives, a quality-duration curve can be plotted for each alternative (Figure 6). Costs of various alternatives can be estimated by usual cost-estimating methods. Effectiveness of alternative developments is indicated by the change in area under the quality-duration curve. The total change is a measure of the total improvement in quality. The choice of alternatives can be made on the basis of net economic benefits and the accompanying environ-

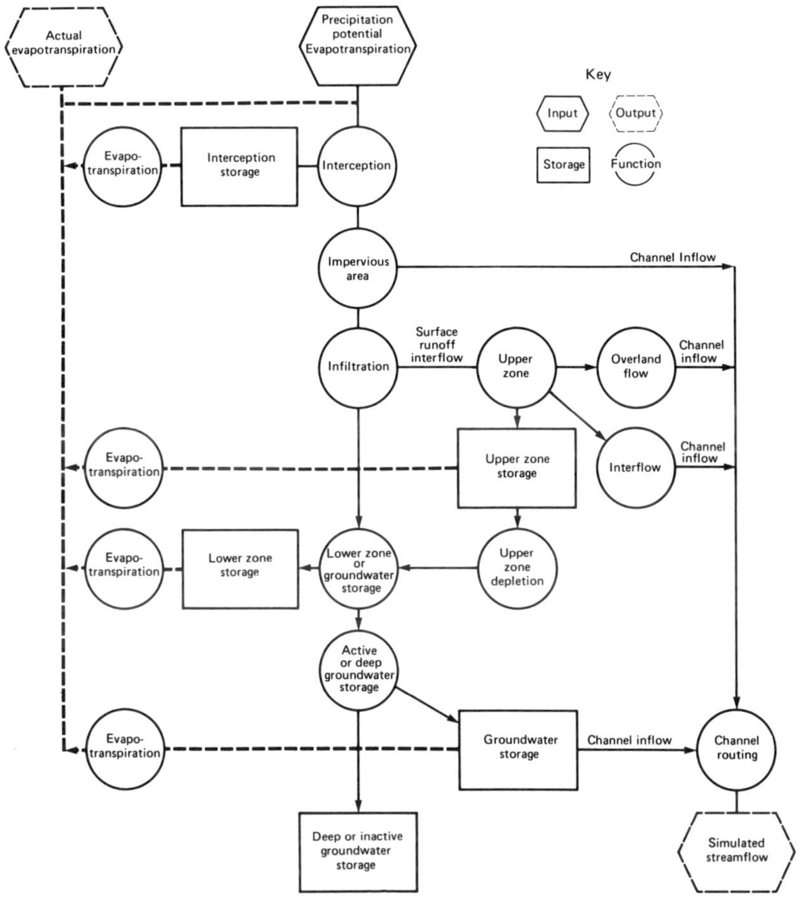

Figure 3. Land segment flowchart.

mental enhancement [17]. Recently, U.S. Environmental Protection Agency (EPA) regulations for industrial water pollution were blocked in federal court in Richmond, Virginia, because the EPA "failed to weight costs against benefits."

Current Studies

Recent work on the HSPF system [18] has emphasized the importance of a fully integrated software system for data management, graphics and simulation of all interfacing hydrologic, hydraulic and water-quality

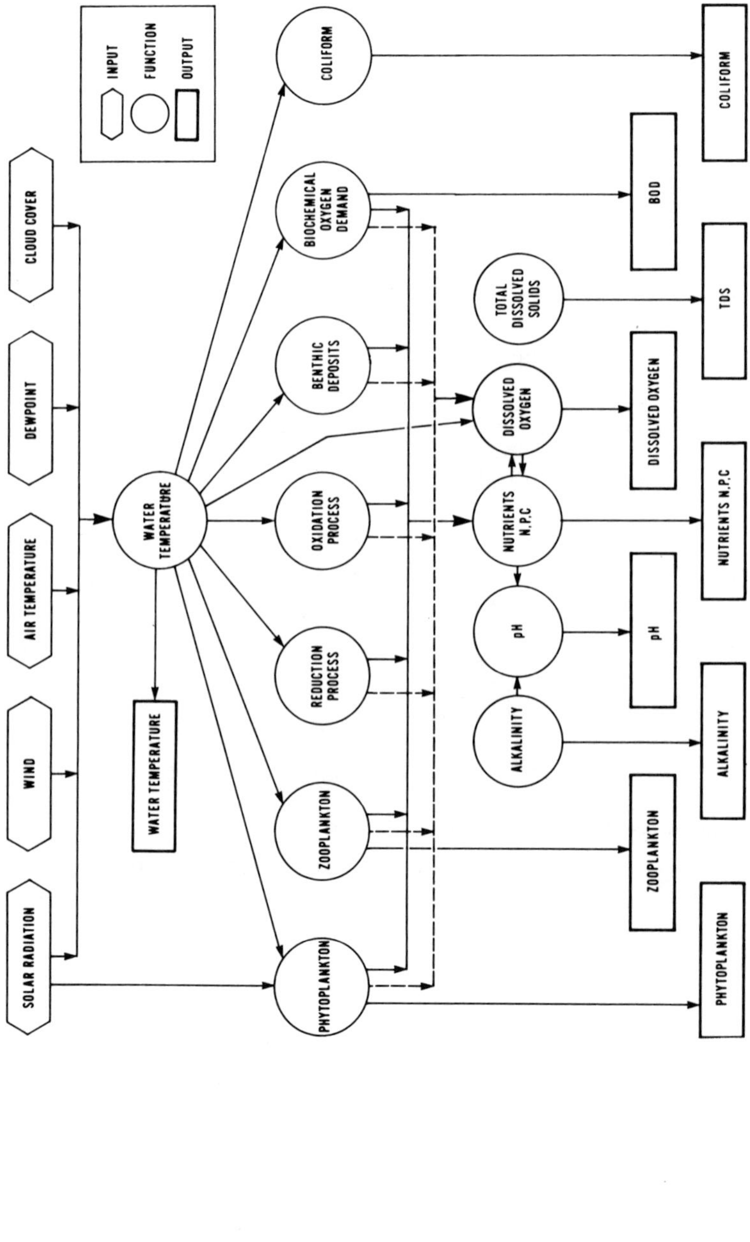

Figure 4. Water quality flowchart.

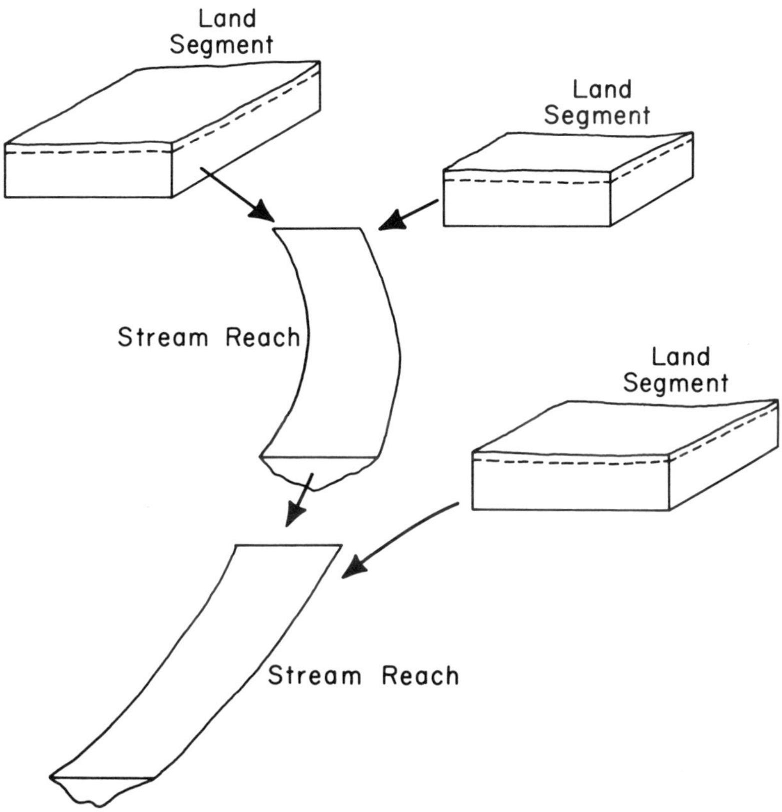

Figure 5. Land segments and stream reaches.

processes. Current research in hydrologic transport modeling is refining and improving individual algorithms for particular processes. Areas of current work include the detailed processes of sediment transport in first-order channels and in larger river systems, and the adsorption/desorption phenomena for compounds carried in streamflow.

USE OF MODELS IN ENVIRONMENTAL MANAGEMENT

We are moving from an era of research on environmental model development into an era of use of these models in environmental management. This is a fundamental change. The activities in research and development of models are:

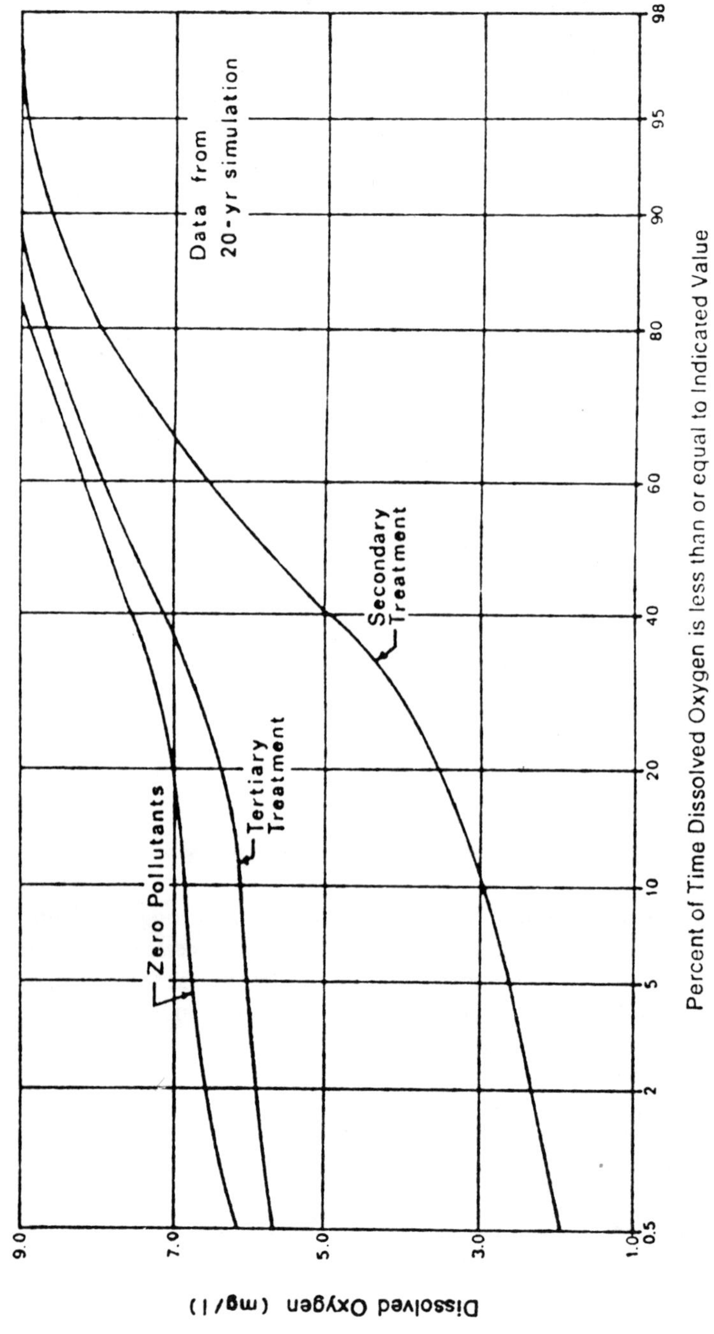

Figure 6. Quality-duration curves. South Platte River, Denver, Colorado (based on HSP, data courtesy Black & Veatch).

1. laboratory investigation of processes;
2. development of accurate modeling algorithms to represent physical processes; and
3. field data collection on experimental watersheds to verify model algorithms.

The application of models in day-to-day environmental management or regulatory activities is quite different. The activities in routine use of modeling tools are:

1. development of accurate model inputs using national meteorologic, hydrologic and geologic data bases;
2. systematic and efficient production of model results at many locations; and
3. political and legal defense of the model results used for planning, including assessment of uncertainty in model results.

In everyday use of models in environmental management, the goal is accurate assessment of environmental processes in many geographic locations. Sensitivity analyses show that sources of error in simulation modeling in order of importance are:

1. uncertain climatic variables;
2. uncertain watershed or pollutant characteristics;
3. uncertain pollutant application rates or rates of entry of pollutants into the environment; and
4. inadequate simulation model algorithms.

Consideration of these "sources of error" dictates spending less time on algorithm development and calibration of individual watersheds, and more time on the mechanics of developing reliable meteorlogic and physical data for model input, systematic model use and record-keeping.

The professional time (%) required to do a typical simulation application is as follows:

1. data gathering (hydrometeorologic data and data on the physical characteristics of watersheds and pollutants): 55%;
2. setting up the computer data base and the initial model runs: 15%;
3. calibrating model behavior against field data: 15%; and
4. producing final model results, reports and meetings: 15%.

Thus, the key to efficient use of models for day-to-day environmental management is data gathering. High-quality data gathering also improves the accuracy of model results, since bias in input data is a major source of error in model analysis.

Systematic Use of Models

The information needed to operate a simulation model consists of meteorologic data, watershed characteristics and initial conditions. These are shown in Figure 7. Efficient use of models occurs when assembly of the basic information required by the models is automated, and information moves into the modeling system with a minimum of human effort. A key operational problem in model use is creating a direct link to existing national environmental data bases and geographic information sources. Sources of information for meteorologic and hydrologic time series are the National Weather Records Center in Asheville, North Carolina, for meteorologic records; the U.S. Geological Survey (USGS) for streamflow, sediment transport and water quality records; and the EPA (Storet system) for water quality and pollutant data. Other sources of information that may be important in some parts of the United States are the U.S. Forest Service, the U.S. Department of Agriculture (USDA), and state and local agencies.

Watershed characteristics are the other major input to a model system. Maps of soils, vegetation, agricultural lands and typical cropping patterns are necessary and are available from the USGS, the Forest Service and the USDA. Maps of land surface elevation and slope are used to derive channel cross sections and lengths. The use of LANDSAT information to derive vegetation and crop patterns for modeling has been studied [19] and needs to be developed further.

When an adequate data base has been established, the remaining tasks in model use are calibration and production of model results for planning or regulatory alternatives. Calibration requires a high level of training and skill. Computer-based graphics systems are very desirable to allow the analyst to explore relationships between input parameters (representing watershed characteristics) and model response. Where data allow, calibrating the model on part of the field data and holding part of the field data aside to verify the model calibration is a good practice. The operation of a calibrated model to produce time series of streamflow or pollutants for planning alternatives is straightforward. The cost of computer resources to carry out comprehensive simulations of large river systems has been dropping steadily, so complex simulations of river systems with and without new reservoirs, treatment plants, land uses or agricultural practices are now feasible.

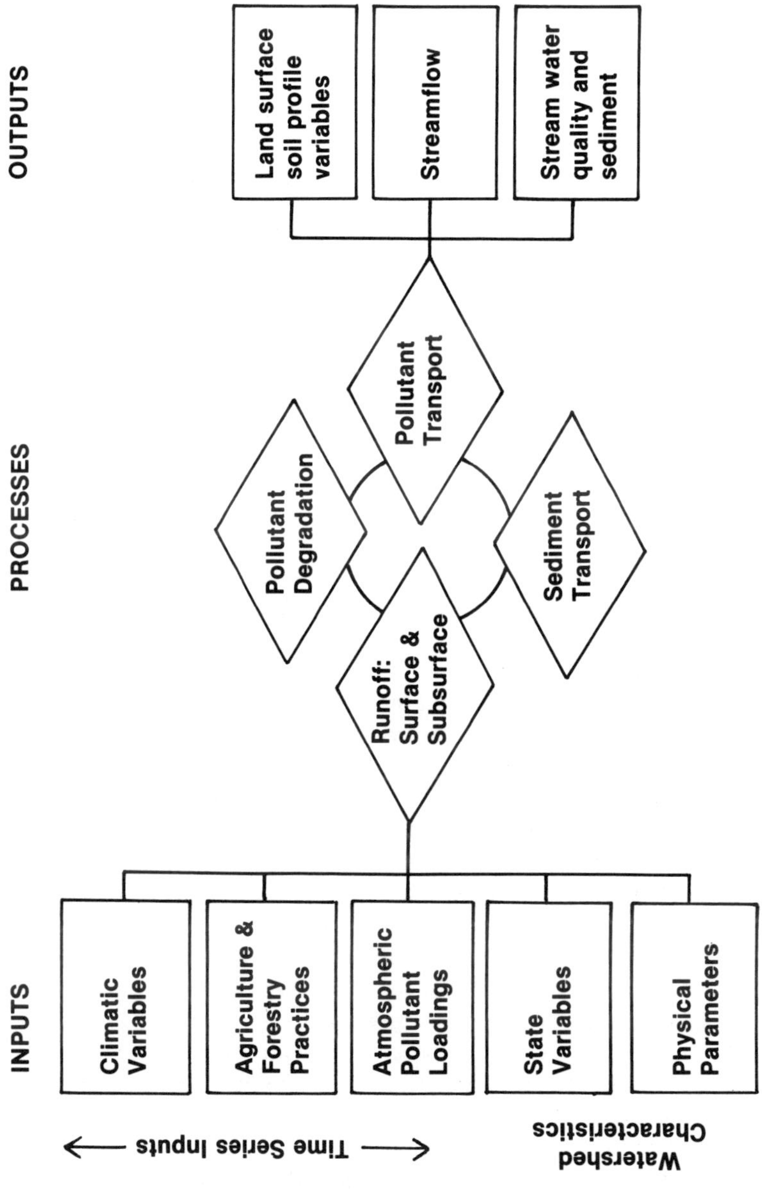

Figure 7. Model input/output.

Legal and Physical Considerations in Large-Scale Model Applications

Regulatory uses of modeling bring models out of the research and development laboratories into the public arena and will result in broader discussion and controversy about models and the conclusions that can be drawn from their use. The legal challenges to models will be based on:

1. scientific validity of the model and its general acceptance and use;
2. adequacy of the data base used for the model and its calibration;
3. validity of assumptions about watershed characteristics that are used in modeling regulatory alternatives; and
4. assessment of errors in model results.

To withstand legal challenges to the regulatory use of models, the model analysis needs to be technically conservative, especially when projections for future conditions are made. The uncertainty in model results must be recognized and estimated.

The key physical consideration in large-scale model use is geographic variability. For example, it is estimated that the mean annual direct runoff is 10 inches in Tennessee and 5 inches in Michigan. Does this mean that the annual sediment yield or the annual movement of pesticides from croplands in Tennessee would be twice that in Michigan? Unfortunately, no. Runoff in Tennessee occurs during the growing season, when pesticides are applied to crops. Runoff in Michigan occurs in the spring snowmelt period, and Michigan summer rainfalls cause almost no surface runoff. The variation in mean annual direct runoff between Tennessee and Michigan is not a good index to the relative movement of pesticides that would occur in Michigan and Tennessee.

Natural geographic variability can be considered in risk assessment. If, for example, an insecticide applied to cotton is to be labeled and sold throughout the cotton growing areas, how could a risk assessment for this compound be made? If the risk assessment is based on the average conditions where the insecticide will be applied, it may be quite safe. However, for 10% or perhaps 5% of the watersheds in which the insecticide is applied, it may cause severe environmental problems. Thus the natural, site-specific variability of processes, as a result of variability in slope erosivity, runoff and vegetation, need to be appreciated and studied. If a risk index of an insecticide were developed and evaluated on a number of watersheds, the variability of risk in different locations could be plotted as shown in Figure 8. When the physical environmental factors that create high risk are understood, pesticide label restrictions or

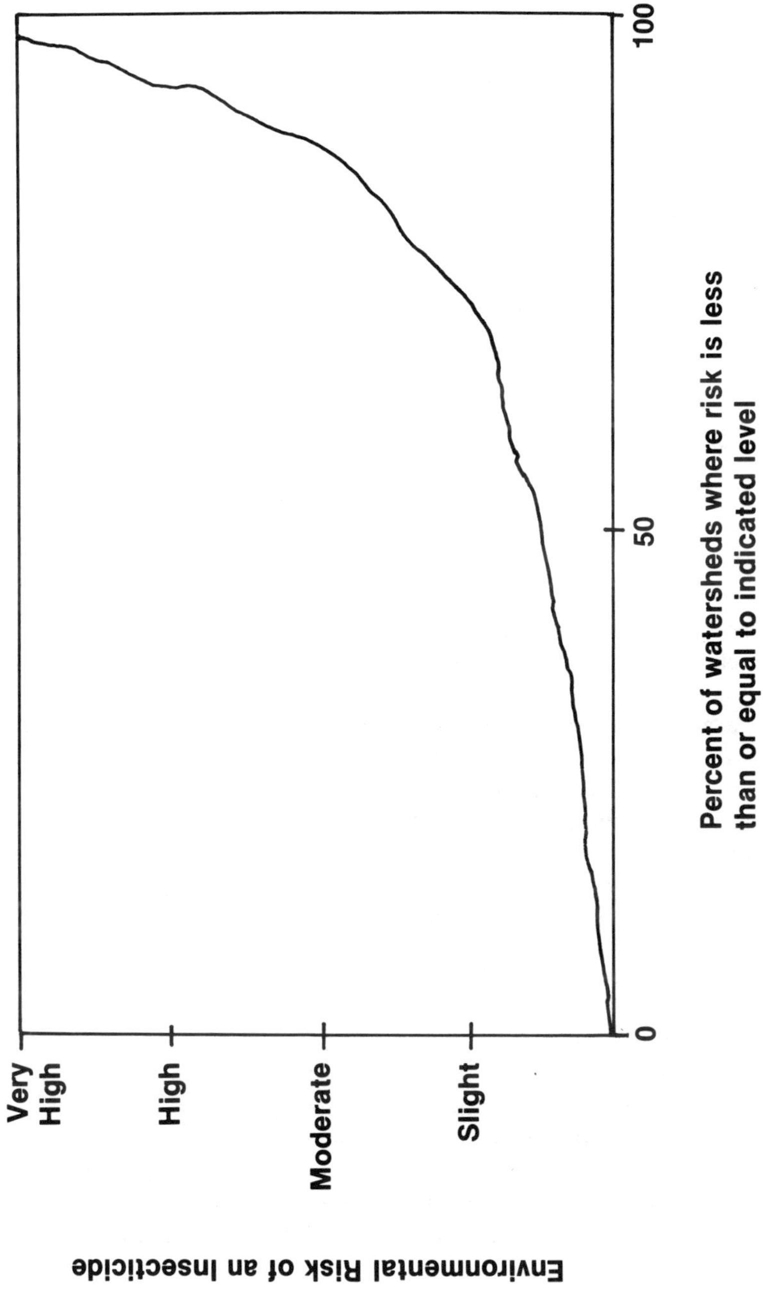

Figure 8. Variability of risk in different areas.

other steps can be taken to prevent environmental damage. The necessary tools exist in geographic data management systems to handle the large quantities of information and will need to be applied to solve problems of physical diversity in environmental regulation.

RECOMMENDATIONS

Large-scale use of models for environmental regulation is just now beginning. Formidable problems of data gathering, calibration, geographic variability and error assessment for intensive use of models have not received much attention. The success of modeling for environmental regulation will depend on creative and effective solutions to these problems. Modeling tools developed in the last 20 years are powerful and effective, and problems in the wide use of these modeling tools tend to be logistic and economic rather than scientific. Further development of model algorithms is desirable and useful, but is not a barrier to the current use of models for environmental regulation.

Preparation of information on climate and watershed characteristics is the major time-consuming element in a model application, accounting for up to two-thirds of the total cost of a project. The costs of using models for environmental regulation can be reduced most dramatically by reducing the human effort that goes into the mechanical tasks of data gathering to prepare for a model run. Automated development of watershed characteristics from LANDSAT information or national digitized agricultural, geologic or topographic data should be developed further. The goal is to interconnect data bases of meteorologic and watershed characteristic information with models, so that a model run could be initiated by a simple request for output at a site.

Human resources are increasing in costs each year, but computer resources are decreasing in costs each year. Reducing costs for environmental analysis will require more and better use of computer resources. Computer resources need to be traded for human resources wherever possible.

Environmental modeling marshals very large amounts of information and produces still larger amounts of information. Graphics systems are important to display results so that nontechnical persons can understand model results and make planning or regulatory decisions. In the last 20 years, unprecedented capability to measure, record, store and analyze information have been developed. These new tools will be used for environmental planning. Our challenge is to use our new tools effectively.

REFERENCES

1. Crawford, N. H. "The Synthesis of Continuous Streamflow Hydrographs on a Digital Computer," PhD Thesis, Stanford University, Stanford, CA (1962).
2. Huff, D. D. "Simulation of the Hydrologic Transport of Radioactive Aerosols," PhD Thesis, Stanford University, Stanford, CA (1968).
3. Anderson, E. A., and N. H. Crawford. "The Synthesis of Continuous Snowmelt Runoff Hydrographs on a Digital Computer," Stanford University, Department of Civil Engineering Technical Report No. 36 (1964).
4. Crawford, N. H., and R. K. Linsley. "Digital Simulation in Hydrology: Stanford Watershed Model IV," Stanford University, Department of Civil Engineering Technical Report No. 39 (1966).
5. "Hydrocomp Simulation Programming Operations Manual, 4th ed.," Hydrocomp International, Inc., Palo Alto, CA (1975).
6. "Water Quality Program Operations Manual," Hydrocomp International, Inc., Palo Alto, CA (1977).
7. Donigian, A. S., Jr., and N. H. Crawford. "Modeling Pesticides and Nutrients on Agricultural Lands," U.S. EPA Environmental Research Laboratory Report 600/13-76-083 (1976).
8. Neger, M. "A Sediment Model on a Digital Computer," Stanford University, Department of Civil Engineering Technical Report No. 76 (1967).
9. Dawdy, D. R., and T. O'Donnell. "Mathematical Models of Catchment Behavior," *J. Hydraul. Div., ASCE* 19(HY4):123-139 (1965).
10. Sugawara, M. "On the Analysis of Runoff Structure about Several Japanese Rivers," *Jap. J. Geophys.* Vol. 2 (1961).
11. Bagley, J. M. "Electronic Analog Modeling of River Basin Hydrology," paper presented at the 5th Western National Meeting of the American Geophysical Union, Dallas, TX, 1965.
12. Rockwood, P. M. "Columbia Basin Streamflow Routing by Computer," *Trans. ASCE* 126(4):32-56 (1961).
13. Holtan, H. N., G. J. Stittner, W. H. Henson and N. C. Lopez. "USDAHL-74 Model of Watershed Hydrology," USDA, Agricultural Research Service Technical Bulletin No. 1518 (1975).
14. Metcalf and Eddy. "Storm Water Management Model," U.S. EPA Water Quality Office Report—11024DOCO4171, Vols. I-IV.
15. "Storage, Treatment, Overflow, Runoff Model (STORM)," U.S. Army Corps of Engineers, Hydrologic Engineering Center, Computer Program 723-58-L7520 (1976).
16. Yaramanoglu, M. "A Finite Element Model for Simulation of Hydrographs from Natural Watersheds," PhD Thesis, University of Maryland, College Park, MD.
17. Water Resource Council. "Principles and Standards for Planning Water and Related Land Resources," *Federal Register* 38(174):24778-24869 (1973).
18. Hydrocomp International, Inc. "User's Manual for the Hydrological Simulation Program—FORTRAN (HSPF)," U.S. EPA Environmental Research Laboratory, Athens, GA (1979).

19. Franz, D. D., and S. M. Lieu. "Evaluation of Remote Sensing Data for Input into Hydrological Simulation Program—FORTRAN (HSPF)," U.S. EPA Environmental Research Laboratory, Athens, GA (1980).

CHAPTER 13

MODELING CHEMICAL REACTIONS AND TRANSPORT IN GROUNDWATER SYSTEMS: A REVIEW

Philip B. Bedient, N. K. Springer, C. J. Cook and Mason B. Tomson

National Center for Groundwater Research
Department of Environmental Science and Engineering
Rice University
Houston, Texas 77001

Prediction of the transport of contaminants in groundwater systems through measurement and modeling has received great interest with the growing trend toward subsurface waste disposal practices. Landfills, chemical waste sites, deep-well injection and radioactive disposal have become increasingly more important from an environmental impact standpoint, along with predicting the fate and significance of associated contamination.

The state-of-the-art in groundwater flow prediction is well advanced in terms of analytical and numerical models under a variety of geometric conditions. However, existing models of mass transport have met with difficulty due largely to parameter estimation of dispersion in the field and to numerical problems with finite-difference computer solutions.

A potentially more difficult problem is the current lack of knowledge regarding quantification of chemical reaction terms. Reactions involving adsorption, ion exchange, degradation and others have been studied in great detail in the laboratory, but few actual field studies have been completed. Modeling such reactions presents the next generation of difficulty to groundwater models and field calibration. Major chemical reaction terms and processes are discussed in detail in this chapter.

With the above problems, there has still been a great advance in the state-of-the-art of groundwater models over the past decade. Many useful approaches have been developed and applied successfully to particular contaminant problems, especially for radionuclides and salinity. Depending on the relative importance of dispersion (spreading caused by heterogeneities) and advection (bulk fluid movement), two approaches have been advanced to predict contaminant transport: (1) advection/dispersion models, which include both mechanisms, and (2) advection models, which only consider flow lines and travel times. Both types of models may include chemical reaction terms.

The next sections present the governing equations of groundwater flow and transport, followed by a presentation of chemical reaction processes. Mathematical models, both analytical and numerical, are discussed in detail, followed by model applications and field validation problems.

GOVERNING GROUNDWATER EQUATIONS

The differential equation for simulating groundwater flow in two dimensions is usually written

$$\frac{\partial}{\partial x_i}\left(T_{ij}\frac{\partial h}{\partial x_j}\right) = S\frac{\partial h}{\partial t} + W \tag{1}$$

where T_{ij} = transmissivity tensor $(L^2/T) = K_{ij}\,b$
K_{ij} = hydraulic conductivity tensor (L/T)
S = storage coefficient
W = source or sink term (L/T)
x_i = cartesian coordinates (L)
h = hydraulic head (L)
b = aquifer thickness (L)

The dispersion equation is usually written

$$\frac{\partial}{\partial x_i}\left(D_{ij}\frac{\partial c}{\partial x_j}\right) - \frac{\partial}{\partial x_i}(cV_i) - \frac{c_o W}{nb} + \sum_{k=1}^{m} R_k = \frac{\partial c}{\partial t} \tag{2}$$

where c = concentration of solute (M/L^3)
V_i = seepage velocity (L/T) averaged in the vertical direction
D_{ij} = coefficient of dispersion (L^2/T)
c_o = solute concentration in source or sink fluid (M/L^3)
R_k = rate of addition or removal of solute $(\pm)(M/L^3T)$
n = porosity

The main difficulty in solving the above equation for concentration is estimating the value of D_{ij}. Scheidegger [1] showed that when molecular diffusion is negligible,

$$D_{ij} = \alpha_{ijmn}\left(\frac{V_m V_n}{V}\right) \tag{3}$$

where for an isotropic medium, all components of α_{ijmn} are zero except for certain ones; and

$$D_{xx} = \alpha_L \frac{V_x^2}{V} + \alpha_T \frac{V_y^2}{V} \text{ in the X direction}$$

$$D_{yy} = \alpha_L \frac{V_y^2}{V} + \alpha_T \frac{V_x^2}{V} \text{ in the Y direction}$$

where $V = \sqrt{V_x^2 + V_y^2}$
α_L = longitudinal dispersivity (L)
α_T = longitudinal dispersivity (L)

The pıoblem of specifying D_{ij} involves determining the velocity distribution from Equation 1, Darcy's law, and the values of α_L and α_T. Next, Equation 2 is solved subject to certain boundary and initial conditions, and resulting concentration plumes in X and Y are compared to actual data. In practice, these equations are usually calibrated to field data by adjusting T_{ij} in the model to yield observed heads and D_{ij} to yield observed concentration patterns. Flow and mass transport equations generally are treated as uncoupled equations in that concentration changes do not affect the flow field; this applies where density differences are unimportant, which is the case for most contaminant problems.

There are two situations where solutions to the above equations can be used:

1. to assess the impact of subsurface waste disposal at proposed sites that have not yet been contaminated; and
2. to assess the impact of already contaminated sites in terms of predicting plume migration and recommending remedial measures

The key to successful application of such models is, of course, to have a reliable measure of hydraulic conductivities and dispersivities in the field. Most efforts have relied on trial-and-error approaches to solving the inverse problem of parameter estimation. This problem is discussed in more detail later in this chapter. Assuming that the flow and mass transport models can be calibrated, the response of the system to future conditions of loading or remedial action can be simulated.

MATHEMATICAL MODELS

Mathematical expressions representing physical processes, such as those described above, must be combined with a mass balance, linked with a flow model, or perhaps even tied by an energy or entropy balance [2]. The resulting set of equations is then solved by any of a number of methods. These various methods generally fall into two categories—analytical and numerical methods.

Analytical Models

Analytical methods of problem solution involve setting up the governing equations and simple boundary conditions such that the system can be solved and the solution may be written in a closed functional form or as an infinite series.

Analytical solutions, restricted to certain equation forms and simple boundary conditions, have a somewhat limited applicability to practical problems. However, their simplicity of use and understanding make them valuable and convenient tools. A large number of analytical flow models and some transport models are available and discussed in the literature [3–6].

The restrictions placed on the governing equations to make possible an analytical solution prevent the incorporation of processes other than dispersion and first-order decay. However, a modified analytical method, called the method of characteristics, which may also employ numerical methods for complete solutions, can help overcome this restriction if transport is dominated by convection. The method of characteristics allows transformation of certain governing hyperbolic partial differential equations (PDE), such as the mass transfer equation without dispersion, into a set of ordinary differential equations [7–9]. In some cases, this simplified problem may be solved analytically; in others, a numerical solution is required [9–11].

The advantage of the analytical model lies in its simplicity and efficiency. Many analytical solutions can be written as easily evaluated closed-form functions that allow easy visualization of the relative importance of the various parameters. In general, the cost of analytical solutions is nominal compared to more expensive numerical schemes.

Application of simple or complex groundwater transport models include problems of groundwater contamination in land disposal sites, hazardous wastes, petroleum spills, radioactive waste disposal and deep-well injection. Cases that have simple geometries, homogeneous sands and measurable input contaminant concentrations, e.g., land application of sewage, might fit a fairly simple analytical model quite well. Suppose one wished to examine solute transport of land-applied wastewater under the hypothetical condition of a rectangular infiltration pond lying above a horizontal infinite aquifer. Knowing or assuming coefficient of dispersivity, flow velocity and decay rates, in two dimensions, one could apply an analytical mass transport model and obtain the numerically evaluated results shown in Figure 1 for a horizontal plane [6].

Numerical Models

Unfortunately, in many practical cases, the governing processes and boundary conditions are too complex for analytical methods. In these cases, the solutions must be approximated by a numerical scheme, usually performed by a digital computer. The basic method is to break up continuous space into cells, approximate the governing partial differential equations by differences between the values of the parameters over the network, and then compute the discrete parcel. Generally, this is accomplished by finite-element or finite-difference schemes.

Interest has recently shifted to the application of finite-element methods [12–15]. Early work used variational principles to formulate integral equations, followed later by Galerkin methods of weighted residuals. Finite-element methods are very flexible in terms of representing complex geometries with variable element shapes. These methods are quite complex geometries with variable element shapes. These methods are quite complex to learn without some formal mathematical training and are also generally very expensive to operate on the computer.

The earlier finite-difference methods operate by dividing space into rectilinear cells along the coordinate axes. Homogeneous values within each cell are represented by values at a single node. Partial differentials can then be approximated by differences (Figure 2) and the resulting set of equations solved by iteration [12,16,17]. Approximating the differentials by a difference requires neglecting remaining terms (designated by R in Figure 2), which results in truncation error. Finite-difference models

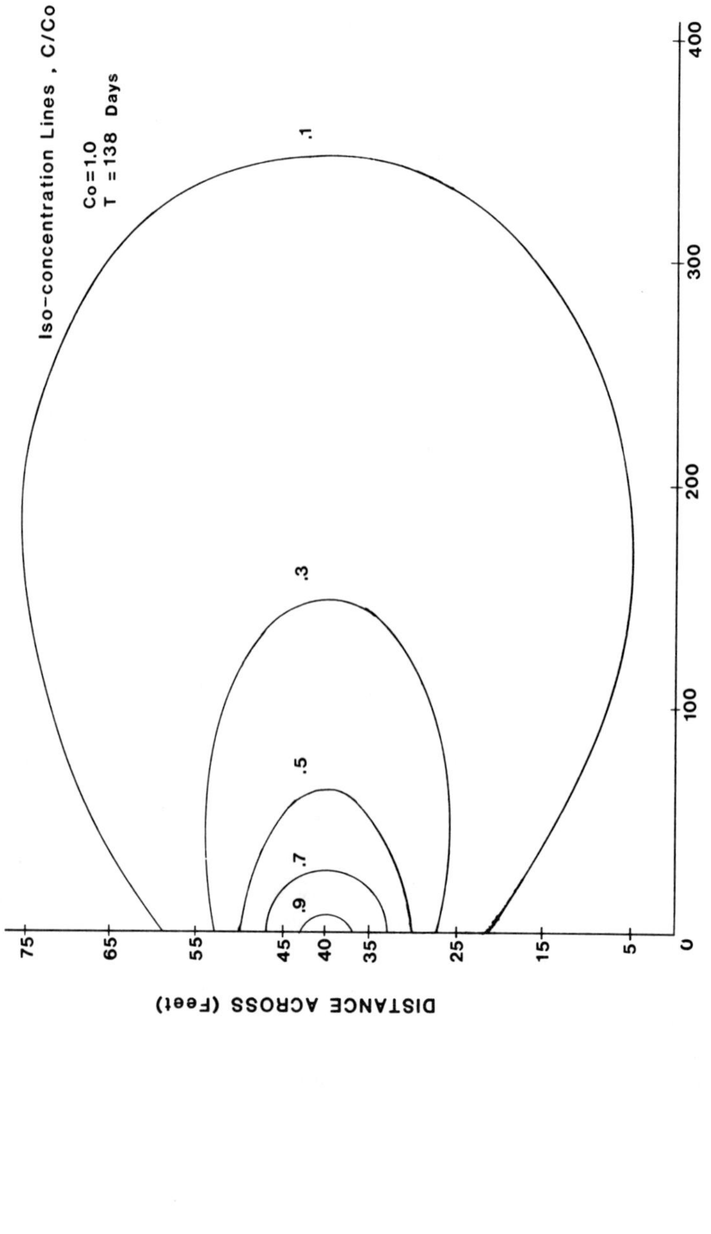

Figure 1. Hypothetical solute transport of land-applied wastewater.

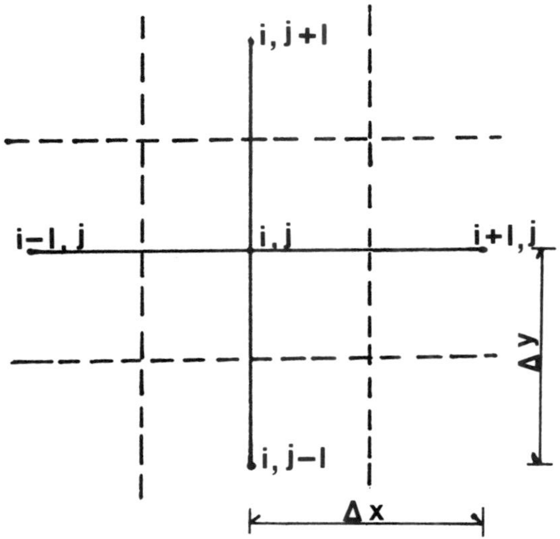

$$U = U(x,y)$$

$$\frac{\partial U}{\partial Y} = \frac{U_{i,j+1} - U_{i,j-1}}{2\,\Delta Y} + R$$

$$\frac{\partial^2 U}{\partial X^2} = \frac{U_{i-1,j} - 2U_{i,j} + U_{i+1,j}}{\Delta X^2} + R$$

Figure 2. Finite-difference models, central-difference scheme.

exist for various types of field problems, for saturated and unsaturated flows, and for transient and constant pollutant sources [8–11,18–25].

Finite-difference models suffer from numerical dispersion, and in some cases it may be of the same order of magnitude as the physical dispersion process. When velocities are large, the equation acts like a hyperbolic PDE, which has generated the use of method of characteristics solutions for transport [9].

One subset of the numerical models is the statistical models. One of these statistical models, the Discrete-Parcel-Random-Walk transport model [26], predicts transport of a parcel by allowing two types of parcel movement. One movement transports the parcel with the convecting fluid. An additional random-walk component simulates dispersion.

The major advantage of numerical methods is their general applicability to all types of problems, both simple and complex. However, they cost more to use than analytical models. In addition, one must be careful that truncation errors do not accumulate as iteration progresses. Numerical methods lack the ease of operation and closed-form solutions possible with analytical methods.

Most case studies of groundwater contaminant transport use numerical models, usually finite-element or finite-difference models, to match data and to predict contaminant behavior. The U.S. Geological Survey (USGS) has supported several numerical groundwater transport simulations at problem sites [9–11,23,27,28]. Many of the USGS models simulate transport of only conservative contaminants, such as chlorides. Two studies [11,23] use models also incorporating radioactive decay and linear adsorption. As a typical example, one of these studies examined the effect of changing waste disposal practices at the National Reactor Testing Station in Idaho. Data on conservative chlorides were matched to verify the model. Then, the model was used to predict future waste patterns for various radionuclides, given either that groundwater disposal was halted or continued. Numerical groundwater transport models have been used at many other sites for various purposes, such as waste behavior prediction, identification of inconsistent or incomplete data, or estimation of monitoring needs [29].

Description of the USGS Groundwater Model

This section presents a more detailed description and application of one of the most widely used groundwater transport models [9]. The USGS model, method of characteristics (MOC), simulates solute transport in flowing groundwater and can be applied to a wide range of one-

and two-dimensional problem types involving steady-state or transient flow. The model computes changes in concentration over time caused by the processes of convective transport, hydrodynamic dispersion and mixing (or dilution) from fluid sources. The model assumes that the solute is nonreactive and that the gradients of fluid density, viscosity and temperature do not affect the velocity distribution. However, the aquifer may be heterogeneous and/or anisotropic.

The computer program solves two simultaneous partial differential equations to simulate transport in the groundwater. The program uses an iterative alternating direction implicit procedure to solve a finite-difference approximation to the groundwater flow equation, and it uses the method of characteristics to solve the solute transport equation. The latter uses a particle-tracking procedure to represent convective transport and a two-step explicit procedure to solve a finite-difference equation that describes the effect of hydrodynamic dispersion, fluid sources and sinks, and divergence of velocity.

The computer program first solves the equation describing the transient two-dimensional flow of a homogeneous fluid through a nonhomogeneous anisotropic aquifer (Equation 1) with the implicit finite-difference approximation presented in Equation 4 [30]. After the head distribution is computed, the velocity of the groundwater flow is computed for each node using an explicit finite-difference form of Darcy's equation:

$$\begin{aligned} T_{xx[i-\frac{1}{2},j]} \frac{h_{i-i,j,k} - h_{i,j,k}}{(\Delta x)^2} + T_{xx[i+\frac{1}{2}\cdot j]} \frac{h_{i+i,j,k} - h_{i,j,k}}{(\Delta x)^2} \\ + T_{yy[i,j-\frac{1}{2}]} \frac{h_{i,i-1,k} - h_{i,j,k}}{(\Delta y)^2} + T_{yy[i,j+\frac{1}{2}]} \frac{h_{i,j+1,k} - h_{i,j,k}}{(\Delta y)^2} \\ = S \frac{h_{i,j,k} - h_{i,j,k-1}}{\Delta t} + \frac{q_{w(i,j)}}{\Delta x \Delta y} - \frac{K_z}{m} [H_{s(i,j)} - h_{i,j,k}] \end{aligned} \tag{4}$$

where i, j, k = indices in the x, y and time dimensions, respectively
$\Delta x, \Delta y, \Delta t$ = increments in the x, y and time dimensions, respectively
q_w = volumetric rate of withdrawal or recharge at the i, j node (L^3/T)

Next, the solute transport equation is solved, which is used to describe the two-dimensional areal transport and dispersion of a given nonreactive dissolved chemical species in flowing groundwater (Equation 2). This equation is solved through a three-step procedure. The first step involves the use of the particle mover that solves for the change in concentration over distance. This in turn is used by a two-step explicit finite-difference approximation to Equation 2. The new nodal concentrations are then calculated at the end of the time period as

$$Ci,j,k = C_{i,j,k}^{*} + \Delta C_{i,j,k} \quad (5)$$

Input parameters for the model include:

1. effective porosity;
2. storage coefficient;
3. longitudinal dispersivity;
4. transmissivity (as a tensor);
5. ratio of transverse to longitudinal dispersivity;
6. ratio of T_{yy} to T_{xx};
7. setting up the time steps, grid spaces and options for output; and
8. initialization of the aquifer with size and shape parameters, locations of sources and sinks, location of observation wells, boundary conditions, background concentrations, potentiometric or water-table levels and stability criteria.

Constant-head and constant-flux boundary conditions are allowed.

The model output can include the head distributions, the X- and Y-direction velocity distributions, the drawdown depths, and the concentration gradients at each time step.

After application, the accuracy of the model was evaluated for two idealized problems for which analytical solutions could be obtained. The agreement was nearly exact for one-dimensional flow, but yielded some numerical dispersion for the case of plane radial flow. Mass-balance errors were generally less than 10%, with largest errors during the first several time steps. The model has been applied to several field examples by the USGS and other groups involved in predicting groundwater pollutant patterns. Obviously, the main limitation of the model is its inability to handle decay and adsorption mechanisms.

Field Determination of Groundwater Parameters

The velocity distribution is needed to quantify the groundwater flow direction and magnitude and the advective transport. Hydraulic heads and conductivities are used routinely in Darcy's equation to calculate V_x or V_y indirectly. Thus knowledge of hydraulic conductivities (K) and system heterogeneities is essential for an understanding of contaminant transport. Trial-and-error adjustment of K until a flow model simulates observed heads can result in several K distributions that give the same head values. Two lines of research presently involve optimization methods to adjust K values or direct field measurements of groundwater velocity [31].

Methods for direct measurement of velocity are given by Fried [32] and Freeze and Cherry [33]. These methods employ a tracer in one or two wells with observations of dilution or breakthrough curves. One problem with such methods is the extremely long times often involved in actual field studies. Use of injection pressures and pumping wells helps to speed up the tracer movement.

The coefficient of dispersion D_{ij} is probably the single most difficult parameter to estimate in the field. It is related to velocity through the dispersivity α presented earlier. The magnitude of α changes, depending on the scale at which measurements are taken [32]. Laboratory values range from 10 to 1 cm, while 10 to 100 m have been obtained in field problems. Dispersivity is usually determined from tracer injections and monitoring concentrations at observation wells. Experimental data are then fitted using solutions to the dispersion equation, thus yielding values for α. In the laboratory, dispersion is measured at a microscopic scale in column studies that yield breakthrough curves.

Fried [32] defines four levels to measure field dispersivities, dependent on distance traveled.

1. local scale: 2–4 m
2. global scale 1: 4–20 m
3. global scale 2: 20–100 m
4. regional scale: >100 m

Different experimental techniques are recommended for each scale. Table 4 in Anderson's [31] excellent review shows the range of dispersivities that have been measured in field studies. Little real progress has been made in developing a sound method for determining spatial variations of dispersivity either by field measures or by fitting the dispersion equation [31].

One promising approach is to quantify spatial variations in dispersivity in terms of variations in hydraulic conductivity, which should lead to a better understanding of the physical process. Gelher, cited by Anderson [31], stochastically analyzed hydraulic conductivity variations to estimate dispersivity, and suggested an equation of the form

$$\alpha_L = 1/3 \frac{l^2 \sigma^2}{\alpha_T} \tag{6}$$

where σ^2 = standard deviation of the log of hydraulic conductivity
l = correlation distance of covariance function

Problems of dispersion estimation can be overcome by using a flow model (advection) to obtain a first approximation of contaminant movement. Flow lines and travel times are used to estimate where contaminants are located and when they will arrive at a given location. This approximation neglects the effects of a spreading front and gives the arrival time of the main zone of contamination. Many examples exist in the literature for both analytical and numerical problems in two dimensions [31]. In particular, radial flow toward a well has been handled in this manner. Simple chemical reactions that can be represented as time lags can also be considered.

The next section indicates the major chemical and biological mechanisms that affect the transport of chemicals in groundwater. Although advection and dispersion are the dominant physical processes that describe bulk movement and spreading of a contaminant, a large number of chemical and biological reactions and transformations are also of great importance. These are presented in some detail in the next section, along with a detailed review of pertinent literature.

CHEMICAL AND BIOLOGICAL PROCESSES

Equilibrium

Sorption of trace-level organics onto high-surface-area clays and soil organic matter is a prime mechanism of trace-level organic removal from groundwater. Many mathematical formulas are used to describe the equilibria of compounds between phases. All such equations must be consistent with the fundamental thermodynamic condition for phase equilibria:

$$G_A^1 = G_A^2 \tag{7}$$

where G_A^1 and G_A^2 = the molar Gibbs' free energy (it can be shown that this is the energy that is available to do useful work) of compound A in phases 1 and 2, respectively. Phases 1 and 2 may be air and water, water and sediment, or water and biota, to mention a few combinations. In general, the molar Gibbs' free energy for a compound A in phase i (G_A^i) can be written as

$$G_A^i = G_A^{oi} + RT\, ln\, C_A^i \gamma_A^i \tag{8}$$

where G_A^i = molar Gibbs' free energy as in Equation 7

G_A^{oi} = numerical constant called "reference state Gibbs' free energy." The numerical value depends on temperature and the units used to calculate concentration
R = gas constant, 1.99 cal/mol-deg
T = temperature (°K)
C_A^i = concentration of A in phase i (mol/l, mg/l, mole fraction, etc.). Since concentration and pressure are directly proportional to one another, C_A^i can be replaced by P_A times a constant that simply changes the numerical value of G_A^{oi} for the gas phase.
γ_A^i = "activity coefficient" of A in phase i

If C_A^i is replaced by P_A, this quantity is called the "fugacity coefficient" and is given the symbol Γ_A. As the concentration of A becomes negligible, two things happen: (1) the volume of A becomes negligible compared to the phase it is in, and (2) the electrostatic, van der Waals, etc., interaction energies between pairs of A molecules become negligible. When both of these conditions are fulfilled, the molecules of A behave as if they were an ideal gas in the medium, and the activity coefficient γ_A^i is exactly 1.00.

The product $C_A^i \gamma_A^i$ is called the "activity" of A in phase i, a_A^i. Similarily, $P_A \Gamma_A$ is called the "fugacity" of A, f_A. In either case, the compound acts and reacts as if its concentration were $C_A^i \gamma_A^i = a_A^i$ or $P_A \Gamma_A = f_A$. The fugacity coefficient, Γ_A, can be taken to be 1.00 for compounds in air. For neutral molecules in most natural waters, the activity coefficients can be calculated to within a few percent by the formula

$$\log_{10\gamma A} = 0.10\text{TDS} \tag{9}$$

where TDS = total dissolved salt content (mol/l)

Similarly, the activity coefficient of an ionic solute A, of charge (+)z or (−)z ($A^{\pm z}$), can be calculated by the Davies equation:

$$\log_{10}\gamma_{A\pm z} = -0.5z^2(\text{TDS}^{1/2}/1 + \text{TDS}^{1/2})^{-1} - 0.3\,\text{TDS} \tag{10}$$

More precisely, the TDS value should be replaced by the ionic strength I, which weights the concentration of multivalent ions by the square of their charge:

$$\text{TDS} \cong \text{I}\,(\text{mol/l}) \equiv 1/2 \sum_{\text{all ions}} (Z^2)(\text{conc.}) \tag{11}$$

For most natural waters, TDS or I can be calculated approximately from the specific conductance:

$$TDS = 1.6 \times 10^{-5}\ \mu mho/cm \quad (12)$$

Figure 3 illustrates the magnitude of the activity coefficient in natural waters of TDS or ionic strength for neutral molecules using Equation 9, and for mono-, di- and trivalent ions using Equation 10.

Although ferric iron and aluminum are often thought of as trivalent ions in natural waters, above about pH 4 both ions are highly hydrolyzed and exist almost exclusively as mono- or divalent species such as $FeOH^{2+}$, $Fe(OH)_2^+$, $Fe(OH)_4^-$, and various polynuclear species. The principal divalent ions in natural waters are Ca^{2+}, Mg^{2+}, Fe^{2+} and SO_4^{2-}. More than 95% of freshwaters have TDS values less than 1000 ppm, which corresponds to activity coefficients of 1.004, 0.881 or 0.602 for neutral molecules, mono- or divalent ions, respectively. Therefore, for practical purposes, Equation 8 can be written for neutral molecules or monovalent ions in terms of concentration only:

$$G_A^i = G_A^{oi} + RT\, ln\, C_A^i \quad (13)$$

For divalent and trivalent ions, an activity coefficient correction should probably be used any time a reaction is involved, as in ion exchange, precipitation, adsorption, redox or hydrolysis.

If the overall salt content of a water system being modeled changes more than a few percent in the course of the model, the activity coefficients calculated from Equations 9 and 10 may change appreciably, as can be seen from Figure 3. For this reason, the ion activity is suggested for use in mass transport models [34,35].

Neutral compounds, such as benzene, naphthalene and dichlorobenzene, partition between all phases of the system being considered. For example, consider the partitioning of benzene (B) between the air phase and an aqueous phase:

$$B_{air} \rightleftharpoons B_{aq} \quad (14)$$

Then, according to Equation 7:

$$G_B^{aq} = G_B^{air} \quad (15a)$$

and from Equation 13:

$$G_B^{air} = G_B^{o,air} + RT\, ln\, P_B \quad (15b)$$

$$G_B^{aq} = G_B^{o,aq} + RT\, ln\, C_B^{aq} \quad (15c)$$

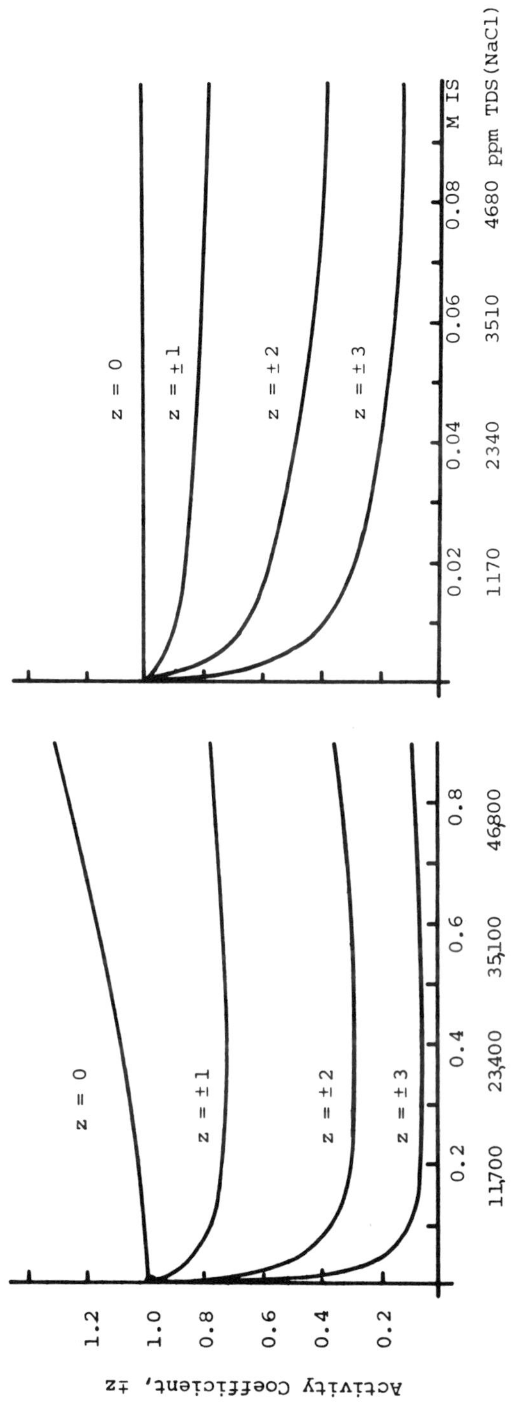

Figure 3. Variation of activity coefficient with ionic strength at two ranges.

or

$$G_B^{o,aq} + RT\,ln\,C_B^{air} = G_B^{o,aq} + RT\,ln\,P_B \quad (15d)$$

or

$$-(G_B^{o,aq} - G_B^{o,air}) = -\Delta G_B^o = RT\,ln\,\frac{C_B^{aq}}{P_B^{air}} \quad (15e)$$

The negative sign in Equation 15e is to conform to the convention of writing "products minus reactants." Since ΔG_B^o is the difference between two numerical constants, the quantity C_B^{aq}/P_B must be a constant K_B at a given temperature. For benzene at 25°C [36]:

$$K_B = \frac{C_B^{aq}}{P_B} = 0.24 \text{ mol/l-atm} \quad (16a)$$

or

$$\Delta G_B^o = -RT\,ln\,K_B = -(1.99 \text{ cal/mol-deg})(298°K)\,ln\,0.24 \quad (16b)$$

$$= +850 \text{ cal/mol}$$

If Equation 16a had been written in terms of benzene concentration (M) in the gas phase C_B^{air} obtained from the ideal gas law, a unitless constant K_B' would have been obtained:

$$K_B' = \frac{C_B^{aq}}{C_B^{air}} = 5.8 \quad (16c)$$

or

$$\Delta G_B^{o\prime} = -1048 \text{ cal/mol} \quad (16d)$$

There are two things to note about Equation 16. First, the reference state Gibbs' free energy value changes, depending on the units used and, second, K_B in Equation 16a is the reciprocal of Henry's law constant (H) in units of atm/*M*. A verbal interpretation of Equation 16a is instructive: hypothetically, if the partial pressure of benzene above water were raised to 1 atm, then the equilibrium concentration in the water would be 0.24 *M*; or if 0.1 atm partial pressure of benzene were put above water, the aqueous concentration would be 0.024 *M*, etc. Similarly, for any other phase, such as biota or sediment, an equilibrium constant K could be derived (as in Equation 16a) and measured correspondingly to a constant air partial pressure of 1.0, or 0.1 atm of benzene. In each case the numerical value of K would correspond to the concentration of benzene in that phase when exposed to a reference benzene partial pressure of 1 atm (i.e., the concentration in that phase per atmosphere). This constant K_B or K was recently given the name "fugacity capacity, Z" [36].

Although it will not be shown here, a parallel treatment starting with Equation 7 can be used to derive the equilibrium expression for the solubility product, ion exchange and octanol/water partition coefficient, to mention a few. It should be noted that a corresponding interpretation of surface capacity or adsorption is only possible for the linear portion of an isotherm where surface concentration is directly proportional to aqueous or air concentration. This fundamental breakdown in correspondence for surface adsorption is a consequence of the fact that surfaces are treated as of finite size and those size limits are often approached in adsorption, whereas Equation 16a implies that unlimited concentrations and pressures are possible. This apparent conflict would not have occurred if mole fraction units were used throughout. (For a further discussion of mole fraction units versus molarity or molality, see Denbigh [37].)

Kinetics

Kinetics is based on a set of empirically derived rate laws. Three rate laws can be used to describe most natural processes:

$$dC_a/dt = -k_0 \quad \text{Zero-order} \tag{17}$$

$$dC_B/dt = -k_1 C_A \quad \text{First-order} \tag{18}$$

$$dC_B/dt = -k_2 C_A C_B \quad \text{Second-order} \tag{19}$$

where k_0, k_1, k_2 = rate constants (mol/1-sec, /sec, 1/mol-sec, respectively)
C_A, C_B = some reacting species

If molecules of A react with some fixed surface (metal catalyst, enzyme surface, soil surface, etc.), two extremes of concentration can be delineated. First, when the concentration of A is small, few of the available sites on the surface are covered, and therefore, the reaction rate dC_A/dt is proportional to the concentration of A (Equation 18). Second, when C_A is so large that essentially all of the surface is saturated with A, the rate is constant (Equation 17). This transition from first- to zero-order kinetics with increasing C_A is illustrated in Figure 4. The combined functional form of the rate law can be written as:

$$dC_A/dt = \frac{-k_o C_A}{k' + C_A} \qquad (20)$$

where $k' = k_0/k_1$

Although Equation 20 is a general form often used in biological research, it probably will seldom be necessary to incorporate this complete form into mathematical models for most natural water systems. Most groundwater systems are expected to exist in the linear first-order portion for most processes of interest, although few quantitative data are available for groundwater systems for processes other than adsorption. Most groundwater transport models use an equilibrium isotherm expression to represent adsorption or ion exchange, but this assumption of "local equilibrium" may not be appropriate [38].

For the purposes of modeling, it is often assumed that second-order reactions (Equation 19) can be approximated by pseudo-first-order reactions in which one of the species is "essentially" constant for the conditions being modeled. For reactions dependent on pH, this is reasonable for highly buffered waters. This assumption is probably better for well-stirred surface waters than for groundwater systems. As groundwater infiltration occurs, bacteria may metabolize available carbon, consuming dissolved oxygen, producing CO_2 and thereby decreasing the pH. Also,

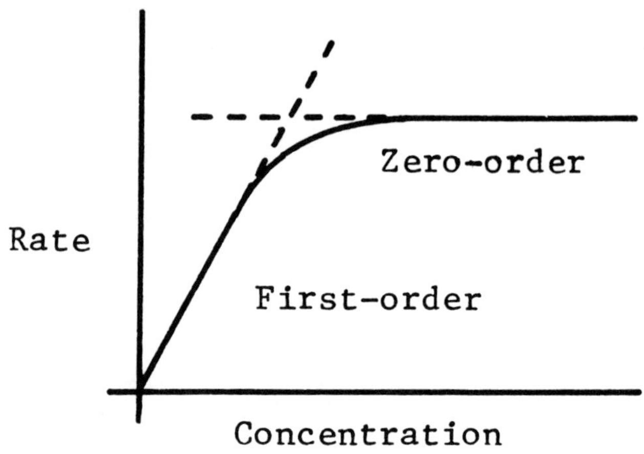

Figure 4. Reaction rate vs concentration for a surface area-limited process.

various soil formations act as strong pH buffers, redox buffers, and sources and sinks for numerous materials in water by surface and ion exchange reactions. The oxidation kinetics of iron (Fe^{2+}) by aqueous oxygen and influenced by pH is one of the few environmental reaction systems to have been studied sufficiently well to be usable for quantitative illustration. The rate law at 20°C is [39]:

$$-\frac{dFe(II)}{dt} = \frac{3 \times 10^{-12}(M/\text{min})[O_{2aq}][Fe^{2+}]}{[H^+]^2} \tag{21}$$

with $[Fe^{2+}]$, $[O_{2aq}]$ and $[H^+]$ in M and time in min. Using Equation 21 and assuming O_2 saturation, it can be shown that at pH 5 it would take 50–100 days to significantly oxidize a few mg/l iron, whereas at pH 7 a majority of the oxidation would be complete in a few minutes. In either case, a significant portion of the O_2 present in solution would be consumed. This rather simple system serves to highlight the caution needed in the normal practice of assuming that second-order reactions can be modeled by pseudo-first-order reactions.

Sorption of Pollutants

Inorganic

To understand the ionic interactions between solution ions, such as Cu^{2+}, Mg^{2+} or Na^+, and soil surfaces, it is useful to picture both the overall ionic properties as in Figure 5 and the specific bonding processes discussed below, e.g., $\equiv$Si–O–Cu^+, which may exist at a silicate surface. As a specific example, one might begin just inside the crystal surface depicted in Figure 5 and move outward, toward the bulk solution. First, a layer of net negative charge at the crystal-solution interface would be encountered. This layer may be due to $\equiv$Si–OH ionization or due to an Al being substituted for a Si in the tetrahedral surface layer, to mention two common explanations. Next, there would probably be a fixed layer of dehydrated adsorbed counterions from solution called the "Stern layer." If the primary force of adsorption is electrostatic, there will probably be fewer positive charges in the Stern layer than negative charges in the solid surface layer. If specific adsorption forces, such as complex formation via covalent bonding, are also involved, the charge on the Stern layer may exceed the surface charge. For sake of illustration, assume the latter case. Next is encountered a rather thick layer (10–1000 Å) of ions and water diffusing under the influence of thermal energy. On the average, the diffuse layer immediately adjacent to the

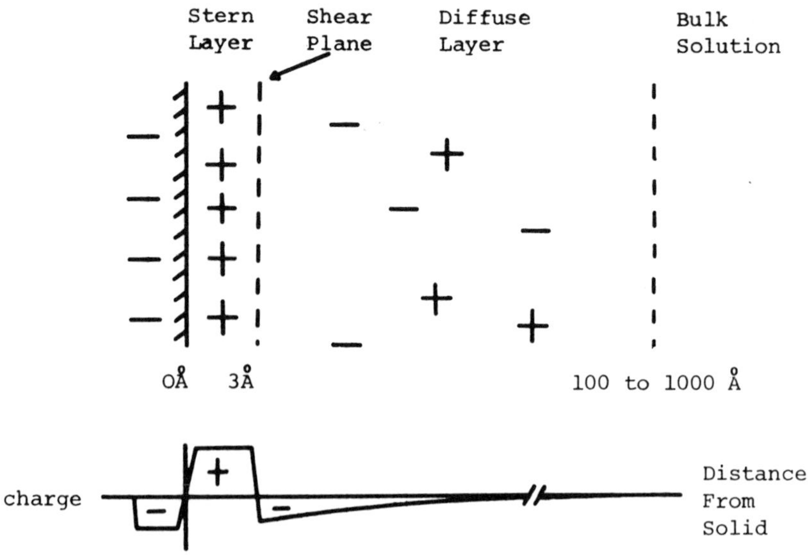

Figure 5. Charge distribution from a solid surface outward.

fixed Stern layer has a net excess of negative ions. This "average" excess of negative charges gradually tapers off as one proceeds toward the bulk solution. The total excess of negative charges in this diffuse layer exactly compensates the excess positive charge of the Stern layer so that overall electroneutrality is obeyed. If the particles move, or if the water moves, the diffuse layer can be mobilized about a "shear plane" that exists approximately at the outer edge of the Stern later, which leads to the classic zeta potential for a given solid. The representative thickness of this diffuse layer varies with ionic charge and with salt concentration (Table I) [40].

In principle, there may be an unlimited number of surfaces for solution ions to interact with, but we shall confine our attention to clays, metal oxides and macromolecular organics, such as humic acid, lipids and proteins. Representative surface reactions are illustrated in Table II. Most reactions involve a R–OH grouping and may be classified as either ionization, complexation or ion exchange. Based on thermodynamics, it should be possible to characterize each of the reactions in Table II by a single equilibrium constant, as would be for Reaction 2 in Table II:

$$K_2 \equiv \frac{\overline{\{\equiv \text{Si–OH}_2^+\}}}{\overline{\{\equiv \text{Si–OH}\}}\{H^+\}} \tag{22}$$

Table I. Thickness of Diffuse Layer for Mono- and Divalent Ions at Various Concentrations

C (mol/1)	Characteristic Thickness [40]	
	Monovalent Ions	Divalent Ions
10^{-5}	1000	500
10^{-3}	100	50
10^{-1}	10	5

Table II. Examples of Surface-Solution Bonding Interactions that Produce Surface Charge

					Reaction
Clays					
$\equiv$Si-OH			$\longrightarrow$	$\equiv$Si-O^- + H^+	1
$\equiv$Si-OH	+	H^+	$\longrightarrow$	$\equiv$Si-OH_2^+	2
$\equiv$Si-OH	+	Mg^{2+}	$\longrightarrow$	$\equiv$Si-OMg^+ + H^+	3
$\equiv$Si-OH	+	Na+	$\longrightarrow$	$\equiv$Si-ONa + H^+	4
2$\equiv$Si-ONa	+	Ca^{2+}	$\longrightarrow$	($\equiv$Si-O)(≡Si-O) > Ca + $2Na^+$	5
Metal Oxides					
$\equiv$Al-OH			$\longrightarrow$	$\equiv$Al-O^- + H^+	6
$\equiv$Al-OH	+	H^+	$\longrightarrow$	$\equiv$Al-OH_2^+	7
$\equiv$Al-OH	+	$CuOH^+$	$\longrightarrow$	$\equiv$Al-O-CuOH + H^+	8
$\equiv$Fe-OH			$\longrightarrow$	$\equiv$Fe-O^- + H^+	9
$\equiv$Fe-OH	+	$H_2PO_4^-$	$\longrightarrow$	$\equiv$Fe-OPO_3H^- + H_2O	10
$\equiv$Fe-OH	+	K^+	$\longrightarrow$	$\equiv$Fe-OK + H^+	11
$\equiv$Fe-OH	+	Li^+	$\longrightarrow$	$\equiv$Fe-OLi + H^+	12
Macroorganics: Humic Acids, Protein, Lipids, etc.					
⊃> OH			$\longrightarrow$	⊃> O^- + H^+	13
⊃> COOH			$\longrightarrow$	⊃> CO_2^- + H^+	14
~ NH_2	+	H^+	$\longrightarrow$	~ NH_3^+	15
⊃> O^-	+	Cu^{2+}	$\longrightarrow$	⊃> O-Cu^+	16

where { } would denote activities and a bar signifies the solid phase.

Similarly, for Reaction 5:

$$K_5 = \frac{\{\overline{(\equiv \text{Si-O})_2\text{Ca}}\}\{\text{Na}^+\}^2}{\{\overline{\equiv \text{Si-O Na}}\}^2\{\text{Ca}^{2+}\}} \tag{23}$$

which is normal ion exchange. As exchange or surface reaction takes place, the tendency for further exchange is altered because these clays, metal oxides and macroorganics are small, and the forces of bonding are often sufficient to physically distort the particle size and shape. This is equivalent to saying that the activity coefficients of the reacting ions in the solid phase change. Often the clays expand and contract on exchange, and no simple activity coefficient expression has yet been derived and accepted to describe how these variations with ion charge and size affect the solid-phase activity coefficients. Therefore, individual ion-distribution–coefficient curves for each clay and set of ions are generally used if the water ionic composition or solid phases are expected to change in the course of the model. These ion distribution curves often take on the general shape illustrated in Figure 6. The curves in Figure 6 show a general trend that the multivalent ions are preferred in the clay phase and even more so at lower solution concentration. Stumm and Morgan [39] give data that support these notions for kaolinite, illite and montmorillonite.

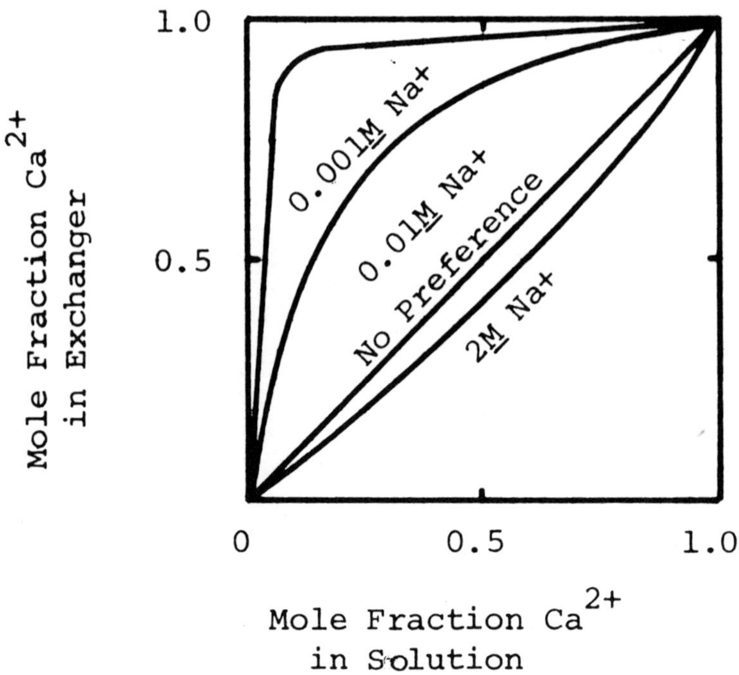

Figure 6. Typical variation of ion exchange selectivity with solution concentration.

Most of the above discussion has been based on equilibrium considerations. For most field studies of groundwater flow with water velocities of 1 ft/day, it may be reasonable to assume that equilibrium, with respect to the solution ions and solids, is established. However, if the velocities become a few feet per day, as is often the case in laboratory soil columns, this assumption of "local equilibrium" may not be valid [41]. If local equilibrium is not established, much more difficult kinetic models of adsorption must be developed and used; again, almost no rate-constant data exist [38].

Organic

In a typical surface water stream or pond, 90–95% or more of the organics in water are humic and fulvic acids, that are generally partial breakdown products of plant matter. The fractions are highly ionic in nature and are resistant to biological metabolism and O_2 oxidation. These materials are possibly major transport vehicles in surface and groundwater. The heavy metals form complexes and ion exchange with the ionic surface of these macromolecules, and neutral organics may adsorb to the organic hydrophobic backbone of the humic acid [42].

The remaining 5% of total organic carbon (TOC) is mostly low-molecular-weight organics, such as acetic, citric or benzoic acids; volatile organics, such as the priority pollutants; and other refractory industrial trace-level organics (TLO), such as antioxidants and plasticizers. TLO are minor components in quantity but are of inordinate importance to groundwater pollution because of their xenobiotic and often toxic nature, and because they generally bioconcentrate in tissue.

In surface waters, volatilization and adsorption are major transport mechanisms. In groundwater, volatilization is probably of minor importance, but adsorption may be a major mechanism of the fate and movement of organics.

Recently, important progress was made toward understanding the adsorption of organics to soils in terms of the percent soil-organic and the water solubility of the TLO solute by Karickhoff et al. [43]. Similar to the equilibrium constants in Equations 16a and 16c, the partition coefficient K_p in terms used by Karickhoff et al. is:

$$K_p = \frac{x}{c} \tag{24}$$

where x = concentration of sorbate on the sediment (ppb, dry wt basis)
c = equilibrium sorbate concentration in solution (ppb)

The adsorption constant can then be normalized for the fraction of organic carbon (OC) in the soil:

$$K_{oc} = K_p/OC \tag{25}$$

For a whole sediment composed of various size fractions by mass f use:

$$K_p = \Sigma K_{p,i} f_i \tag{26}$$

or, using Equation 25 to calculate $K_{p,i}$:

$$K_p = \Sigma K_{oc,i} f_i (OC)_i \tag{27}$$

In practice, Equation 27 can be separated into two terms, one for the fines ($\leq 50\ \mu m$) and one for the sand. The K_{oc} for sand has been shown in practice to be approximately 0.20 of K_{oc} for the fines, or

$$K_p = (K_{oc} fOC)_{fines} + 0.2(K_{oc} f'OC)_{sand} \tag{28}$$

where $f' = 1 - f$. Since $0.2(OC)f'_{sand}$ is generally small compared with $f(OC)_{fines}$, for a given soil, K_p can be written as simply proportional to the K_{oc} partition coefficient:

$$K_p \cong K_{oc} fOC \tag{29}$$

Additionally, Karickhoff et al. experimentally showed that the K_{oc} was directly proportional to the normal octanol/water partition coefficient K_{ow}:

$$K_{oc} = 0.63\, K_{ow} \quad (r^2 = 0.96) \tag{30}$$

Finally, the K_{oc} is related to the water solubility, S:

$$K_{oc} = 2.75\, S^{-0.54} \tag{31}$$

where S = water solubility (mole fraction)

In log-log form, these equations become:

$$\log K_{oc} = 1.00 \log K_{ow} - 0.21\, (r^2 = 1.00) \tag{32}$$

$$\log K_{oc} = -0.54 \log S + 0.44 (r^2 = 0.94) \tag{33}$$

The relative velocities of a contaminant v_c to water v_o in soil of bulk specific gravity ρ_b and porosity n is [33]:

$$\frac{v_o}{v_c} = 1 + \frac{\rho}{n} K_p \tag{34}$$

where K_p = partition coefficient, as in Equation 29

Substituting Equations 29 and 31 into 34, and neglecting the one on the right hand side, Equation 34 becomes:

$$\frac{v_o}{v_c} = \frac{\rho}{n} 2.75\, S^{-0.54}\, fOC \tag{35a}$$

or

$$\log \frac{v_o}{v_c} = \log(2.75\, \rho fOC/n) - 0.54 \log S \tag{35b}$$

For any given soil, Equation 35b predicts a log relationship between the relative retardation for a set of organic compounds and their water solubilities. If such predictions are proven correct in field studies, they would be a great asset to groundwater modeling.

Research at Rice University is presently being carried out to investigate the hypothesis that the work of Karickhoff et al. or a similar theory will describe the interaction between TLO and dissolved organic carbon (DOC) in water. Interaction in this case would account for facilitated transport. That is, the water column may actually be a three-phase system—soil, water and DOC.

In this case, using Karickhoff's theory, $K_p = K_{oc}$, since OC = 1.0 is assumed. Preliminary results using phenanthrene as a test compound (solubility = 1.29 ppm) yielded $K_{p,exp} = 10.0 \times 10^4$, which agrees reasonably with $K_{p,theor} = 1.4 \times 10^4$.

The reversibility of these interactions with DOC is important, as this mechanism of transport may act as a supply of free organics to aquifer systems, which may otherwise be unaffected in reasonable time due to adsorption and retardation. The work of Karickhoff et al. with soils showed the binding of pyrene and methoxychlor to be reversible, and work at Rice University indicates that the association between DOC and phenanthrene is also reversible. Similar work using DDT and a surface water with a TOC of 32 ppm yielded data from which $K_{p,exp} = 10^{4.51}$ can be calculated, which compares reasonably with $K_{p,theor} = 10^{6.0}$, from

Equation 30 [44]. Further research using other compounds and various humic and fulvic acid cuts is underway by the authors to determine if a correlation such as Equation 32 or 33 exists to describe this type of organic-mediated transport. If it is shown that DOC mobilizes TLO in groundwater, it will have far-reaching consequences to the modeling of the distribution of organics in natural waters.

Microbiological Processes

Removal or transformation of organic pollutants from the subsurface environment can occur through the action of microorganisms that may be attached to the soil or contained in the void spaces. The soil environment is undoubtedly the most complex of microbial habitats and contains within variances of time and space a plethora of microorganisms capable of chemical transformations of most organic pollutants. An excellent review of environmental factors affecting distribution and activity of bacteria in surface soils has been presented by Stotzky [45]. Microbial interactions in the surface environment have been studied much more extensively than those in the subsurface. However, bacteria have been found throughout extensive depths in different locations, along with varying organic carbon concentrations. A recent review of bacteriological surveys of groundwater used for drinking water supplies in the United States found that 9–51% of the samples contained coliform bacteria [46]. The compatibility of the subsurface for microbial growth and metabolism has been addressed by McNabb [47], who concluded that subsurface conditions may not preclude the existence of active microbial flora. It is not unreasonable to presume that, despite their low numbers, subsurface microbes could possess an adequate fraction of metabolic activities needed to significantly reduce the levels of organic compounds migrating through the subsurface soil profile.

Approaches to microbial kinetics generally are based on Monod bacterial kinetics, in which the rate of substrate utilization is proportional to the concentration of the microorganisms present and is a function of substrate concentration [48]:

$$\frac{-d[S]}{dt} = \frac{k\,[X][S]}{y([K_s] + [S])} \tag{36}$$

where [S] = substrate concentration
k = maximum utilization rate for the substrate per unit mass of bacteria

[X] = concentration of bacteria
$[K_s]$ = half-velocity coefficient for the substrate
y = yield coefficient = d[X]/d[S]

At typical groundwater pollutant concentrations, [S] is normally less than K_s, which ranges typically from 0.1 to 10.0 mg/l [47]. In this case, the above equation reduces to the familiar form of second-order kinetics:

$$\frac{-d[S]}{dt} = \frac{k}{yK_s}[X][S] = K'[X][S] \qquad (37)$$

Second-order rate constants have been measured [48] and found to be first-order in both substrate and population size when substrate concentrations are kept low. The mean values were highly independent of site.

Another important microbial concept concerning groundwater organic utilization rates is that of threshold at low organic concentrations. Boethling and Alexander [49] have speculated on this, but data are inconclusive.

Structure-activity relationships have long been recognized in chemistry. Relationships between rate constants and molecular structure may prove useful as a predictive tool. Correlation, for instance, between second-order microbial rate constants and second-order hydrolysis constants appears good over orders of magnitude.

Biodegradation has been modeled by biofilm kinetics [50], in which the concentration of the substrate available to the microorganism on an inert surface is a function of the concentration gradient across the biofilm/water interface and can be described by Fick's first law of diffusion:

$$J_z = D\,\frac{d[S_z]}{dZ} \qquad (38)$$

where J_z = diffusional flux (M/L^2T)
D = diffusion coefficient (L^2/T)
Z = direction normal to the surface (x + y taken as zero)

In this way, microbial utilization is seen as a process of mass transportation from the bulk fluid and biodegradation within the biofilm.

The above discussion does not account for the effect of other processes, such as multiple carbon sources, temperature, competitive substrate utilization or cometabolism, but these are certainly factors for which many questions must be answered before biodegradation can be adequately modeled.

SUMMARY AND CONCLUSION

Many groundwater flow and transport models have been developed and applied to field problems to predict migration patterns of contaminants. However, accuracy of predictive ability is limited because of uncertainties in estimating field parameters, such as hydraulic conductivity, velocity and dispersivity. In addition, numerical errors can mask computer solutions and produce another major source of error.

Yet, with these inherent calibration problems, models of groundwater behavior still offer the most reliable approach to prediction and ultimate management of waste disposal problems. Analytical models offer a useful first approximation to flow directions and arrival times under homogeneous conditions. Limited field data can be used to partially calibrate the model and at least provide estimates of gross impacts. Advection models are probably adequate for cases in which nontoxic chemicals are involved and general levels of accuracy are required or advection is known to be dominant.

Advection/dispersion models, although more accurate, require extensive field data on hydraulic conductivity and dispersivity; data are presently unavailable for most waste sites, except where large sums have been allocated for monitoring wells and water quality sampling. Heterogeneities within an aquifer determine to a large extent the magnitude of dispersion, and dispersivity is a function of variations in hydraulic conductivity. More research is need in theoretical development and field sampling to define better the effects of dispersion. The lack of field data is particularly clear since the ability to solve governing equations is many years ahead of the physical and geochemical description of aquifer systems. Thus, models have helped point the way toward correcting the deficiencies. These corrections need to be implemented in field data and sampling methods.

The USGS MOC model is probably one of the best-documented and most widely used advection/dispersion models. Galerkin finite-element models are typically most accurate, but are much more expensive to operate. Given the lack of available data on input parameters, choice of a model is probably the least of the problems at this time.

Chemical reaction processes for low-molecular-weight compounds are well understood for some equilibrium and kinetic processes, such as hydrolysis, complexation and acid-base equilibria. However, constants associated with heterogeneous reactions between colloids, high-molecular-weight materials (i.e., fulvic and humic acids) and clays are not well known because of the difficulty of defining structures of these compounds. The adsorption of low-molecular-weight neutral organics with high-molecular-weight collodial soil organic matter is at present an area of active research. Heterogeneous reactions between aqueous and gas or solid phases, such as volatilization, precipitation and dissolution, are also generally well known.

Although certain reaction rates are known for adsorption, ion exchange or degradation, their inclusion into standard groundwater models and application to field studies has been extremely limited. Ongoing research efforts at the U.S. Environmental Protection Agency and Rice University have attempted to define adsorption coefficients for selected organics and to include the adsorption mechanism in transport models. Again, the greatest obstacle at this time is a severe lack of appropriate field and laboratory data.

REFERENCES

1. Scheidegger, A. E. "General Theory of Dispersion in Porous Media," *J. Geophys. Res.* 66(20):3273–3278 (1961).
2. Mauersberger, P. "On the Theoretical Basis of Modelling the Quality of Surface and Subsurface Waters," Publ. No. 125, International Association of Hydrological Sciences (1978).
3. Al-Niami, A. N. S., and K. R. Rushton. "Radial Dispersion to an Abstraction Well," *J. Hydrol.* 39(3/4):287–300 (1978).
4. Hunt, B. "Dispersive Sources in Uniform Ground-Water Flow," *J. Hydraul. Div., ASCE* 104(HY1):75–85 (1978).
5. Walton, W. C. "Progress in Analytical Groundwater Modeling," *J. Hydrol.* 43:149–159 (1979).
6. Cleary, R. W., and M. J. Ungs. "Groundwater Pollution and Hydrology: Mathematical Models and Computer Programs," Princeton University, Water Resources Program Report No. 78–WR–15.
7. Charbeneau, R. J. "Groundwater Contaminant Transport with Adsorption and Ion Exchange Chemistry: Method of Characteristics for the Case without Dispersion," *Water Resources Res.* (submitted).
8. Chaudhry, M. H. *Applied Hydraulic Transients* (New York: Van Nostrand Reinhold Co., 1979).
9. Konikow, L. F., and J. D. Bredehoeft. "Computer Model of Two-Dimensional Solute Transport and Dispersion in Ground Water," in *Automated Data Processing and Computations, Techniques of Water Resources Investigations of the U.S. Geological Survey, Book 7: Automated Data Processing and Computations* (1978).
10. Bredehoeft, J. D., and G. F. Pinder. "Mass Transport in Flowing Groundwater," *Water Resources Res.* 9:194–209 (1973).
11. Robertson, J. B. "Digital Modeling of Radioactive and Chemical Waste Transport in the Snake River Plain Aquifer at the National Reactor Testing Station, Idaho," USGS Water Resources Div. Report No. IDO–22054 (1974).
12. Faust, C. R., and J. W. Mercer. "Ground-Water Modeling: Numerical Models," *Ground Water* 18(4):395–409 (1980).

13. Gray, W. G. "The Finite Element Method in Groundwater Transport," Advances in Groundwater Hydrology, Z. A. Saleem, Ed. (Minneapolis, MN: American Water Research Institute, 1976), pp. 130–143.
14. Narashiman, T. N., S. P. Neuman and Witherspoon. "Finite Element Method for Subsurface Hydrology Using a Mixed Explicit-Implicit Scheme," *Water Resources Res.* 24(5):863–877 (1978).
15. Varoglu, E., and W. D. L. Finn. "A Finite Element Method for the Diffusion-Convection Equation with Constant Coefficients," *Adv. Water Resources* 1(6):337–343 (1978).
16. Carnahan, B., H. A. Luther and J. O. Wilkes. *Applied Numerical Methods* (New York: John Wiley & Sons, Inc., 1969).
17. Prickett, T. A. "Modeling Techniques for Groundwater Evaluation," in *Advances in Hydroscience, Vol. 10,* V. T. Chow, Ed. (New York: Academic Press, Inc., 1975), pp. 1–143.
18. Codell, R. B., and D. L. Schreiber. "NRC Models for Evaluating the Transport of Radionuclides in Groundwater," Proceedings of Symposium Low-Level Radioactive Waste Management, USGS and National Regulatory Commission, Atlanta, GA (1979).
19. Gorelick, S. M., T. Remson and R. W. Cottle. "Management Model of a Groundwater System with a Transient Pollutant Source," *Water Resources Res.* 15(5):1243–1249 (1979).
20. Gureghian, A. B., D. S. Ward and R. W. Cleary. "Simultaneous Transport of Water and Reacting Solutes Through Multilayered Soils under Transient Unsaturated Flow Conditions, *J. Hydrol.* 41(3–4):253–278 (1979).
21. Konikow, L. F. "Modeling Chloride Movement in the Alluvial Aquifer at the Rocky Mountain Arsenal, Colorado," USGS Water Supply Paper No. 2044 (1977).
22. Perez, A. I., et al. "A Water Quality Model for a Conjunctive Surface-Groundwater System," U.S. EPA Office of Research and Development, Socioeconomic Environmental Studies Series EPA 600/S–74–013 (1974).
23. Robertson, J. B. "Numerical Modeling of Radioactive Solute Transport from Waste-Seepage Ponds at the Idaho National Engineering Laboratory," USGS Open File Report 76–717, IDO–22057 (1977).
24. Satter, A., et al. "Chemical Transport in Porous Media with Dispersion and Rate-Controlled Adsorption," *Soc. Petrol. Eng. J.* 20(3):129–138 (1980).
25. "GWSIM-II Groundwater Simulation Program: Program Documentation and User's Manual," Texas Department of Water Resources, Data Collection and Evaluation Section, Austin, TX (1978).
26. Ahlstrom, S. W., and H. P. Foote. "Transport Modeling in the Environment Using the Discrete-Parcel-Random-Walk Approach," NTIS Publication No. BNWLSA–5528 (1976), pp. 1–4.
27. Konikow, L. F., and J. D. Bredehoeft. "Modeling Flow and Chemical Quality Changes in an Irrigated Stream-Aquifer System," *Water Resources Res.* 10(3):546–562 (1974).
28. Robson, S. A. "Application of Digital Profile Modelling Techniques to Groundwater Solute Transport at Barstow, California," USGS Water Supply Paper No. 2050 (1978), pp. 1–28.
29. Remson, I., S. M. Gorelick and J. F. Fliegner. "Computer Models in Groundwater Exploration," *Ground Water* 18(5):447–451 (1980).

30. Helfferich, F. *Ion Exchange* (New York: McGraw-Hill Book Co., 1962).
31. Anderson, M. P. "Using Models to Simulate the Movement of Contaminants Through Ground Water Flow Systems," *CRC Crit. Rev. Environ. Control* 9(2):97–156 (1979).
32. Fried, J. J. *Ground Water Pollution* (Amsterdam: Elsevier Scientific Publishing Co., 1975).
33. Freeze, R. A., and J. A. Cherry. *Groundwater* (Englewood Cliffs, NJ: Prentice-Hall, Inc., 1979).
34. Benson, S. W. *The Foundations of Chemical Kinetics* (New York: McGraw-Hill Book Co., 1971).
35. Kazmierczak, T. F., M. B. Tomson and G. H. Nancollas. "The Crystal Growth of Calcium Carbonate, a Controlled Composition Kinetic Study," *J. Phys. Chem.* (in press).
36. Mackay, D., and S. Patterson. "Calculating Fugacity," *Environ. Sci. Technol.* 15:1006–1014 (1981).
37. Denbigh, K. G. *The Principles of Chemical Equilibrium: with Applications in Chemistry and Chemical Engineering,* 3rd ed. (Cambridge: Cambridge University Press, 1971).
38. Cameron, D. R., and A. Klute. "Convective-Dispersive Solute Transport with a Combined Equilibrium and Kinetic Adsorption Model," *Water Resources Res.* 13:183–188 (1977).
39. Stumm, W., and J. J. Morgan. *Aquatic Chemistry,* 2nd ed. (New York: Wiley-Interscience, 1981).
40. Moore, W. J. *Physical Chemistry,* 4th ed. (Englewood Cliffs, NJ: Prentice-Hall, Inc., 1972), p. 512.
41. Janes, R. V., and J. Rubin. "Applicability of the Local Equilibrium Assumption to Transport Through Soils of Solutes Affected by Ion Exchange," in *Chemical Modeling in Aquatic Systems,* E. A. Jenne, Ed., American Chemical Society Symposium Series 93 (Washington, DC: American Chemical Society, 1979), pp. 225–235.
42. Gjessing, E. T. *Physical and Chemical Characteristics of Aquatic Humus* (Ann Arbor, MI: Ann Arbor Science Publishers, Inc., 1976).
43. Karickhoff, S. W., D. S. Brown and T. A. Scott. "Sorption of Hydrophobic Pollutants on Natural Sediments," *Water Res.* 13:241–248 (1980).
44. Poirrier, M. A., et al. "Adsorption and Concentration of Dissolved C-14 DDT by Coloring Colloids in Surface Water," *Environ. Sci. Technol.* 6:1033–1035 (1972).
45. Stotzky, G. "Activity, Ecology, and Population Dynamics of Microorganisms in Soil," in *Microbial Ecology,* A. Laskin and H. Lechevalier, Eds. (Cleveland: CRC Press, Inc., 1974).
46. Dunlap, W. J., and J. F. McNabb. "Subsurface Biological Activity in Relation to Groundwater Pollution," U.S. EPA Report 660/2-73-014 (1973).
47. McNabb, J. F. "Current Developments in Assessing the Role of Subsurface Biological Activity in Ground Water Pollution," Proceedings of Conference on Ground Water Quality, Measurement, Prediction, and Protection, Water Research Centre, Medmenham, Great Britain, (1976).
48. Tinsley, I. J. *Chemical Concepts in Pollutant Behavior* (New York: Wiley-Interscience, 1979).
49. Boethling, R. S., and M. Alexander. "Effect of Concentration of Organic Chemicals on Their Biodegradation by Natural Microbial Communities," *Appl. Environ. Microbiol.* 37:1211–1216 (1980).

50. McCarty, P. L., B. E. Rittmann and M. Reinhard. "Processes Affecting the Movement and Fate of Trace Organics in the Subsurface Environment," in *Proceedings of Symposium,* T. Asano and P. V. Roberts, Eds., California State Polytechnic University, Pomona, CA, September 5-7, 1979.

CHAPTER 14

COMPLEX EFFLUENT FATE MODELING

Harold L. Bergman and Joseph S. Meyer

Fish Physiology and Toxicology Lab
Department of Zoology and Physiology
University of Wyoming
Laramie, Wyoming 82071

The models that have been devised to predict the fate of chemicals in the aquatic environment typically deal with one chemical compound at a time. These models, including both mathematical (rate) and physical (microcosm) models, have been reviewed previously [1,2]. In this book, the emphasis has been on mathematical contaminant fate models.

The emphasis of all of these fate models on individual compounds no doubt stems from early concerns about the persistence and bioaccumulation of individual pollutants, such as certain pesticides and heavy metals. More recently, passage of the Toxic Substances Control Act (TSCA), with its emphasis on production and use of new chemicals in commerce, further stimulated development, perfection and application of single-compound environmental fate models.

On the other hand, some of the earliest mathematical fate models used for the aquatic environment could be termed "complex effluent" models. For instance, sanitary engineers have used models to predict biochemical oxygen demand (BOD) downstream from municipal sewage discharges, and hydrologists have mathematically modeled the fate of total dissolved solids (TDS) levels associated with saline discharges such as those from irrigated agriculture. However, BOD and TDS, termed "conventional" pollutants, only represent general properties of these complex effluents. Such models typically do not deal with the fate of potentially toxic compounds, such as chlorinated organics or heavy metals, which would be expected to make up only a small fraction of the modeled BOD or TDS parameters.

Because these single BOD and TDS models are not really different conceptually from single-compound models, they do not exemplify the complex-effluent modeling problem we wish to discuss in this chapter. Rather, we will address an even more complicated complex-effluent problem—modeling the fate of potentially hazardous chemical constituents found in complex-effluent mixtures. To further define this problem and our concept of a complex effluent, we suggest that an effluent is truly complex when one or both of the following conditions apply:

1. A number of potentially hazardous chemical constituents are present (inorganic, organic, organometallic, radioactive—any or all), and all constituents cannot be identified within reasonable time or cost.
2. A reasonable possibility exists that the environmental fate of a number of constituents could be affected markedly by complex kinetic interactions (e.g., synergism).

The main objective of this chapter is to stimulate thought and discussion about how one might model the fate of constituents from such a complex effluent. First, however, we must ask what is expected of any modeling strategy applied to this problem. Traditionally, the objective of single-compound fate models, whether they are mathematical (rate) or physical (microcosm) models, has been to predict environmental concentrations of pollutants [1]. This should also be the ultimate objective of complex-effluent models, but these models must be somewhat different since we do not always know the identities of all compounds for which environmental concentrations must be estimated. Thus, we suggest that the objectives of a complex-effluent modeling strategy should be:

1. to identify constituents posing an environmental fate problem; and
2. to predict expected exposure concentration (EEC) of each constituent posing a problem.

In the remaining sections of this chapter, we illustrate the modeling problem posed by complex-effluent mixtures by providing some examples of complex effluents, reviewing mathematical and physical modeling strategies currently being developed for complex effluents, and discussing the trade-offs between mathematical and physical models for complex-effluent mixtures.

THE PROBLEM

Complex-Effluent Examples

Although many surface-water discharges might be classified as complex effluents, several stand out as posing especially difficult problems for identifying hazardous constituents and predicting their fate in aquatic environments. Examples include coal-fired power plant effluents, chemical manufacturing discharges, oil spills, mine effluents, waste-dump leachates and synfuel process waters.

Because synfuel process waters present such a perplexing problem and because our own work over the past few years has been to identify fate and effect problems that might arise from their release, we will describe several synfuel process waters as examples of complex effluents. We must caution that we use the term "effluent" rather loosely here since the synfuel industry is just now being developed in the United States, and significant amounts of raw or treated process waters are not now being discharged. Moreover, no one proposes to discharge the kinds of raw process waters described below directly to surface waters. However, accidental releases of raw process waters will occur and, if only for that reason, potential contaminant fate and effect problems need to be evaluated for the raw waters. More importantly, by identifying constituents posing the greatest fate and effects hazards, these constituents can be targeted for removal during treatment and emphasized in effluent monitoring plans.

There are many different processes for each of the main synfuel technologies, which include coal gasification, coal liquefaction, oil shale retorting and tar sands recovery. Furthermore, each process has a number of contaminated process water streams. Oil shale retorting, for example, produces retort water, condenser water, cooling tower and boiler blowdown, mine water, domestic waste, raw shale leachate, and spent shale leachate [3]. Because of the wide variety of process streams, the chemical makeup of synfuel process waters varies considerably.

Table I [4,5] shows the range of values for selected water-quality parameters in a number of oil shale retort waters and underground coal gasification (UCG) condenser waters. Both process water types are characterized as highly alkaline, high-pH waters with very high organic carbon content. In the case of oil shale retort water, high concentrations of several heavy metals are also present. Figure 1 illustrates typical chromatograms from reverse-phase, high-performance liquid chromatography (HPLC) gradient elutions of an oil shale retort water and a UCG

Table I. Selected Water Quality Characteristics of Oil Shale Retort Waters and UCG Condenser Waters [4,5]

Parameter	Range of Values (mg/l)[a]	
	Oil Shale Retort Waters[b]	UCG Condenser Waters[c]
Alkalinity	1,500–23,000	11,200–12,200
Carbon		
Bicarbonate (HCO_3^-)	0–36,000	5,800–10,000
Carbonate (CO_3^{2-})	0–24,000	2,300–4,500
Total Organic (TOC)	1,000–39,000	8,900
Chemical Oxygen Demand	270–140,000	7,900
Chloride	210–5,800	100–950
Conductance (siemens)	6,500–64,000	16,100
Cyanide	0.001–2.6	0.05–0.13
Elements		
Aluminum	0.03–17	0.05–1.1
Arsenic	0.05–43	0.03–0.08
Boron	0–100	0.3
Calcium	1.9–300	2.7–3.7
Magnesium	0.1–170	0.9–3.8
Potassium	2.0–60	0.5–1.0
Selenium	0.005–3.1	0.26–0.35
Sodium	7.0–5,300	7.5–39
Zinc	0.01–400	
Fluoride	0–270	17
Hardness (as $CaCO_3$)	11–1,200	12–25
Nitrogen		
Ammonia (total NH_3-N)	160–24,000	3,400–16,400
Nitrate (NO_3-N)	0.04	0.1
Total Kjeldahl (TKN)	1,000–28,000	3,800–5,000
pH	3.5–9.5[d]	8.8–9.1
Solids		
Dissolved	3–111,000	210–1,220
Suspended	16–100	2–54
Total	580–111,000	210–1,270
Sulfate (SO_4^{2-})	3.3–17,000	16–300

[a] Except pH and conductance.
[b] Range of values for 1–12 different oil shale retort waters.
[c] Range of values for 1–3 different UCG condenser waters.
[d] pH values for retort water produced during retorting of Western oil shales were all alkaline; one pH value of 3.5 for a retort water produced from a Michigan oil shale.

condenser water. In reverse-phase HPLC gradient elutions, relatively polar and low-molecular-weight organics, such as phenolics, elute earlier than relatively nonpolar and high-molecular-weight organics, such as polynuclear aromatic hydrocarbons (PAH). As suggested by these chro-

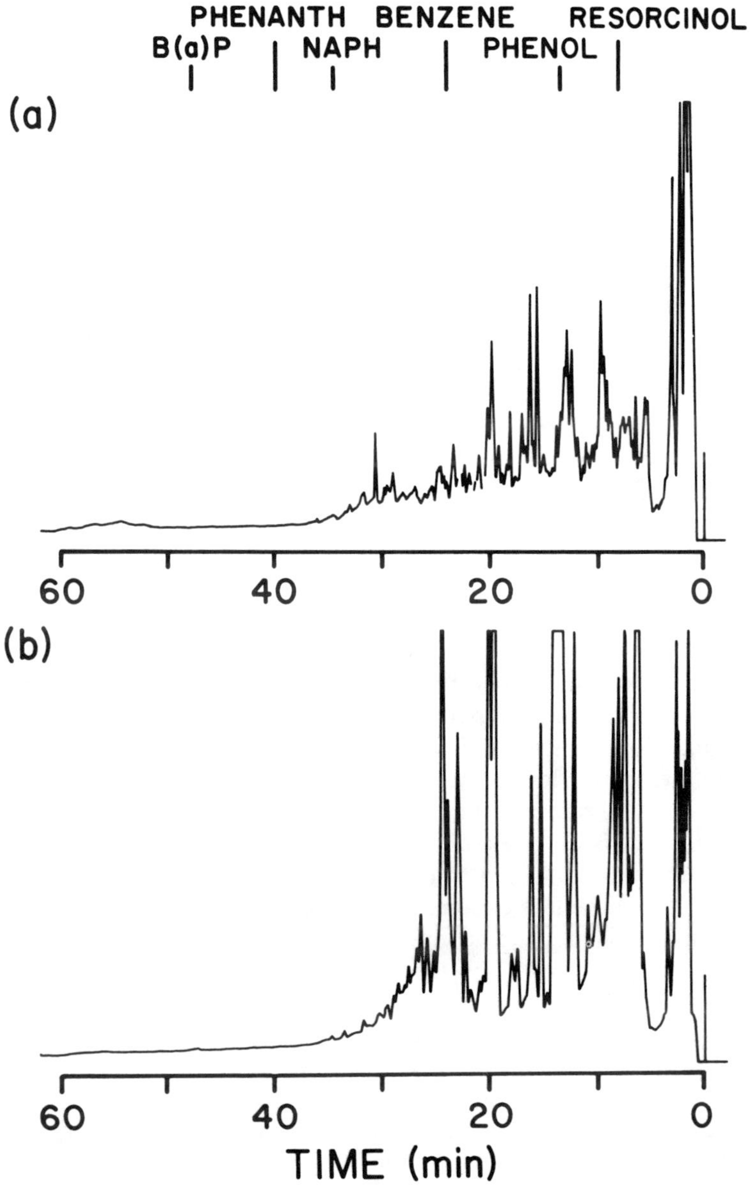

Figure 1. Reverse-phase HPLC gradient chromatograms of (a) Occidental 6 oil shale retort water and (b) Hanna 4B-01W underground coal gasification condenser water. Retentions of 6 reference compounds [resorcinol, phenol, benzene, napthalene, phenanthrene and benzo [a] pyrene (B[a]P)] are indicated above the chromatograms.

matograms and by work of other investigators [6–9], coal gasification and oil shale retorting produce process waters with many compounds at high concentrations and almost innumerable compounds at very low concentrations. Oil shale process waters contain a large number of organic compound classes such as carboxylic acids, aldehydes, ketones, quinolines, nitriles, pyridines and thiophenes [6,10–12], but coal gasification condenser waters contain very high concentrations of phenolic compound classes [8,11,13]. Both types of process waters may contain trace quantities of high-molecular-weight PAH [11].

Without question, synfuel process waters are exceedingly complex chemically. Furthermore, the organic composition of these waters, along with other energy-related liquid and solid wastes, such as those from coal-fired power plants and oil refineries, are very poorly characterized [9].

Modeling Problems

Baughman and Lassiter [1] suggested that "the complexity of ecosystems makes prediction of environmental fate a sobering task, a task often viewed as hopelessly difficult." They were, of course, referring to the difficulties of predicting the fate of individual compounds. They and others have, despite these difficulties, been quite successful in modeling the fate of individual compounds in the aquatic environment. However, the almost overwhelming complexity of some effluent mixtures adds another set of difficulties to the task of contaminant fate modeling. Again, the task is sobering and could be viewed as hopelessly difficult.

An effluent is complex when one cannot reasonably analyze for all potentially hazardous constituents and/or when there is a reasonable possibility that the fate of some constituents could be affected markedly by complex kinetic interactions such as synergism. Although a few investigators [9,11] have now identified a large number of constituents in several synfuel process waters, many compounds have not been identified. Many of these are nonvolatile compounds or are present at low (μg/l and lower) concentrations. Although more effort will likely be expended over the next few years on extensive organic characterization of several typical synfuel process wastes, the cost of these studies will be very high. Because of prohibitive time and cost requirements, such extensive characterizations will probably not be routine for some time. Until these analyses are available routinely, we are faced with the problem of not knowing the identity of many constituents in such complex effluents.

The problem of complex kinetic interactions among chemicals complicates traditional mathematical fate models, and consequently, such interactions are usually ignored. In any case, mathematical models that included a number of interaction terms would no doubt be too complicated for routine use in regulatory decisions. Yet, synergistic and antagonistic interactions among chemicals could affect contaminant fate in important ways. For example, a number of physical and chemical processes that influence fate of chemicals (e.g., water solubility, sorption and precipitation) could be influenced by chemical interactions, suggesting the need for higher-order kinetic terms in mathematical models. However, biological processes probably contribute the most variability to environmental fate calculations, and part of the cause may be chemical interactions that are not accounted for in the model.

Such interactions are known to affect bioaccumulation. For instance, Matsumura [14] reported a weak antagonistic effect of polychlorinated biphenyls (PCB) on DDT accumulation in fathead minnows, whereas DDT accumulation decreased with increasing PCB concentrations. Reinert et al. [15] reported a similar relationship between DDT and dieldrin. Their results, adapted here and summarized in Figure 2, show reduced accumulation of dieldrin in yearling lake trout when the fish were exposed simultaneously to DDT. From this and other work, we

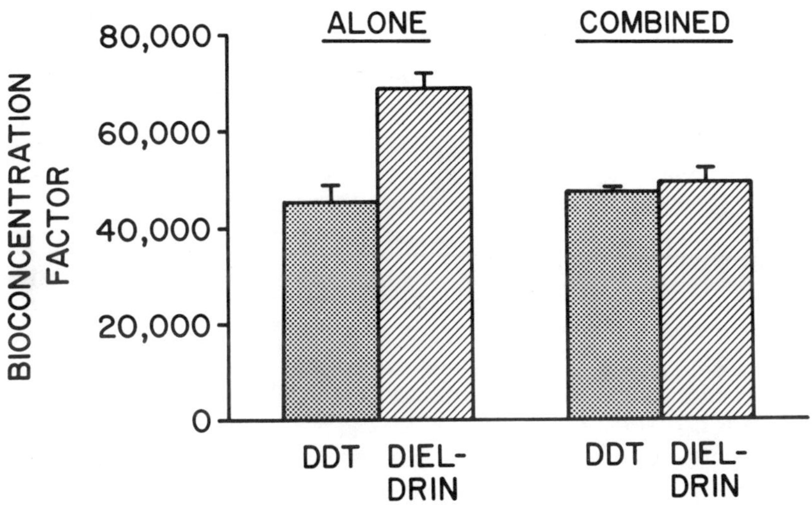

Figure 2. Bioconcentration factors in yearling lake trout exposed for 152 days to *p,p′*-DDT and dieldrin in single-compound and combined exposures; values are means with line indicating one standard error of the mean (adapted from Reinert et al. [15]).

know that an organic compound may influence the uptake and/or depuration of other organics. Both antagonistic and synergistic interactions are possible, since enzyme systems can be induced or inhibited by different compounds. Cytochrome P-450 and mixed-function oxidase systems can be induced by one organic, resulting in enhanced metabolism of a second organic [16]. On the other hand, other organic compounds and some heavy metals can inhibit these same enzyme systems.

Interactions such as these can also affect bacterial metabolism of organic compounds, but preferential metabolism could be even more important in considering microbial degradation of organics. The importance of this phenomenon is suggested by the results reported by Meyer et al. [17] and shown in Figure 3. In single-compound, laboratory shake-flask experiments, phenol and naphthalene were both metabolized rapidly by bacteria seeded from an oil-refinery settling pond [Figure 3a]. However, when phenol and naphthalene were subjected together to the same inoculum, naphthalene was not degraded until phenol disappeared (Figure 3b). We obtained similar results with phenol and benzene mixtures, where phenol again was preferentially metabolized.

The kinds of interactions we have discussed might not be critical in many single-compound models of complex effluents. The effect of one compound on bioaccumulation of a second may not be important if the change is no larger than an order of magnitude, as is the case in the DDT-dieldrin interaction shown in Figure 2. Furthermore, the phenomenon of preferred microbial substrates demonstrated in Figure 3 may be only a laboratory curiosity. On the other hand, in short-term exposures to complex mixtures, such as during accidental spills, the preferred substrate phenomenon may be important for predicting concentrations of mixture constituents. With longer-term exposure, such as in chronic discharges and activated-sludge ponds, the bacterial population may acclimate and simultaneously metabolize all but the most recalcitrant organics. In any case, before we progress with contaminant fate modeling and model validation, we suggest that complex kinetic interactions should be borne in mind for complex effluents and possibly the discharge of individual compounds to a complex receiving water.

In considering the problems of modeling the fate of complex-effluent constituents, we have discussed the implications of the cost of completely characterizing complex effluents and the possibility of complex kinetic interactions affecting the fate of complex-effluent constituents. We also suggest that as effluent complexity increases, chemical characterization costs and potential for compound interactions increase dramatically (Figure 4).

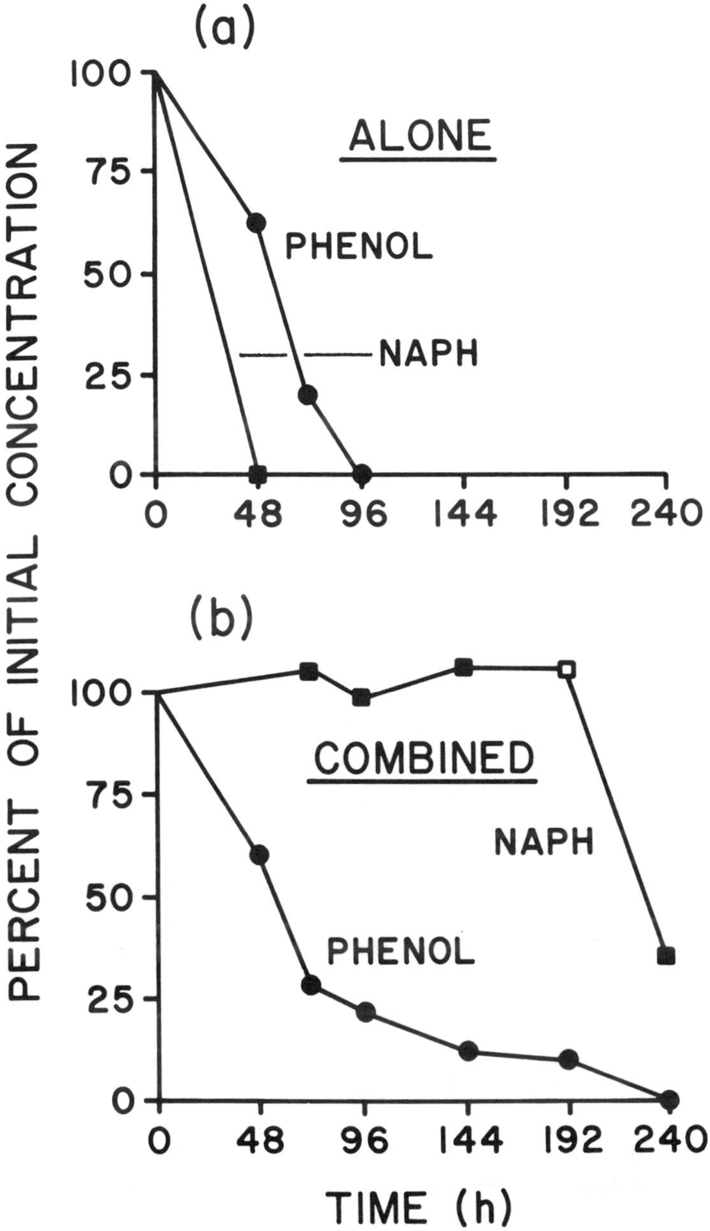

Figure 3. Degradation of phenol and naphthalene by oil-refinery settling-pond bacteria in (a) separate shake flasks and (b) the same shake flask.

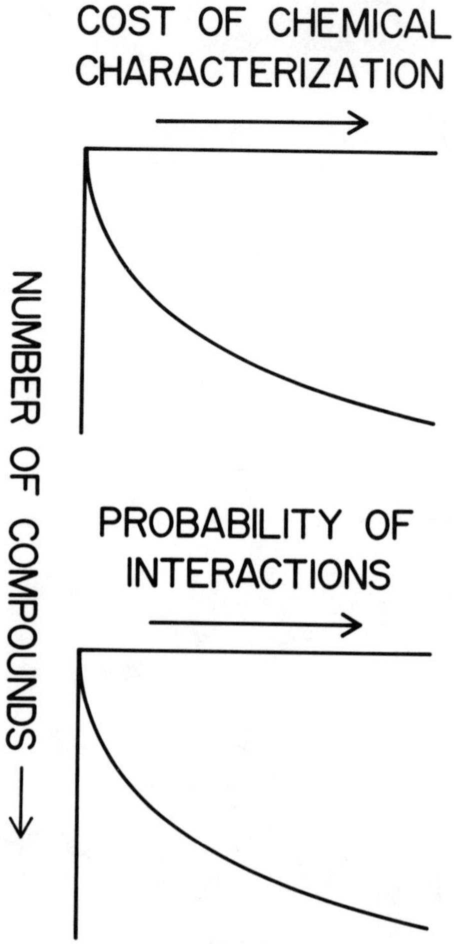

Figure 4. Implications of increased effluent complexity in analytical costs and potential compound interactions.

MATHEMATICAL MODELING APPROACHES

Compared to single-compound applications, few mathematical models of complex effluents have been designed. Among these complex-effluent models, the following three approaches are used: (1) model general properties of the effluent; (2) model chemical classes by choosing representative compounds; and (3) model one compound at a time. Proceeding from the general property models to the specific compound approaches, individual compound fates are more predictable. However, input information requirements (e.g., complete chemical characterization, rate constants for many compounds) rapidly increase with the more specific models and often can limit the objectives and applications of the models.

General-property models include the "conventional" pollutant models for BOD and TDS already discussed. Dissolved oxygen reductions and salinity increases can be predicted with acceptable accuracy, but fate information about problem compounds can not be extracted from these models. Other recent examples of general-property models are the oil-spill models developed by the National Oceanographic and Atmospheric Administration (NOAA) and Oak Ridge National Laboratory (ORNL). In the NOAA models, only physical transport of an oil layer at sea is considered [8]. Primary objectives include predicting movement, landfall sites and vertical distribution of a spill unit. Mass transport into the water column is not modeled. On the other hand, the model developed by Yeh at ORNL addresses potential oil spills in riverine systems [19]. This model is really more than a general-property model, because it incorporates both the differential dispersion of water and oil and the dissolution of organic compounds from the oil to the water.

Modeling by compound class provides a more specific anlaysis of chemical compound contributions. In this approach, important chemical classes in an effluent are identified, and characteristic compounds from each class are chosen for the model. In essence, these are generic models, because they can be applied to a variety of similar complex effluents. Individual compounds and concentrations may vary greatly between the effluents, since only the similarity of compound classes is important. This modeling approach is used to provide only a first approximation and a comparative estimate of environmental fate. Two models being developed at ORNL employ this approach for potential synfuel product spills and effluents. The first expands the Fates of Aromatics Model (FOAM), which was developed at the Savannah River Ecology Laboratory to predict PAH fates in a model stream ecosystem [20]. The expanded model includes several organic compound classes (phenols, anilines, etc.) that represent organics dissolving from oil spills into the water column [21]. Eventually this will interface with the model of Yeh at ORNL to predict both dissolution and subsequent environmental fate of compounds in the water column.

The second ORNL environmental fate model is part of a risk-assessment strategy described by Moghissi and Foley [22]. In that strategy, risk analysis units (RAU), which represent broad organic and inorganic compound classes, are selected to model coal conversion and oil shale process streams. The total concentration of compounds within a RAU is multiplied by the individual risk of that RAU and summed over all RAU to obtain the overall risk. Environmental fate of each RAU model compound is mathematically modeled in the exposure assessment segment of the risk-analysis process, although the algorithms for these calculations are not yet well defined.

Finally, we are aware of only three one-compound-at-a-time complex-effluent modeling strategies. First, nuclear reactor effluent fates in aquatic and terrestrial environments have traditionally been modeled by assuming additive contributions from each radionuclide to a total radiation dose [23]. This is, in comparison with complex organic effluents, a much easier problem to model. For instance, individual compound concentrations are easier to determine in experiments used to estimate rate constants, and bacterial transformations of radionuclides are considered negligible.

Second, groundwater models are often applied to single-compound aqueous streams. Only recently have modelers considered complex effluents in response to problems presented by hazardous waste dumps and solid disposal problems in the emerging synfuel industries. However, chemical reactions and biological transformations pose tremendous modeling challenges [24], and for the most part the groundwater models that deal with complex mixtures address mixture constituents one at a time.

Our final example of a compound-by-compound model is the U.S. Environmental Protection Agency (EPA) Grosse Ile studies reported by Swain [25]. Although PCB were first modeled as a representative chemical category for persistent Great Lakes pollutants, the model is now being revised to consider heavy metal complex effluents [26]. In both cases, the modeled compounds are considered conservative. Bacterial degradation is assumed minimal, and all contributions by individual effluent release are considered additive.

The principal problems with these complex-effluent mathematical fate models are:

1. The identities of potentially hazardous chemicals are not known (general-property model).
2. The time and cost for chemical characterization are very great and, depending on the level of effluent complexity, all constituents will probably not be identified (compound-class and single-compound models).
3. Possible complex kinetic interactions are not accounted for (general-property, compound-class and single-compound models).

PHYSICAL MODELING APPROACHES

Physical (microcosm) models have been devised and used by Metcalf et al. [27], as well as others, to evaluate persistence and bioaccumulation of many inorganic and organic chemicals. A wide variety of physical models have been used, including very simple models incorporating only one or two ecosystem components (e.g., photodegradation and air-water-sediment partitioning) or very complex microcosm models (e.g., artificial streams) that are meant to mimic the ecosystem in question. These physical modeling approaches to contaminant fate problems have been reviewed and evaluated [2,28,29], and some of the general observations were:

1. Fate determinations in complex microcosms (model ecosystems) can have a high variance.
2. Model ecosystems are not as flexible or as cost-effective as mathematical models.
3. Bioaccumulation potentials of chemicals can be predicted adequately with microcosms.
4. Simple microcosms that model one ecosystem process (e.g., biodegradation or volatility) are valuable for determining rate constants.
5. Model ecosystems appear to be useful for validating mathematical fate models.

As with mathematical models, physical models have been applied almost exclusively to single compounds, such as pesticides and heavy metals. However, physical models could be very useful tools in evaluating contaminant fate problems that might arise with discharge of a complex effluent. In fact, these models could be particularly helpful to: (1) identify effluent constituents that pose the most significant fate problems even when all of the effluent constituents are not known; and (2) integrate and thus account for any complex kinetic interactions that might occur among constituents in the complex effluent or among a discharged compound and constituents in a complex receiving water. In the preceding section, we suggested that these are two principal weaknesses of mathematical modeling strategies for complex-effluent constituents.

There are two alternative physical modeling strategies that are being used to identify complex-effluent fate problems: (1) complex-ecosystem models, and (2) a physical modeling strategy that dissects ecosystem processes into a series of single-process laboratory tests. In the first approach, the effluent mixture could be added to a complex ecosystem model and those compounds persisting and/or bioaccumulating could be identified. ORNL is now using this approach for synfuel product spills [21].

We are using the second approach in our own studies to identify contaminant fate problems in complex synfuel process streams. We call this approach a physical "unit process" model because it is designed as a series of experiments, each of which represents a unit (or units) of the ecosystem in the same way that unit processes are sequenced in a chemical engineering process. Although the key experiments have been attempted in pilot runs, not all of those that we will discuss have yet been attempted. The basic philosophy underlying the procedure and key elements of the procedure have been described [4] and consist of the following:

1. devise "unit process" lab tests for volatility, photochemical conversion, biodegradation, etc.;
2. allow the physical, chemical and biological properties of each constituent in a complex effluent to rule its behavior in "unit process" tests; and
3. identify problem constituents, e.g., those expected to persist or bioaccumulate based on unit process tests.

The tests that we have so far planned to include are shown in Figure 5. Of those shown, the three physical tests will be attempted over the next year. The object in each of these planned "screening" tests is to identify only those organics in a synfuel process water that are highly volatile, extracted in high concentration in a series of organic solvents or not sorbed in sediment columns.

The two biological tests shown in Figure 5, biodegradation and bioaccumulation, have both been attempted in a series of pilot tests. The biodegradation tests [5] were conducted to identify potentially persistent organics in three different oil shale retort waters. Dilutions of each retort water were subjected to a microbial inoculum from a local river, and another series of dilutions was inoculated with a microbial population from an oil refinery settling pond. The experimental flasks, with essential inorganic nutrients added, were then placed on a rotary shaker for varying periods up to four months. The flask contents were analyzed before and after the incubation period using HPLC gradient elutions. Those organics remaining undegraded were then collected preparatively off the HPLC and retained for gas chromatography/mass spectrometry (GC/MS) analysis. A bioaccumulation test was conducted by exposing rainbow trout to an oil shale retort water at concentrations that

P-C-B Property	Test
—PHYSICAL—	
•Vapor Pressure	•Headspace GC/MS
•Part.-Water Sol.	•Solvent Ext., GC/MS
•Sorption	•Sediment Col., GC/MS
— CHEMICAL / PHYSICAL —	
•Photo. & Chem. Rxn's	•Dilute & age in sun, GC/MS
— BIOLOGICAL —	
•Biodegradation	•Shake Flask, HPLC, GC/MS
•Bioaccumulation	•Expose, Extract, HPLC, GC/MS

Figure 5. Physical "unit process" tests being developed and evaluated for identification by persistent or bioaccumulated organics from synfuel process water ("P-C-B property" means physical, chemical and biological properties).

bracketed the early-life-stage no-observable effect concentration. Tissue extracts of control and exposed fish were prepared using an exhaustive steam distillation procedure [30] and analyzed using HPLC gradient elutions. Those peaks appearing in the exposed fish but not in the controls were collected for later GC/MS identification.

A full evaluation of the overall physical unit process modeling procedure for complex effluents will have to await completion of more of the work that is now underway. All GC/MS work is still in progress to identify compounds remaining after microbial degradation of the retort waters and those compounds accumulated in fish after exposure to retort water. However, four major peaks remained in the HPLC elutions after four months of exposure to microbes, and one major peak that was not present in control fish extracts was found in the HPLC elutions of fish exposed for three weeks. These observations, although preliminary and not fully evaluated, suggest that the biodegradation and bioaccumulation testing and analysis procedures should work.

An obvious disadvantage with the proposed physical modeling procedure we have described will be that expected exposure concentrations cannot be determined without first determining rate constants for compounds identified as potential problems, followed by traditional mathematical fate modeling.

TRADE-OFFS

To help in comparing mathematical and physical modeling approaches for predicting environmental fate of complex-effluent constituents, Figure 6 summarizes the steps necessary in three of the approaches we have discussed: (1) single-compound mathematical models; (2) chemical-class mathematical models; and (3) physical models. The principal difference in mathematical and physical modeling is that full chemical characterization is needed as a first step in mathematical models. Physical models, on the other hand, attempt to identify problem compounds (those posing a persistence or bioaccumulation hazard) based on the results of microcosm or unit process screening experiments. In the end, a mathematical modeling step is necessary in all approaches to predict expected exposure concentrations for all compounds (single-compound models), representative compounds (chemical-class models) or problem compounds (physical models).

To obtain physical, chemical and biological rate constants (e.g., volatility, hydrolysis and biodegradation) for the mathematical modeling step, the procedure might be different for the different approaches, as suggested in Figure 6. For very complex mixtures, the large number of compounds may make actual determination of rate constants impractical in the single-compound mathematical modeling approach, necessitating the prediction of many rate constants from structure-activity relationships such as those reviewed by Hansch [31] and Martin [32]. However, with chemical-class mathematical models or physical models, the number of compounds for which rate constants are needed is reduced. Thus, laboratory determination of the necessary rate constants may be feasible.

Although mathematical and physical modeling approaches are still being developed for complex effluents, and none of the various approaches has been adequately tested, we can draw some conclusions about the trade-offs between these approaches, even if only on logical grounds. The strictly mathematical modeling approaches are acknowledged to be reliable and cost-effective for predicting the fate of individual contaminants [2,28,29]. However, for highly complex effluents, mathematical models rely on extensive chemical characterization of the effluent. Based on our arguments about the probable inability to identify all constituents of complex mixtures and the extensive analytical time

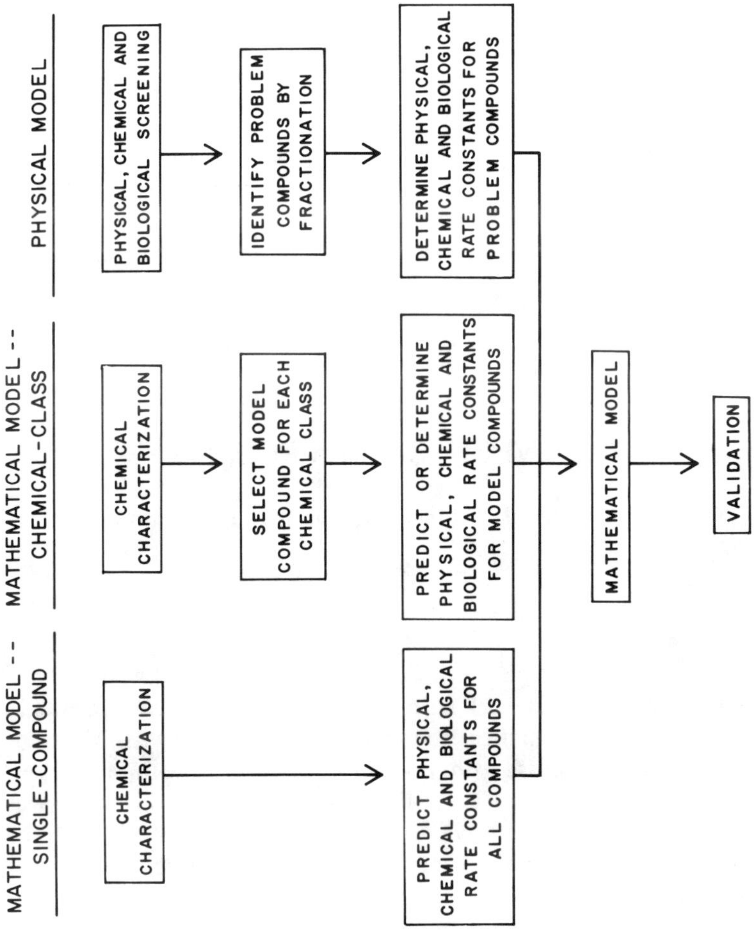

Figure 6. Alternative modeling approaches for complex mixtures.

and cost requirements for complete chemical characterization, we believe that reliability and cost-effectiveness of mathematical modeling become lower and lower with increasing numbers of compounds in complex effluents. At some level of complexity, then, we argue that physical modeling approaches become more reliable and cost-effective than mathematical models for identifying compounds that may pose a persistence or bioaccumulation problem. This view is summarized in Figure 7.

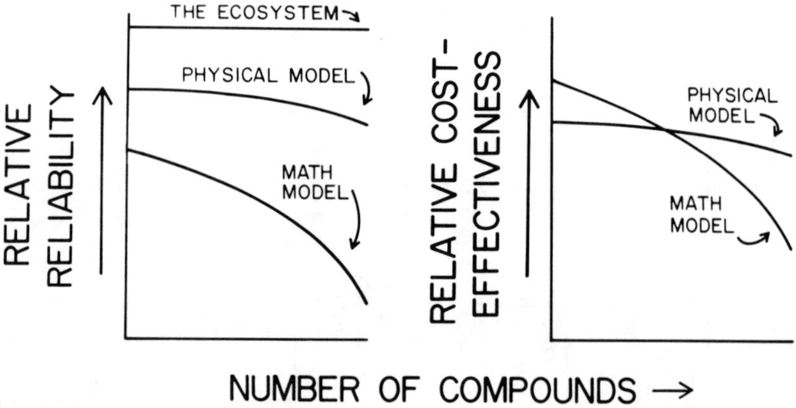

Figure 7. Relative trade-offs between physical and mathematical models as affected by effluent complexity.

ACKNOWLEDGMENTS

We thank S. Herbes, S. Bartell, G. Torgrimson and W. Swain for discussing their unpublished work with us, so that current developments in complex-effluent mathematical modeling could be reviewed. Many of the ideas in this paper arose from invaluable discussions with M. Parker, B. Steadman, G. Linder and T. La Point. Preparation of this paper was supported in part by Cooperative Agreement No. CR808671010 with the Environmental Research Laboratory—Duluth, U.S. Environmental Protection Agency.

REFERENCES

1. Baughman, G. L., and R. R. Lassiter. "Prediction of Environmental Pollutant Concentration," in *Estimating the Hazard of Chemical Substances*

to Aquatic Life, ASTM Technical Publication 657, J. Cairns, Jr., K. L. Dickson and A. W. Maki, Eds. (Philadelphia: American Society for Testing and Materials, 1978), pp. 35–54.

2. Branson, D. R. "Predicting the Fate of Chemicals in the Aquatic Environment from Laboratory Data," in *Estimating the Hazard of Chemical Substances to Aquatic Life,* ASTM Technical Publication 657, J. Cairns, Jr., K. L. Dickson and A. W. Maki, Eds. (Philadelphia: American Society for Testing and Materials, 1978), pp. 55–70.
3. Fox, J. P., and T. E. Phillips. "Wastewater Treatment in the Oil Shale Industry," in *Oil Shale: The Environmental Challenges,* K. K. Petersen, Ed. (Golden, CO: Colorado School of Mines Press, 1981), pp. 253–284.
4. Bergman, H. L., G. M. DeGraeve, J. S. Meyer, M. D. Marcus and D. L. Geiger. "Aquatic Ecosystem Hazard Assessment of Underground Coal Gasification Process Waters," 20th Hanford Life Sciences Symposium on Coal Conversion and the Environment, Richland, WA, October 19–23, 1981.
5. Marcus, M. D., and H. L. Bergman, Eds. "Effects of Aqueous Effluents from *in situ* Fossil Fuel Processing Technologies on Aquatic Systems," annual progress report to the U.S. Dept. of Energy and the U.S. EPA, Contract No. DE-AS20-79-LC01761 (1980).
6. Felix, W. D., D. S. Farrier and R. E. Poulson. "High Performance Liquid Chromatographic Characterization of Oil Shale Retort Waters," in Proceedings from the Second Pacific Chemical Engineering Congress, Denver, CO (1977), pp. 480–485.
7. Marcus, M. D., H. L. Bergman, D. A. Stoller, S. K. Brown and P. A. Molholt. "Bibliography of Aquatic Ecosystem Effects, Analytical Methods and Treatment Technologies for Organic Compounds in Advanced Fossil-Fuel Processing Effluents," U.S. EPA Ecological Research Series 600/3-80-026 (1980).
8. Mattox, C. F., and M. J. Humenick. "Organic Groundwater Contaminants from UCG," *In Situ* 4(2):129–151 (1980).
9. Pellizzari, E. D., and N. P. Castillo. "Organic Constituents of Waters Associated with Coal and Oil-Shale Conversion," in *Synthetic Fossil Fuel Technology: Potential Health and Environmental Effects* (Ann Arbor, MI: Ann Arbor Science Publishers, 1980), pp. 57–70.
10. Anderson, A. D., M. E. Lebsack, G. M. DeGraeve, D. S. Farrier and H. L. Bergman. "Toxicity of an *in situ* Oil Shale Process Water to Rainbow Trout and Fathead Minnows," *Arch. Environ. Contam. Toxicol.* 9:171–179 (1980).
11. Pellizzari, E. D. "Identification of Components of Energy-Related Wastes and Effluents," U.S. EPA Interagency Energy-Environment Research and Development Report 600/7-78-004 (1978).
12. Stuber, H. A., and J. A. Leenheer. "Fractionation of Organic Solutes in Oil Shale Retort Waters for Sorption Studies on Processed Shale," in *New Analytical Methods Applied to Fuels and Fuel Processing and Environmental Aspects of Fossil Fuel Processing,* (Washington, DC: American Chemical Society, 1978), pp. 165–174.
13. DeGraeve, G. M., D. L. Geiger, J. S. Meyer and H. L. Bergman. "Acute and Embryo-Larval Toxicity of Phenolic Compounds to Aquatic Biota," *Arch. Environ. Contam. Toxicol.* 9:557–568 (1980).

14. Matsumura, F. "Absorption, Accumulation, and Elimination of Pesticides by Aquatic Organisms," in *Pesticides in the Aquatic Environment,* M. A. Q. Khan, Ed. (New York: Academic Press, Inc., 1977), pp. 77–105.
15. Reinert, R. E., L. J. Stone and H. L. Bergman. "Dieldrin and DDT: Accumulation from Water and Food by Lake Trout (*Salvelinus namaycush*) in the Laboratory," Proceedings of the 17th Conference on Great Lakes Research (1974), pp. 52–58.
16. Gillette, J. R. "Effects of Induction of Cytochrome P-450 Enzymes on the Concentration of Foreign Compounds and Their Metabolites and on the Toxicological Effects of these Compounds," *Drug Metab. Rev.* 10(1):59–87 (1979).
17. Meyer, J. S., H. L. Bergman and M. D. Marcus. "Inhibitory Interactions of Aromatic Organics during Microbial Metabolism," paper presented at the Annual Meeting of the Ecological Society of America, Stillwater, OK, August 12–17, 1979.
18. Torgrimson, G., National Oceanographic and Atmospheric Adm., Seattle, WA. Personal communication (1981).
19. Herbes, S., Oak Ridge National Laboratory, Oak Ridge, TN. Personal communication (1981).
20. Bartell, S. M., P. F. Landrum, J. P. Giesy and G. Leversee. "Simulated Transport of Polycyclic Aromatic Hydrocarbons in Artificial Streams," paper presented at the International Symposium on Energy and Ecological Modeling, Louisville, KY, April 20–23, 1981.
21. Bartell, S., Oak Ridge National Laboratory, Oak Ridge, TN. Personal communication (1981).
22. Moghissi, A. A., and G. J. Foley. "Assessment of Environmental Impact of Coal Conversion and Oil Shale Technologies," paper presented at the International Symposium on Health Impacts of Different Sources of Energy, Nashville, TN, June 22–26, 1981.
23. "Calculation of Annual Doses to Man from Routine Releases of Reactor Effluents for the Purpose of Evaluating Compliance with 10 CFR Part 50, Appendix I," U.S. Nuclear Regulatory Commission, Regulatory Guide 1.109 (1977).
24. Anderson, M. P. "Using Models to Simulate the Movement of Contaminants Through Groundwater Flow Systems," *CRC Crit. Rev. Environ. Control* 9(2):97–156 (1979).
25. Swain, W. R. "An Ecosystem Approach to the Toxicology of Residue Forming Xenobiotic Organic Substances in the Great Lakes," Environmental Studies Board of the National Research Council, National Academy of Science (1980).
26. Swain, W. R., U.S. EPA Grosse Ile Large Lakes Research Station. Personal communication.
27. Metcalf, R. L., G. K. Sangha and I. P. Kapoor. "Model Ecosystem for the Evaluation of Pesticide Biodegradability and Ecological Magnification," *Environ. Sci. Technol.* 8:709–713 (1971).
28. Johnson, H. E., V. Zitko and F. Matsumura. "Synopsis of Discussion Session: Case Studies," in *Biotransformation and Fate of Chemicals in the Aquatic Environment,* A. W. Maki, K. L. Dickson and J. Cairns, Jr., Eds. (Washington, DC: American Society for Microbiology, 1980), pp. 136–139.

29. Matsumura, F. "Use of a Microcosm Approach to Assess Pesticide Biodegradability," in *Biotransformation and Fate of Chemicals in the Aquatic Environment,* A. W. Maki, K. L. Dickson and J. Cairns, Jr., Eds. (Washington, DC: American Society for Microbiology, 1980), pp. 126–136.
30. Veith, G. D., and L. M. Kiwus. "An Exhaustive Steam-Distillation and Solvent-Extraction Unit for Pesticides and Industrial Chemicals," *Bull. Environ. Contam. Toxicol.* 17(6):631–636 (1977).
31. Hansch, C. "A Quantitative Approach to Biochemical Structure-Activity Relationships," *Acc. Chem. Res.* 2:232–239 (1969).
32. Martin, Y. C. "A Practitioner's Perspective of the Role of Quantitative Structure-Activity Analysis in Medicinal Chemistry," *J. Med. Chem.* 24(3):229–237 (1981).

CHAPTER 15

SYNOPSIS OF DISCUSSION SESSION: STATE-OF-THE-ART AND CASE STUDY EXAMPLES

G. Loewengart
Session Chairman

P. B. Bedient, W. E. Cooper, N. H. Crawford and L. Mulkey

A critical part of the risk assessment of chemicals released into the aquatic environment is prediction of the transport and fate of chemicals both spatially and temporally in the various environmental compartments (e.g., water column, aquifer, sediment bed, suspended solids and biota). Mathematical modeling of chemicals in the aquatic environment may be used to predict the behavior and fate of new chemicals before their release or to assist in the evaluation of questions related to past or current releases.

In a historical perspective, modeling "activity" has increased rapidly in hydrology, ecology and environmental studies over the last 10 years. There is no doubt that modeling will increase in the future: the legions of 10-year-olds who run computers today at home and in school will not hesitate to express concepts and ideas in models when they reach universities. Modeling will be done and the "model" or "do not model" issue is moot. The issue is how to do modeling effectively.

A suite of models is available to describe processes in various aquatic environments, including lakes, rivers, estuaries, coastal regions, groundwater, surface runoff and complex effluents. In each case, the resulting estimates are based both on laboratory and environmental measurements and on underlying assumptions basic to each model. The corrections and nature of these assumptions (theories) and the accuracy of the input data in describing field events determine to a great extent the reliability and resolution of the model's predictions.

Previous workshops in this series [1–3] investigated data requirements (i.e., physicochemical properties, rate constants, equilibrium coefficients and properties of the environments) for making useful predictions of the fate and concentration of chemicals released to the aquatic environment. In this section, case studies were presented describing how this information is used in various environmental systems to make reasonably accurate and therefore useful estimates and predictions of chemical fate and transport.

Because of criticism and misunderstanding of the current use of mathematical models, particularly with their predictability and resolution, it is important to consider the alternatives. In the extreme, these alternatives could be not to evaluate at all the quantitative fate of chemicals over space and time or, at the other extreme, to perform complex field studies for many chemicals and for many sites of release. Obviously, the use of mathematical fate models is preferable to either of these extremes, better addresses current needs and makes the best use of limited resources. The critical issue, then, becomes to gain a broad appreciation of models' limitations and to apply modeling results within the confidence boundaries of their accuracy.

In this respect, complex, detailed models require more input and often require a higher degree of accuracy of the input data than do simple models used mainly to rank chemicals and focus on areas of general concern. Also, the degree of validation should reflect the degree of resolution necessary. However, not all inputs are equally precise, and experience of round-robin testing of biological effects and environmental fate tests by the American Society for Testing and Materials (ASTM) and the U.S. Environmental Protection Agency (EPA) has shown a surprisingly large variation of supposedly "standard" tests, including simple measurement of physical properties. Variations of biodegradation tests will be more uncertain. Also, with respect to the environmental input data, it was the general opinion of the authors in this section that, for validation purposes, it is important that environmental data be coupled sequentially, if possible, for best input use. Further, if laboratory data are being used, these tests should mimic the environment for critical parameters as closely as possible.

Field experiments to validate and assess the overall accuracy of models are needed on well-defined systems. For example, selection of sites with single, known contaminants with low volatilization, low degradation and known adsorption could be used to evaluate field calibration. The objective is to limit the number of unknown coefficients to the extent possible and add them into the application sites one at a time or in small groups. For instance,

1. adsorption only—site 1;
2. adsorption + biodegradation—site 2; and
3. volatilization + biodegradation—site 3.

The following topics provide the focus of this synopsis: theories (underlying assumptions and principles), mathematical approach, available models, and their uses, degrees of validation, limitations and areas for additional research.

These are discussed and compared for different models so that an understanding of their proper use and role in the hazard assessment process may be obtained and, equally important, to help focus on the areas where further work is needed to improve or simply the predictions made.

GENERAL COMMENTS

Modeling the transport and fate of toxic substances requires modeling the basic physical and biological processes of transport of materials in water. This includes modeling of rainfall, runoff, soil moisture, groundwater recharge and movement, streamflow, sediment movement, biota, lakes, reservoirs, and estuaries. These physical elements and processes are inherently dynamic and spatially variable. Precipitation causes runoff into streams and runoff causes erosion of sediments. Water temperatures in rivers and lakes respond to air temperature and solar radiation. Phytoplankton change their activities as water temperature changes. All of the climatic variables are time-dependent and include diurnal and seasonal cycles.

The effects of spatial variability are important when considering properties, vegetation density, topography and the loading of substances on the land surface. Soil erosion has been found to vary by several orders of magnitude in short distances because of changes in soil properties and vegetation. Moderate changes in aquatic environments from point to point on a stream can have dramatic effects on biota.

Given the physical diversity of the land and aquatic systems and the complex interrelated processes important to transport and fate of toxic substances, a variety of approaches to modeling have evolved. To reduce the scope of the problem, spatial and/or time variability of process can be eliminated from a model. The major types of models that result are shown in Table I.

Success of modeling to date varies among processes and among physical environments. A very subjective judgment about levels of success ranging from excellent to poor is given in Table II. The success index is a level-of-confidence index for model predictions given the current state-of-the-art. Low levels of confidence may result from difficulty in one or more of the following:

1. establishing needed model inputs (e.g., the physical system of groundwater models often lacks permeability data for the entire flow field);
2. lack of data to calibrate model behavior or inaccurate data resulting from measurement problems (e.g., algae and trace contaminants are difficult to sample); and
3. incomplete or uncertain model algorithms for complex processes (biota).

The state-of-the-art will likely improve in the future for all processes, but the overall relative ranking among processes may not change. It is not likely that all biologic process algorithms will improve to give results comparable to algorithms for channel flow hydraulics.

The applications of each of the model types in Table I will increase in the future. Model applications are becoming more and more practical for two reasons: (1) greatly expanded field data collection in recent years has provided needed model inputs, and (2) computer resources for data management and computations are decreasing in cost.

Validation

Validation of a model is a process of investigating the degree of usefulness, given the set of expectations associated with a particular user group. A mechanistic model whose components represent basic processes and whose utility is to explore unexperienced situations is validated by rigorously testing the initial assumptions, the completeness of the component array and the logic of the couplings. Experimentation is performed with individual components and subsystem clusters of interactions. An integrated combination of laboratory and field experimentation is invariably needed to develop and demonstrate a useful model.

Models can be developed from empirical relationships without including information on the cause-and-effect relationships. In general, confidence in these management tools comes from repeated experience with

Table I. Comparison of Types of Models

Model Type	Spatial Variability	Time Variability	Advantages/Limitations
Unit World	No	No	Assumes fully mixed air, water soil/sediments and biota. Concentrations are calculated given constant system inputs and losses. Losses are to stratosphere or buried sediments. Calculations are quick and simple. The assumed fully mixed environment only gives information about relative distribution of compounds in major compartments.
Idealized Environment	Partial	Select variables only	Models improve the "unit world" concept by introducing some spatial or time variability. An example is one-dimensional modeling of water quality variables in a stream at a fixed flowrate and water temperature. The limitation of these models is failure to represent the actual frequency of events. Specific conditions can be represented that could occur, but information on the likelihood or frequency of the result is not produced.
Comprehensive, Lumped-Parameter Models	Yes, limited by model elements or compartments	Yes	This approach expands selected environment models by allowing spatial and time variability. They produce results similar to selected environment models, but do so continuously in time and space, answering "what happens?" "how frequently does it happen?" and "at what intensity?"

Table I, continued

Model Type	Spatial Variability	Time Variability	Advantages/Limitations
Comprehensive Distributed-Parameter Models	Yes	Yes	Models include spatial variability with "finite element" representation of soils and runoff paths. Data on environmental properties for each element are needed.

Table II. Confidence of Input Data Determined for Different Physical Environments

Physical Environment	Water Movement	Water Temperature, Conservative Compounds	Nutrients	Erosion, Sediment Transport Processes	Adsorption/ Desorption	Volatilization/ Physical Degradation	Biotransformations
Soil Surface Unsaturated Zone	G[a]	F	F	F/P	F	F	P
Saturated Zone	G	G	F	NA	F	NA	P
Stream/ Channels	E	E	F	F	F	F	P
Reservoirs/ Lakes	G	G	G	F	P	F	F
Estuaries	F	G	F	F	P	F	P

[a] G = good; F = fair; P = poor; NA = not applicable.

empirical case studies. If validation is empirically determined by correspondence between predictions and observations, then quality performance is ensured by not exceeding the limits of the empirical relationships. Confidence arises from the logical consistency of the macroscopic structure and the relevancy of the outputs.

Mechanistic Models

The processes of model development and validation should be iterative activities designed into any research plan. Proposals for model development should include laboratory and field experimentation as an integrated program with the formulation of the mathematical structure.

Experimental tests of a model should be designed to challenge its validity. They should not be designed to demonstrate its correctness. Validity is obtained by the convergence of specific predictors with direct observations obtained under controlled experiments where external factors are known.

STATE-OF-THE-ART

Groundwater

Groundwater flow and chemical transport models have been developed and applied to selected problems in soil columns and field studies. Although soil column models are reasonably well advanced for predicting contaminant migration, accuracy of prediction for many field problems is limited because of heterogeneities in lateral and vertical geological strata. Models can closely approximate, but not duplicate, the environment. A useful dynamic model will provide a good representation of spatial and temporal events.

The magnitude of field dispersion is a function of variations in hydraulic conductivity K, which is the driving force for hydraulic flows in porous media. Mathematical models of the advective/dispersive transport mechanisms in one and two dimensions have been developed, but there is a severe lack of physical and geochemical data to support these efforts.

Problems of dispersion measurement can be overcome by using an advective model to obtain a first approximation of contaminant movement. Flow lines and travel times are calculated from hydraulic conductivities to yield the response of the main zone of contamination. These types of models are useful in level 1 assessments, in which the system is assumed homogeneous and advection-dominated. Simple chemical reactions, which can be represented as time lags, can be included. Once the flow field is known, models that utilize the "particle mover" analogy can be used to predict the fate of contaminants.

Advection/dispersion models have been used where spreading of contaminants is dominant. If the distribution of heterogeneities could be described, the magnitude of dispersion could be estimated, based on statistical analyses. However, in most cases of interest, a severe lack of field data has precluded calibration and verification of such models. One notable exception includes some of the excellent applications of models to predict radionuclide transport in porous media developed during the past decade. Modeling of adsorbing organics requires the development of extensive field data bases, such as those associated with the nuclear radiation industry.

The effort to include chemical reaction terms into many of these models is just now beginning. Simple adsorption and first-order biological degradation have been understood for years in surface water quality, and these mechanisms are being incorporated for groundwater systems. While laboratory kinetic rate constants are reasonably well known for adsorption or degradation, their application to field studies has been quite limited. Therefore, the greatest research need is for more extensive field monitoring efforts to generate coefficients necessary to calibrate models even to a first approximation.

Runoff Modeling

Rainfall, evaporation, infiltration, transpiration, runoff into streams, water storage in lakes and reservoirs, and flow through estuaries into the oceans comprise the hydrologic cycle. Moving water transports dissolved solids, sediments and compounds that adsorb onto sediments. Runoff erodes continents and creates deep marine sedimentary basins. These basins are lifted over geologic time to form new eroding land masses. The hydrologic cycle is highly variable in time. Seasonal and diurnal cycles provide a background for chaotic time sequences of floods and droughts, which are sufficiently random that they defy prediction. Erosion and movement of sediments and dissolved materials caused by water flow are also chaotic and unpredictable. Sediment erosion from a farm in one storm might exceed the total sediment movement from the farm in the previous 10 years. The hydrologic cycle is a critical link in multimedia analyses of the transport and fate of toxics in the environment. Airborne contaminants may reach the land surface as they fall, or they may combine with precipitation. For centuries, humans have associated rainfall with the cleansing renewal of the earth. The phrase "acid rain" has now entered the language.

Modeling the hydrologic cycle is advanced and successful. Hydrologic models that operate using continuous precipitation, temperature, potential evapotranspiration and other meteorologic data produce continuous snow depths, soil moistures, surface and subsurface flows, and streamflows. Hydrologic transport algorithms are linked to the hydrologic models and produce time series of sediment transport and the mass transport of pollutants.

The widely used hydrologic or runoff models are physically deterministic models based on continuous water-balance accounting. The associated hydrologic transport models also use physically deterministic mass-balance algorithms. Hydrologic modeling has replaced statistical and graphical hydrologic techniques in sensitive applications that require superior accuracy, such as flood forecasting and flow forecasting for power generation.

Models of sediment, water temperature, dissolved solids, oxygen and other transport processes vary in reliability. Water temperature is an example of a process that is based on well known physical principles, and water temperature is measured routinely in the field. These processes are modeled very successfully. Phytoplankton is an example of a process that is less well understood, and for which field measurements are infrequent.

Rivers and Lakes

Modeling efforts for flowing freshwater systems have been focused along two major lines of activity. Transport and fate model emphasizing hydraulic movement and sediment interactions have attracted the attention of many investigators interested in the movements and distributions of toxic materials. Ecologists, on the other hand, have focused on nutrient- and energy-processing models that relate directly to primary and secondary productivity. Both efforts have produced a number of models that work and can provide analytical and predictive capability to appropriate user groups. The hydraulic and sediment models are most useful in estimating ambient concentrations of materials over temporal and spatial gradients. The ecological models are essential for estimating ecological effects and environmental risk assessment. The material transport and transfer models can utilize deterministic functions and parameters in determining temporal and spatial gradients as long as they deal with chemical concentrations that do not initiate state-dependent ecological feedbacks that alter the basic modes of behavior. The ecological models usually include dynamic biological feedbacks that require dynamic modeling techniques. Whereas models like EPA's Exposure Analysis Modeling System (EXAMS) and the "allochthonous leaf input" [4] are most generally utilized, there are no generic models that can uniformly fulfill the needs of all user groups across all stream types. System-specific alternatives will still be needed.

It is convenient in making statements about the "state-of-the-art" for a given modeling exercise (in this case transport and fate modeling in lakes and streams) to deal with the strengths, weaknesses and controversies associated with "standard" models (or model types) in place. One must also further define "state-of-the-art." Two state-of-the-art levels usually exist—development and application. Clearly models exist that are developed but have little or no application history. To date, application is too limited to warrant much discussion. Indeed, sufficient field data do not exist to make meaningful statements about model performance in an application setting. Thus, the state-of-the-art here applies to the research and development of models only.

Strengths

The major strength of current models is their ability to utilize well defined, chemically based fate process information in first-order rate expressions for most of the known processes (hydrolysis, photolysis and oxidation). The models also use, to the maximum extent possible, conventional equilibrium data for neutral organics (however, some controversial issues remain here). Polar organics and inorganics are less well understood and utilized. There is no reason to expect that existing models cannot fully utilize known theory in this area, but little effort appears to have been made to do so. This is probably related more to the need for such model components than to state-of-the-art limitations. Volatilization is modeled in a way that is consistent with accepted mass transfer processes. Dispersion processes within the water column are used when necessary, and the approaches are those used in general water quality modeling.

Weaknesses

The greatest weaknesses of current models relate to sediment processes—both sediment/pollutant interactions and sediment transport dynamics. Transport dynamics per se are much better understood, but are not fully utilized in most models. This is probably justified, given the uncertainty associated with sediment/pollutant interaction. The major weaknesses include the exchange of sediment-laden pollutants with interstitial waters, diffusive exchange between interstitial water and overlying water, and exchange of bed sediment-laden pollutants and overlying sediment/water mixtures. Finally, the degradation of bed sediment pollutants is very uncertain and hence a recognized weakness.

Controversies

The major controversies (legitimate differences of opinion) relate to two general areas: (1) interpretation of basic fate processes, and (2) the role of dynamic vs steady-state hydraulics and other model processes. Some models (e.g., EXAMS) include fate processes designed around second-order kinetics. This requires estimation of second-order rate constants. The alternative is first-order rates measured with in situ media or laboratory media analogs. The relative measurement difficulties, costs and ease of application are at issue. A second issue is the importance of equilibrium. The issue of steady-state vs dynamic model constriction may be significant for large lake systems, but systems with short residence times and time-variable loads may require more nearly dynamic approaches.

Complex Effluents

As a result of limited environmental data on synfuel process streams and effluents, as well as other complex mixtures (e.g., shale oil, leachates from waste disposal sites), severe regulatory restrictions are being considered. The result of regulation potentially causes new operational, habitat and water habitat problems (such as increased salinity because of greater water consumption, but restricted water release). A proposed hazard evaluation procedure using a physical rather than a mathematical model was described by Bergman [4], who discusses the identification of persistent or bioaccumulative chemicals in complex mixtures.

The procedure does not provide an in-stream exposure concentration of any chemicals identified as potential problems. These compounds may require mathematical modeling or other estimation techniques on an individual basis. This evaluation procedure is an adaptation of the principles developed at the previous Pellston workshops for single compounds. Although there are several levels of testing incorporated in the scheme, each with associated decision criteria, bioaccumulation and biodegradation tests emerge as critical studies. The procedure attempts to reduce the analysis from the large number of chemicals that may initially be present to focus on those of greatest environmental concern. Acute and chronic toxicity would be addressed with the aggregate mixture, whereas bioconcentration and biodegradation would deal with individual or a limited number of chemicals. A high level of quantitative chemical analysis is required to support this approach. Nevertheless, the total level of effort, both biological and analytical, would be far lower than that required for an evaluation on a compound-by-compound basis. Further, this procedure incorporates the complex interactions that may occur, which would render less reliable compound-by-compound evaluation, summation and generation of rate constants.

This approach is an early attempt to address the problem of evaluation of complex mixtures. With the continued development of synfuels and the implementation of the Resource Conservation and Recovery Act (RCRA), it can be expected to become one of the more important issues of the 1980s.

AREAS OF CONCERN AND RECOMMENDATIONS FOR FUTURE RESEARCH

Throughout the workshop, several areas of inadequacy or uncertainty were expressed in almost every session. These areas are listed below. Other, more specific, recommendations are discussed by model type.

General Concerns and Recommendations

User Sophistication

There are different uses and limitations of models. There are also some fundamental differences in the variety of models; as a result, data needs and application vary widely. The models used for chemical ranking are sufficiently developed and easy to use. More complex models used in hazard assessment require more experienced and sophisticated users. These models must be approached with great caution to provide the correct input and expert interaction required.

Biotransformation Rate Constants

It was the consensus of the workshop that one of the sources of greatest uncertainty in model prediction resulted from errors in the biotransformation rate constants. This was also the subject of great debate at the previous Pellston workshop. At that time [3], it was proposed that second-order rate constants for biotransformation reactions would provide the required accuracy. Further, second-order kinetics would normalize the rate constants by microbial population and would therefore have general application. Several cases were described in this book wherein, after an attempt was made to use second-order rate constants, it was found that first-order rate constants provided a better fit for the data. This is an important and controversial subject that remains to be clarified.

Sediment Decay

There is a need to determine sediment-decay processes and rates. It was shown in this book that if degradation of a chemical in sediments occurs, it could have dramatic results on the modeling prediction. We do not currently know how to determine the rate constants that describe these processes.

Ecological Oscillations

Most fate models attempt to predict "steady-state" conditions (some older models, like runoff models, are dynamic). However, the environment is dynamic, oscillating around and constantly striving to obtain equilibrium or a steady-state condition. The frequency and magnitude of this oscillation is rarely well defined and may exceed the capability of steady-state models to predict environmental concentrations with confidence. A corollary of this is that once the frequency and amplitude of these oscillations have been measured, an estimate can be made of the resolution required of the model (i.e., the confidence intervals of the predicted values).

A related issue is the requirement of models to use average values. The uncertainty with which the model predicts concentrations in specific segments or subsegments in question is an inherent concern and not always well defined.

Model Specific Research Needs

Groundwater Models

A current impediment to more widespread use of groundwater models for transport and fate of toxics is the uncertainty in spatial properties for groundwater systems. Development of comprehensive field data sets to define such properties (e.g., hydraulic conductivity) should be a high-priority research effort. Sorption and microbial degradation are currently included in model research efforts and should be further refined. Hydrolysis, volatilization and chemical precipitation are probably important processes that have not been used to date, but should be incorporated.

River and Lake Models

Transport-and-fate and ecological models for rivers and lakes have developed to the point where merging them appears worthwhile. Research to accomplish this could lead to highly refined hazard assessment models. Transport-and-fate models should be further developed to include time-variable processes to enable frequency analysis of expected behavior. In addition to field data for fate model validation, controlled field experiments should be performed to investigate ecological functional-group behavioral response to toxic chemicals. These experiments and associated time-variable models will enable investigation of ecosystem resilience to toxic chemical perturbations. Finally, the orthogonality (i.e., additivity and independence) of community sensitivity to stress induced by multiple exposures should be investigated.

River and lake models are currently limited by our understanding of the basic fate and transport processes. Specifically, research needs include investigation of the influence of sediment/water concentrations on equilibrium and kinetic sorption processes and investigation of the role of bioturbation (e.g., worms) on the benthic boundary layer, mixing zone and associated mass transfer. A major transport question is the study of the mechanics of sediment transport (e.g., settling and resuspension) in lakes/impoundments given the apparent influence of sediment on pollutant transport. Finally, comprehensive field data sets must be generated for model calibration and validation.

Complex Effluents

The whole procedure discussed for modeling complex effluents has only been proposed and minimally tested in several laboratories. Methodology needs to be refined and then validated. Other procedures, including mathematical modeling, should also be considered as potentially useful.

Runoff

Hydrologic modeling in areas with poor rain gauge coverage or thunderstorm rainfall is difficult. Techniques for rainfall estimation in these climates, by meteorologic simulations or stochastic analysis, should be explored.

The spatial distribution of topography, vegetation and soil properties control both runoff and sediment erosion. Techniques for representing spatial variability should be explored. Runoff may stem from limited specific sites. "Source-area" runoff concepts should be further developed, since these areas are key to the hydrologic transport of pollutants.

Analytical (i.e., mathematical) techniques for direct solution for water movement in thc unsaturated zone should be explored and developed for applications to routine hydrologic modeling. These techniques may provide more physically based algorithms for the movement of water, which are useful in pollutant transport analysis. For example, infiltration is currently handled very empirically. Better physical understanding could then improve the model also.

In summary, development of models will surely continue, but investigations and improvements of the modeling subparts need to be performed and some new approaches tried to raise the state-of-the-art rather than just refining what is currently available.

REFERENCES

1. Cairns, J., Jr., K. L. Dickson and A. W. Maki, Eds. *Estimating the Hazard of Chemical Substances to Aquatic Life,* ASTM Technical Publication 657 (Philadelphia: American Society for Testing and Materials, 1978).
2. Dickson, K. L., A. W. Maki and J. Cairns, Jr., Eds. *Analyzing the Hazard Evaluation Process* (Bethesda, MD: American Fisheries Society, 1979).
3. Maki, A. W., K. L. Dickson and J. Cairns, Jr., Eds. *Biotransformation and Fate of Chemicals in the Aquatic Environment* (Washington, DC: American Society for Microbiology, 1980).
4. Bergman, H. L., and J. S. Meyer. "Complex Effluent Fate Modeling," Chapter 14, this volume.
5. Boling, R. H., Jr., E. D. Goodman, J. A. VanSickle, J. O. Zimmer, K. W. Cummings, R. C. Peterson and S. R. Reice. "Toward a Model of Detritus Processing in a Woodland Stream," *Ecology* 56:141–151 (1975).

SECTION 4

VALIDATION AND APPLICATION TESTING

CHAPTER 16

TESTING MODELS OF THE FATE OF CHEMICALS IN AQUATIC ENVIRONMENTS

Ray R. Lassiter
Institute of Ecology
University of Georgia
Athens, Georgia 30602

It is generally accepted that models should be tested before they are put to serious application. The way in which models ought to be tested, however, is not widely agreed on, nor even frequently discussed. I shall present the view of models and model testing that underlay and motivated testing the Exposure Analysis Modeling System (EXAMS). In predicting the fate of chemicals in the aquatic environment, both empirical and theoretical models have been used. Empirical models are discussed by others in this book, and I shall consider them only to aid in discussing theoretical models. Both classes of models, in the final analysis, are tools that, in the present context, are used to predict some aspect of a chemical's fate in the aquatic environment. In the past, both empirical and theoretical models have been needed in this field, and there is reason to believe that it will continue to be so well into the future.

Different kinds of models and different uses of models need to be tested differently. One needs to test whether a model is adequate for its purpose. As a basis for considering what aspects need to be tested for models of the environmental fate of chemicals, I shall present, briefly, my perception of the way that theoretical models relate to theory, and what is necessary for them to be applied in practice. A brief and very general description of empirical models, as applied in environmental science, should help to clarify the differences by contrast between empirical and theoretical models.

EMPIRICAL MODELS

Caswell [1] refers to empirical models as predictive models. These models are entirely utilitarian; their sole purpose is to produce useful predictions. Therefore, one cannot validly test or even question their theoretical basis. Only the adequacy of the results can be criticized. A regression model is one of the simplest cases. In environmental science, the various calibrated models are empirical and predictive. Calibration (fitting) involves varying thc paramctcrs of the model until the differences between the predictions and observations satisfy some criterion, e.g., the minimum sum of squares. Such models might employ terms intended to reflect underlying theory, but if they depend on calibration for their predictive power, they are empirical. Testing these models entails establishing their adequacy to satisfy the purpose for which they were created. Whether such models are adequate in a given instance depends heavily on the statistical characteristics of the calibration data set. The description of empirical models as predictive must be understood in a special way. There is no intent for such a model to be used to make predictions outside the range for which it was calibrated. Indeed, there is no basis for credibility of such a prediction. Therefore, an empirical model's prediction is in some sense more akin to an interpolation. Of course, there may be many specific predictions within the range of calibration that have never been observed and/or which could not be deduced without the model. In fact, the utility of an empirical model lies in its ability to make such predictions.

THEORETICAL MODELS

Theoretical models are distinguished from empirical models by the way in which their parameters receive values. A parameter value of a theoretical model is associated with a term that represents a process. Its value is estimated in some physical reference to that process, that is, in an experiment or other task in which measurements are made on relevant aspects of the process. Never is a contrast made between predictions by the model and observations of what is predicted for the purpose of establishing a value of any parameter. Therefore, predictions of a theoretical model are not limited by the range of observations of a data set. They are limited by the generality of the terms that represent the processes and by uncertainty associated with estimates of the process parameters. Of course, they also are potentially limited by mistakes and omissions made when the model was constructed. Apart from these potential limitations, no bounds are placed on the range of predictions that a theoretical model can legitimately make. It is, therefore, these limitations that need to be examined when theoretical models are tested.

Theoretical models can be utilitarian, or, as Caswell [1] puts it, their purpose can be "to gain insight into how the system operates." I prefer to distinguish between model and theory, and would modify his statement to read, "to gain insight into how theory operates," or perhaps "... into how theory implies that the system operates." Theoretical models are supposed to be conceptually correct by their presumed correspondence to underlying theory. A theory is a body of concepts and relationships explaining some aspect of the world. These concepts and relationships can be used to make observable predictions about aspects of the world that are not explicit in the theory. Some of the concepts and relationships of a particular theory originate in other theories. We generally use "theory" to refer generically to the body of concepts, relationships and implications associated with theories. Models themselves, however, are not theories. A specific, theoretical model is one of many possible instruments by which theory can be translated into specific cases (predictions). A single equation or logical relationship can be considered a model if it is used to obtain a result implied by theory. At the other end of the spectrum is the large set of coupled or otherwise interconnected equations and logic too complicated for paper and pencil manipulation and, therefore, manipulated by computer. Currently, particularly in environmental science and related fields, the latter is called a model, whereas a single equation or logical relationship often is not. Every such expression of theory, however, legitimately can be called a theoretical model.

Overton's [2] view of the modeling process is that it consists of searching for appropriate model components to accomplish two objectives. First, the model predictions must match observed values ("prescribed behavior"). Second, it must incorporate sufficient theory to permit extrapolation. Thus, models can interact with theories in two directions. That is, they can be based on theory; ultimately, however, theory can also be based on a model, though in an indirect way. A theory is not necessarily complete; yet a model, if it is to make concrete predictions, must be logically complete. Where the basis for a model does not exist in underlying theory, the modeler is obliged to improvise. The improvisation can take the form of an empirical submodel, or, if warranted, the modeler can complete the model by using a conjectural, theoretical construct to substitute for the nonexistent theory. In keeping with Popper's [3,4] theory of the growth of scientific knowledge, such a conjecture, if submitted to repeated experiments designed to refute it should it be false, can become accepted as part of the underlying theory. In environmental science, the underlying theory consists of relevant portions of the theories of physics, chemistry, biology, geology, meteorology and perhaps many other disciplines.

TESTING THEORETICAL MODELS

The essence of testing a model is a comparison of its predictions to observations of the phenomena that it predicted, followed by analysis, conclusions about the model and, if weaknesses are detected, conjectures for revising the model to overcome the weaknesses. The intent of testing, as already alluded to, is to subject the model to the severest tests conceivable, to expose any deficiencies or fallacies as soon as possible. Of equal importance to exposing errors is the subsequent revision of the model with the intent that the revised model will also be subjected to equally critical tests. Popper [4] refers to it as essentially a method of trial and error, or more explicitly, "conjecture and refutation." The method of science, however, differs from simple trial and error by conscious efforts to uncover errors in the testing stage so as to avoid their turning up inopportunely in the use stage. Popper has effectively laid to rest any notion that a theory (and therefore the basis for a theoretical model) can be validated in a positive sense. One can show that a model's predictions were accurate in a given instance or several instances, but one can never show that a model or its underlying theory is correct, in general. At best, the credibility or degree of belief that it will predict accurately for all future cases can be improved. As Polya [5] puts it, verification in a singular case makes the general case "more plausible."

EXPOSURE ANALYSIS MODELING SYSTEM

EXAMS was constructed to make evaluative analyses of the fate of organic chemicals in aquatic environments [6]. To make an evaluative analysis of a chemical to be released to the environment, we visualized calculating its expected fate using environment representations that span the characteristics of the real environments that the chemical is expected to enter. These environment representations, of course, do not correspond to specific sites. Analyses of this sort can be useful only if calculations are accurate for an environment as described. That is, if an environment did exist exactly as described in the evaluative analysis, predicted concentrations should match the observed ones. Otherwise, predictions would be misleading and therefore of negative value to one charged with making a regulatory decision. In general, the environments used in evaluative analyses are fictitious, and it is a hypothetical question to ask whether predictions would match observations on the system. However, some existing environments, notably laboratory ecosystems, can be characterized in the manner required by EXAMS, and therefore, they can be used to provide observations against which to compare the model's predictions. Accordingly, we used laboratory ecosystems maintained with several different environments to make the tests.

Our first concern was to test whether EXAMS was theoretically sound. If the model perfectly described an environment and the transport and transformations of a chemical in it, predictions for fate of that chemical would perfectly match the chemical's actual fate. In testing EXAMS for its conceptual validity, we used this idea as a pattern. We hypothesized that the model's structure and theoretical content were correct, and supplied the model with the best available parameters. The best available parameters were those that corresponded most closely to the sites and times where we made the observations on chemical fate. Similarly we used specific chemistry parameters for the test chemical, methyl parathion. For biological decomposition, a separate second-order rate coefficient was estimated and used for each of the experimental environments. By this means, the parameterized model was, as nearly as possible, a specific representation of the test systems. Thus, any differences between model and observation should be attributable to fallacies in the model's structure or theoretical content. We formulated the null hypothesis to reflect these ideas: "there is no difference between predicted and observed concentrations when the model is parameterized with environment-specific parameters." Results of this test were analyzed by linear regression of observed on predicted concentrations. The null hypothesis implies a linear relationship with intercept of zero and slope of one. Figure 1 is a display of the data plotted along with this hypothetical line. Student's t was used to test whether slope and intercept differ from their hypothesized values. These results are given in Table I. No differences can be detected, and therefore the null hypotheses were not rejected. As a result of this experimental test, the model has increased in plausibility.

In outline, this approach to testing consists of the following steps:

1. design the content of the model test;
2. devise a statement of no difference between model predictions and observations;
3. use regression of observations on predictions and test working-level null hypotheses of intercept = 0 and slope = 1; and
4. interpret tests and draw conclusions.

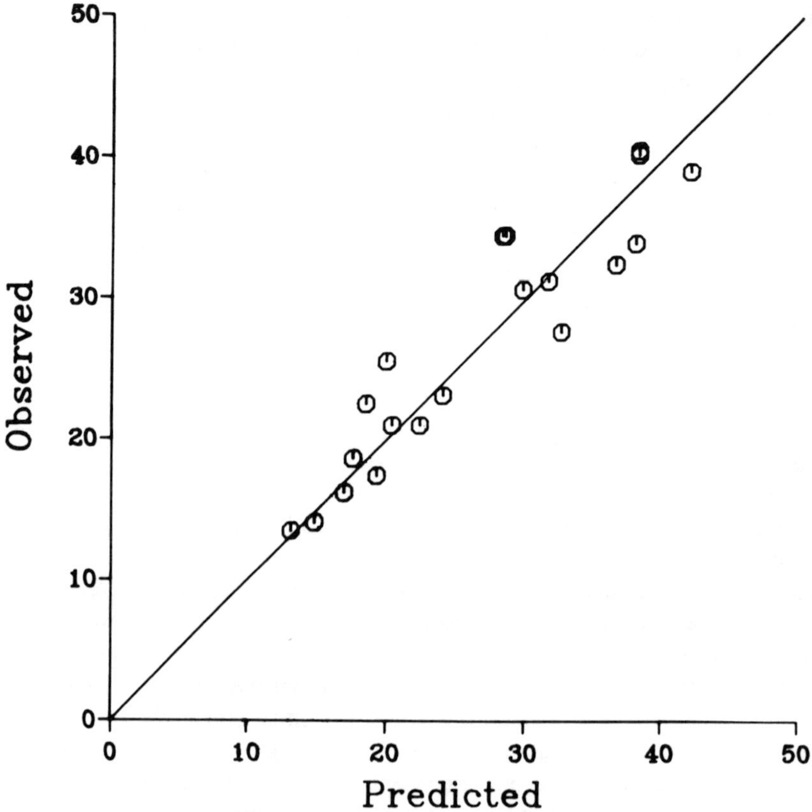

Figure 1. Observed vs predicted concentrations plotted with the line of no difference. Data are from the experiment to test concept validity.

Designing the content of the model is the step requiring the greatest comprehension of the problem at hand. In testing EXAMS, we established content of the test by the assumptions that we used to make the predictions. In the test described above, we used the most specific parameters that we had available, under the assumption that the model, if conceptually sound, would thereby provide the most accurate predictions. In another test, we used the same experiment with different assumptions (more observations qualified) to test the model's utility.

Some aspects of model utility will need to be checked with reference to the specific application. In general, though, no user will have access to site- and situation-specific parameters with which to run the model. To the extent that situation-specific parameters are necessary to provide

Table I. Test of Intercept = 0 and Slope = 1 from Regression of Observed on Predicted Concentration for Validity of Concept

		Parameter Value	t	p>t
n = 20	Intercept	2.75	1.18	0.25
$r^2 = 0.87$	Slope	0.91	1.10	0.29

useful results in practice, the utility of the model will be limited. This suggests that one useful test of utility is to remove any situation-specific parameters from the model and test the hypothesis of no difference using situation-independent parameterization. We know of no evidence to suggest that abiotic chemistry processes vary as a function of environment. That is, these rate coefficients and equilibrium constants are considered to be situation-independent. It is frequently suggested, however, that biological decomposition rate coefficients vary with environment, and accordingly in the test for concept, we used situation-specific values for these rate coefficients as already discussed. To test utility, however, we obtained a rate coefficient applying to all environments by averaging the situation-specific rate coefficients over environments. The null hypothesis was: "there is no difference between predicted and observed concentrations when a single biological decomposition rate coefficient is used for all environments."

Regression analysis was used as in the outline above. Therefore, the null hypothesis was translated into working null hypotheses, as in the test for concept, that intercept = 0 and slope = 1. A plot of observed vs predicted concentrations is shown in Figure 2, and results of the test are presented in Table II. There is little doubt that the null hypotheses should be rejected and the conclusion should be drawn that the predictions are detectably different from the observations.

The model has been accepted as conceptually sound, but rejected for use with environmentally invariant biological decomposition rate coefficients. To interpret this rejection in terms of model utility, however, additional factors are relevant. What is desired as a result of the testing is to be able to establish bounds on the situations for which EXAMS is applicable. We can say from this single test that the biological decomposition rate coefficient is not universally invariant with respect to environment. On the other hand, it is not prudent (and certainly not logically justifiable) to extend the interpretation much beyond that. Our test system was heavily dominated by aufwuchs, and we have not tested the

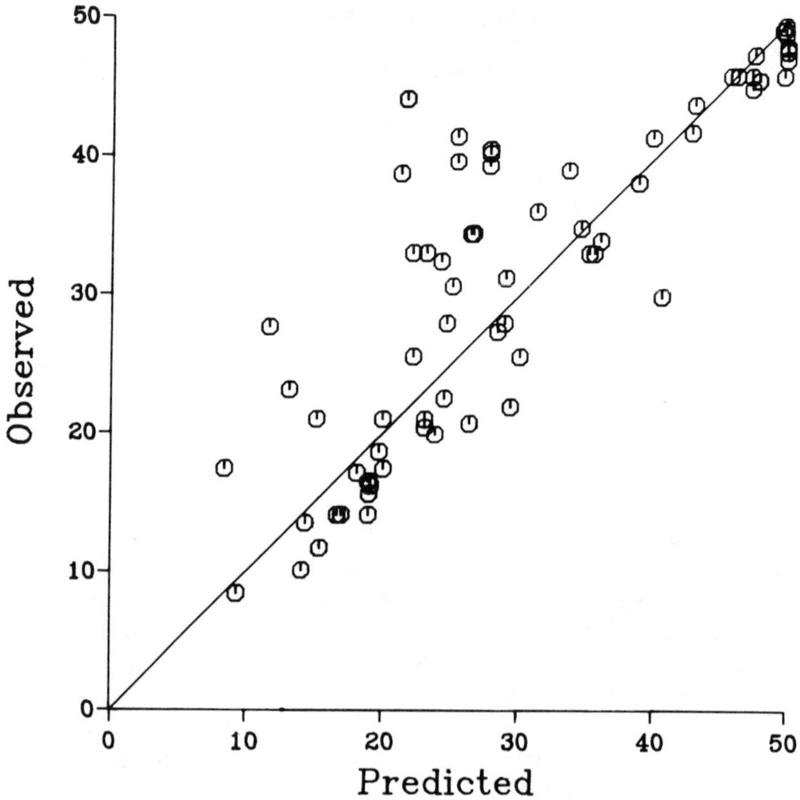

Figure 2. Observed vs predicted concentrations plotted with the line of no difference. Data are from the experiment to test situation independence of the biological decomposition rate coefficient.

invariance of biological decomposition rate coefficients in other classes of systems. There is also evidence that test results were influenced by major shifts in the biological community prior to the last sampling period. Paris et al. [7] studied second-order biological decomposition rates in planktonic systems. They studied three chemicals in more than 40 water bodies. Overall they found a coefficient of variation of about 15%. It is difficult to say just how much variation in model predictions would result from rate coefficients varying over a range equivalent to a coefficient of variation of 65%. However, it will probably be substantially less than the variation noted using rate coefficients from aufwuchs communities as in our tests.

Two recommendations are appropriate based on the above considera-

Table II. Tests of Intercept = 0 and Slope = 1 from Regression of Observed on Predicted Concentrations for Invariance of Biological Decomposition Parameter

		Parameter Value	t	p>t
n = 71	Intercept	7.10	3.40	<0.01
$r^2 = 0.71$	Slope	0.83	2.67	0.01

tions. First, EXAMS needs to be tested more along these same lines using different chemicals and environments. Second, when one uses EXAMS to make analyses for systems dominated by aufwuchs, caution should be exercised. Of particular concern are chemicals that are slow to degrade and whose major pathway of degradation is microbial decomposition. In any event, one can exercise the simple expedient of doing two or more analyses to obtain a range of predicted values when using parameters with associated uncertainty.

The potential limitation of the utility of EXAMS for analyzing the fate of chemicals in systems dominated by aufwuchs naturally led to questions of how to overcome the limitation. Such questions are the beginnings of the process that can eventually lead to conjectures for revising the model. Popper [3] argues that new hypotheses cannot be originated mechanically or analytically, but are purely conjectural. It is reasonable, however, to pursue analyses with the intent and expectation of stimulating ideas. A serious conjecture for revising the model can have its beginning in just such an idea. Differences between observations and predictions conceivably can be associated with factors of the test experiment in ways that suggest model improvement. For EXAMS we analyzed the differences using the predictions made under the assumption of invariance of the biological decomposition parameter. Table III contains the analysis of variance for these data.

Several interesting avenues of discussion are possible based on the results presented in Table III. Although these are discussed elsewhere [8], the clear association of continuously stirred tank reactors (CSTR), maintained as different environments with differences is of particular interest here. We examined this association more closely using Duncan's multiple range test (Table IV). As a result of this test, it is apparent that predictions are poor only for CSTR 3, 4 and 5. Each CSTR received nutrient supplements. We speculated that bacteria without the capability to decompose methyl parathion were disproportionately stimulated to grow in these systems. Their high densities led to overinflated estimates of net

Table III. Association of Treatment Factors with Variation in Differences Between Predicted and Observed Concentrations of Methyl Parathion Resulting from Use of Averaged Biological Decomposition Rate Coefficients to Test Their Invariance

Source	df	F	P>F
Total	70		
Rep	1	25.53	<0.01
CSTR	7	4.11	0.04
Rep × CSTR	7	67.06	<0.01
Date	3	27.92	0.01
Rep × Date	3	115.31	<0.01
CSTR × Date	21	4.25	<0.01
Rep × CSTR × Date	21	27.14	<0.01
Error	7		

Table IV. Duncan Multiple Range Tests Comparing Differences (z) Between Predicted and Observed Resulting from Use of Average Biological Decomposition Rate Coefficients (Averaged over Sampling Dates and Arranged by CSTR)

CSTR	3	4	5	2	6	7	8	1
n	10	10	9	8	8	8	8	10
mean x	9.30	5.82	3.69	−0.68	−0.90	−1.00	−1.28	−1.58

microbial decomposition rates. As a result, the predictions of EXAMS were low. We conjecture that the microbial decomposition rate coefficient for aufwuchs-dominated communities will decrease with increasing degree of eutrophication. At this point this is a qualitative conjecture because neither the term "degree of eutrophication" nor the function relating the rate coefficient to it have been defined. Although needing refinement, it is a conjecture that can lead to additions to underlying theory as well as improvement in the performance of the theoretical model.

As a result of the hypothesis test, one either rejects or accepts (strictly, fails to reject) the hypothesis. As is explained in elementary statistical texts, one can make two kinds of right and two kinds of wrong decisions.

Or one can reject a true hypothesis (type I error), or accept a false one (type II error). The probability of making a type I error is controlled by the selected rejection probability. Too often scant attention is paid to the probability of a type II error. Where the null hypothesis is equivalent to a claim that the model is correct, Balci and Sargent [9] refer to the type I error as model-builder's risk and to the type II as model-user's risk. Typically in environmental science, it is prudent to minimize model-user's risk. That is, accepting a fallacious model is accompanied by too great a potential loss. A user of a faulty model could unwittingly mislead decision-makers to make a decision that would result in environmental damage or significant economic loss.

Before the experiment, the probability of a type II error can be controlled, while holding the probability of type I error at the selected value, by planning for appropriate sample size (number of observations) and adequate replication. Failing to do so introduces the risk that a costly experiment will be run, from which it is not possible to detect a faulty model. Once the experiment is complete, the probability of a type II error can only be reduced by increasing the probability of a type I error. Unfortunately, many studies in environmental science suffer in the analysis stage from lack of forethought in the planning stage. For example, if an experiment is carried out without the design incorporating control over type II error, acceptance of the null hypothesis leaves unanswered a question of the likelihood that a type II error has been made. An optimal design will apportion resources to achieve a powerful test of the hypothesis. It will account for factors in the experiment that could otherwise introduce variation that would reduce the power of the test. Power is the complement to type II error; that is, it is the probability of rejecting the null hypothesis when it is false. The particular experimental design, the underlying assumptions and experimental error together determine the power. It is important to adhere closely to the design, so that the power in practice is equivalent to the designed power. Often as a practical matter, however, some assumptions are not met. It is useful, in these cases, to consider the effect of not adhering to the assumptions, some being more critical to the power than others. It is also enlightening to consider the ability to detect deviations of various magnitudes from the null hypothesis, using estimated error variance for the experiment and assuming that all assumptions of the design are met. In testing EXAMS, we examined the power to detect deviations from the relationship of intercept = 0 and slope = 1 between predicted and observed concentrations. As an example, estimated contours of power for the test of concept validity are given in Figure 3. These contours are calculated with probability of type I error at 0.05. One can read the probability of rejecting the

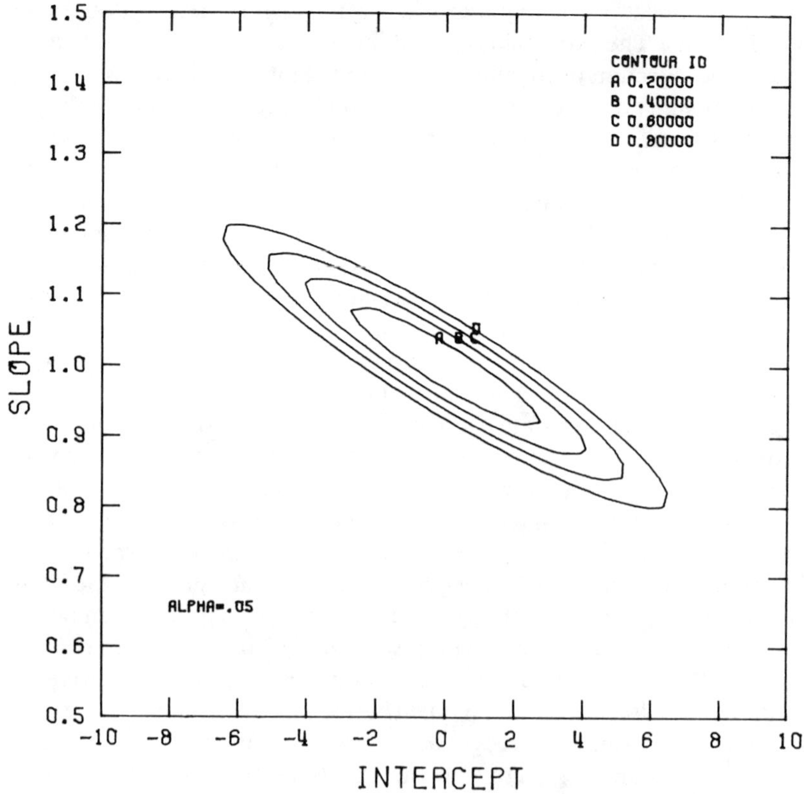

Figure 3. Estimated power contours for values of the slope and intercept. For null hypothesis of intercept = 0 and slope = 1, the curves represent the probability of rejecting the hypothesis should actual intercept and slope be other values. Probability of a type I error is set at 0.05. Note that at intercept = 0 and slope = 1 the probability of rejection is 0.05, the probability of type I error.

hypothesis when the true intercept and slope take any value. Exploration of contours of this sort can be of great value in deciding whether to reject, when the choice is not entirely clear, based on other considerations.

SUMMARY

Testing models of the environmental fate of chemicals ultimately must provide a basis for judging their adequacy for a given purpose. One purpose for which models commonly need to be used is to aid in assessing the probable consequences of releases of new chemicals. In such cases, it is evident that there are no direct measurements of their fate. Theoretical models can be used for this purpose because they do not depend on prior observation for their ability to predict. They depend on the existence of relevant theory, and they must be a faithful representation of that theory. Various factors of their design, the planned uses and the nature of the environment influence their utility. The way that these factors influence the model's adequacy can be discerned only through critical testing and interpretation. In formulating, indeed in recognizing, the tests that need to be made, it is important to comprehend how models relate to theory and of the necessity for that relationship. It is necessary, but not sufficient, that a theoretical model be a faithful representation of theory if it is to be used to make predictions of the sort useful in assessments of chemicals. It suffices that a model be adequate for the purpose to which it is put [2]. Adequacy can only be specified by the user, but at least the model must make use of available information to provide relevant information that does not mislead. In the final analysis, the user must judge whether the model is adequate, although general criteria of usability are often discernible by the developer. Such was the question of invariance of biological decomposition rate coefficient tested for EXAMS.

Figure 4 is an illustration of a basic pattern, as I perceive it, that is followed at any level of testing. A relevant hypothesis about the model is

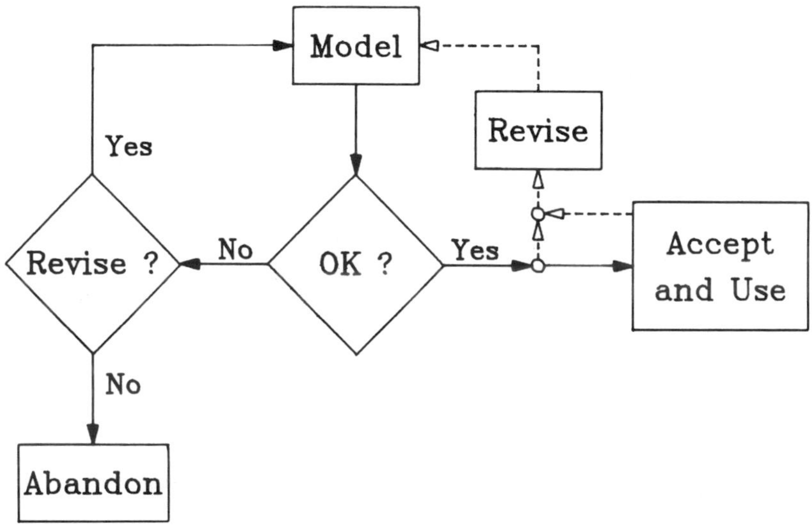

Figure 4. The general pattern of activities in testing a model. Block labeled "model" implies some testable hypothesis about a model. Block labeled "OK?" implies a decision based on test results. Dashed lines indicate potential decision to recycle through the revision and testing without having rejected a hypothesis.

formulated and subjected to critical tests. Depending on the outcome, one can perceive whether the hypothesis should be rejected. If so, it can be revised and subjected to further testing, or it can be abandoned. If not, it can be accepted, or it can be subjected to still further testing, possibly suggested from results of its use. In Figure 5, this pattern is repeated in the context of the three levels of testing as presented in this chapter. For models that depend on theory for their capability to predict, tests are required of theory, of the fidelity with which models represent theory and at the utility level. For models that depend on an empirical fit to a data set for their predictive capability, it is necessary only to test at the level of utility. It is likely that empirical models will be tested for their utility. In the environmental sciences, however, there is a continuing need to test theoretical models for their faithfulness to theory and, espe-

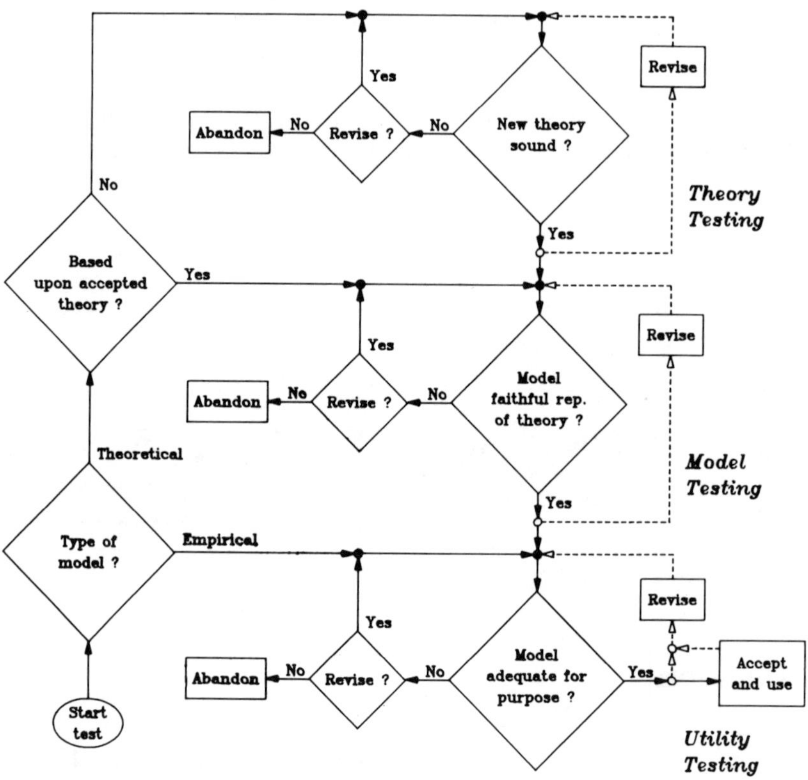

Figure 5. The general structure of model testing. Note the three levels. Note also short circuits to "accept and use" for both empirical models and theoretical models.

cially, to recognize constructs that represent new hypotheses and which, therefore, require testing at the theory level.

REFERENCES

1. Caswell, H. "The Validation Problem," in *Systems Analysis and Simulation in Ecology,* B. C. Patten, Ed. (New York: Academic Press, Inc., 1976), pp. 313–325.
2. Overton, W. S. "A Strategy of Model Construction," in *Ecosystem Modeling in Theory and Practice,* C. A. S. Hall and J. W. Day, Jr., Eds. (New York: John Wiley & Sons, Inc., 1977), pp. 50–73.
3. Popper, K. R. *The Logic of Scientific Discovery* (New York: Harper and Row, 1959).
4. Popper, K. R. *Conjectures and Refutations, the Growth of Scientific Knowledge* (New York: Harper and Row, 1962).
5. Polya, G. *Mathematics and Plausible Reasoning, Vol. II, Patterns of Plausible Inference* (Princeton, NJ: Princeton University Press, 1954).
6. Burns, L. A., D. M. Cline and R. R. Lassiter. "Exposure Analysis Modeling System, EXAMS: User Manual and Documentation," U.S. EPA Environmental Research Laboratory, Athens, GA (1981).
7. Paris, D. F., W. C. Steen, G. L. Baughman and J. T. Barnett, Jr. "Second-Order Model to Predict Microbial Degradation of Organic Compounds in Natural Waters," *Appl. Environ. Microbiol.* 4:603–609 (1981).
8. Lassiter, R. R., R. Parrish and L. A. Burns. "Predictability of the Fate of Chemicals in Aquatic Environments" (in preparation).
9. Balci, O., and R. G. Sargent. "A Methodology for Cost-Risk Analysis in the Statistical Validation of Simulation Models," *Comm. ACM* 24(11):190–197 (1981).

CHAPTER 17

FIELD VALIDATION AND ERROR ANALYSIS OF CHEMICAL FATE MODELS

Anthony S. Donigian
Anderson-Nichols & Co., Inc.
Palo Alto, California 94303

Just as the 1970s was the decade of the emergence of water quality and general environmental concerns, the 1980s will examine the problem, and hopefully some solutions, of toxic chemicals in the environment; their sources, fate, transport and effects will be the focus of many research and development efforts in the current decade. Mathematical models of chemical fate in the environment began appearing in the technical literature in the past five years, and have quickly moved from the research arena to the threshold of accepted methods for assessing risk and exposure to toxic chemicals and for evaluating regulatory and control mechanisms. However, this transfer of new technology to common usage is not without problems. Lack of sufficient, reliable chemical data bases for field validation or verification of chemical fate models is a significant deterrent to expanded use of these models for exposure and control assessments. Moreover, well-defined and accepted procedures, methods and measures of model verification do not exist.

This chapter presents an overview of concepts that should be considered in any validation or verification effort of chemical fate models and in any model application to problems of chemical fate and persistence in the environment. The primary focus, limited largely by the author's experience, is on models of the physical, chemical and biological processes that affect the fate and transport of chemicals in terrestrial and aquatic systems; however, many concepts will also be relevant to other models, such as air and groundwater models. Validation, or verification, will be discussed in terms of common procedures currently used and of the ultimate goal or purpose of model validation. The emphasis in this chapter will be on error analysis, i.e., investigation of the source of discrepancies (errors) or differences between model results and observed data. Various categories of errors are discussed and examples are presented.

VALIDATION/VERIFICATION

Although the literature includes some distinction between the two terms, validation and verification, in this chapter they are used interchangeably to refer to the process of comparing model predictions to observed field data on chemical fate and transport. The purpose of the verification process is to assess the ability of the particular model to represent accurately, under a range of conditions, the important physical (e.g., hydrologic, hydrodynamic and sediment transport), chemical and biological processes that it purports to simulate. Lacking a deterministic understanding of many natural processes and their spatial variation, the verification process is usually preceded by a calibration procedure whereby model parameters (e.g., infiltration rates, channel roughness coefficients, partition coefficients and decay rates) are adjusted to minimize differences between simulated and observed values. The calibration process is required, to varying degrees, for essentially all currently operational models to refine selected model parameters for the specific system being simulated. For physically based models, an expected range of values for each parameter is usually known, and the adjustments are confined to this expected range. As noted above, the calibration process attempts to account for (1) spatial variations not represented by the model formulation; (2) functional dependencies of parameters that are either nonquantifiable, unknown and/or not included in the model algorithms; or (3) extrapolation of laboratory measurements of parameters to natural field conditions.

Split-data testing is often recommended and used in model calibration and validation. This type of testing involves dividing the available data into two portions, one of which is used to calibrate the model and, after the model user is satisfied with calibration results, the second portion is used to verify (or validate) both the model representation of the system and the calibrated parameters. In essence, verification is an independent test of how well the model (with its calibrated parameters) is representing the important processes occurring in the natural system.

Split-data testing has been used commonly in hydrologic modeling because of the general acceptance of the test and the long records (covering many years) of flow data that are often available. Ideally, for water quality, this type of testing should also be used, but data, especially chemical fate data, are often lacking. When data are available, the length of record is usually limited. Consequently, the range of conditions under which the model is being calibrated and verified is also limited, producing an inadequate test of model capabilities. Under limited-data conditions, to divide the available data further, i.e., into calibration and verification data sets, can lead to inadequate calibration and thus produce misleading results during verification. If the purpose of the validation effort is to develop sufficient confidence in the model to use the model to evaluate control measures or changed conditions, it may be adequate to calibrate on the entire set of limited data, since the parameters remain constant throughout the simulation period for the continuous simulation models of concern in this chapter.

For example, the input decay rate remains constant throughout the simulation, but its actual value at any time may be adjusted for temperature and other environmental conditions. In an extreme case, if 20 years of chemical concentration data were available and a chemical fate model could be calibrated to model adequately (based on some measure or test) all the data, there would be no need to further verify or validate the model since it has demonstrated that it can represent the range of conditions over 20 years of data with a calibrated parameter set.

Similarly, if only 12 months of data were available, calibration and verification could be initially performed on the two separate 6 months of data as a test of the model; however, prior to using the model to analyze changed conditions, the parameters that provide the best representation of the entire 12 months should be determined by additional calibration. This would ensure that the model simulations are the most accurate possible over the range of conditions reflected in the available data.

ANALYSIS AND CATEGORIES OF ERRORS

In any model calibration or verification/validation effort, the model user is continually faced with the need to analyze and explain differences between observed data and model predictions. This requires assessments of the accuracy and validity of observed model input data, parameter values, system representation and observed output data. Figure 1 schematically shows the model and the natural system with regard to inputs, outputs and sources of error. Clearly, there are possible errors associated with each of the items/topics noted above and shown in Figure 1, i.e., input, parameters, system representation and output. Each of these error sources or categories is discussed below, along with examples derived from a recent application of the EPA Hydrological Simulation Program—FORTRAN (HSPF) [1] to model the fate and transport of alachlor in an agricultural watershed in Iowa [2].

Input Errors

Errors in model input often constitute one of the most significant causes of discrepancies between observed data and model predictions. As shown in Figure 1, the natural system receives the "true" input, whereas the model receives the "observed" input as detected by some measurement method or device. Whenever a measurement is made, a possible

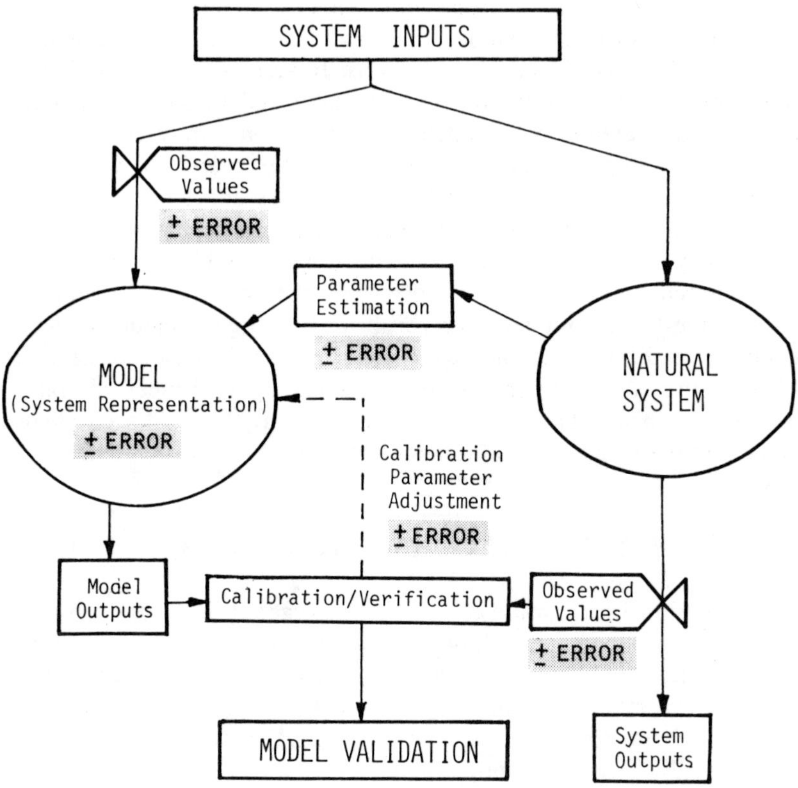

Figure 1. Model vs natural system: inputs, outputs and errors.

source of error is introduced. System inputs usually vary continuously both in space and time, whereas measurements are usually point values, or averages of multiple point values, for a particular time or accumulated over a period of time. Although continuous measurement devices exist, errors are still possible, and essentially all models would require transformation of a continuous record into discrete time and space scales acceptable to the model formulation.

A classic example of input errors in watershed hydrologic modeling is the use of point rainfall measurements to estimate the effective rainfall over a watershed area. Since rainfall is the driving time series in hydrologic modeling, any errors in the input rainfall, i.e., misrepresentation of effective watershed rainfall, will likely propagate through the simulation and thus have a critical effect on model calibration and verification. The hydrologic calibration and verification of HSPF to Four Mile Creek, a 20-mi^2 agricultural watershed in central Iowa, illustrates the importance of analyzing and evaluating the effects of input rainfall errors. Figure 2 shows the map of Four Mile Creek, including data collection sites and segmentation of the watershed area and stream system. Table I presents the results of the flow calibration and verification, with the individual columns representing the steps followed to analyze the effects of non-representative rainfall.

A "pervious land segment" (PLS) (noted in Table I) is a portion of the watershed with (assumed) relatively uniform climatic, hydrologic, soils and land-use characteristics. In Four Mile Creek, four precipitation gauges (Nos. 31, 33, 34 and 35 in Figure 2) were used to approximate the spatial variation in rainfall; the relative shading in Figure 2 delineates the areas represented by each gauge. Within each rainfall area, three major categories of land use were identified—soybeans, corn, pasture/other—producing a total of 12 PLS (i.e., 4 rain gauges × 3 land uses) as noted in Column 1. Soil characteristics were assumed to be relatively uniform throughout the watershed.

In Four Mile Creek, 30 months of detailed rainfall data at four sites within the watershed were available for calibration, while the rainfall data available for verification were limited to a single rain gauge located 10 km southeast of the watershed boundary. Column 1 (Table I) shows the recorded annual runoff and daily flow statistics for both the calibration and verification periods; note that the stated accuracy of this data can vary by 10–15% (or greater) from the actual values (these "output" errors will be discussed later). Column 2 shows the results of the final calibration using 12 PLS and 4 rain gauges. The calibrated annual runoff is within 10–15% of recorded values, and both the mean daily flows (0.186 m^3/sec (cms) simulated vs 0.172 cms observed) and peak daily flows (4.98 cms simulated vs 5.72 cms observed) are closely simulated.

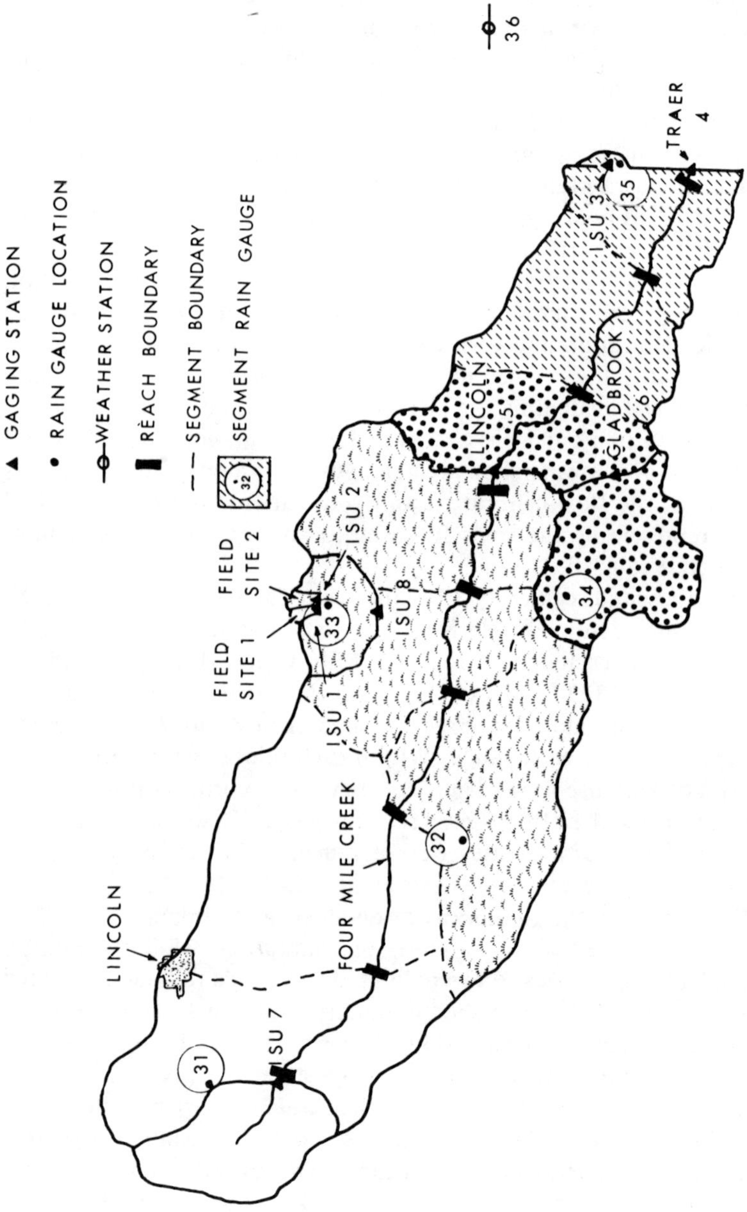

Figure 2. Watershed segments and channel reach boundaries on Four Mile Creek, IA.

Table I. Flow Calibration and Verification Results for the Traer Gauge, Four Mile Creek, IA[a]

	Recorded 1	Final Calibration 2	Runs with Traer Gauge Rainfall: 12 PLS/1 Rain Gauge 3	Runs with Traer Gauge Rainfall: 3 PLS/1 Rain Gauge 4	Verification Run (with Traer Rainfall) 5
Annual Runoff (mm)					
4/76–9/76	76.2	68.6	80.0	87.1	
10/76–9/77	11.4	15.7	24.6	24.4	
10/77–9/78	182.0	205.0	287.0	287.8	
Total	269.6	289.3	391.7	399.3	
1/69–9/69	395.0				334.8
10/69–9/70	144.0				159.8
10/70–9/71	287.0				224.8
10/71–9/72	198.0				234.4
10/72–9/73	458.0				307.1
10/73–9/74	288.8				259.6
Total	1771.0				1520.4
Daily Flow Statistics (cms)					
Mean	0.172/0.4933[b]	0.186	0.252	0.252[c]	0.4247
Standard Deviation	0.376/1.011	0.348	0.530	0.537[c]	0.9115
Maximum	5.27/12.0	4.98	7.90	8.04	14.5
Minimum	0/0.0057	0	0	0	0

Table I, continued

	Recorded 1	Final Calibration 2	Runs with Traer Gauge Rainfall: 12 PLS/1 Rain Gauge 3	Runs with Traer Gauge Rainfall: 3 PLS/1 Rain Gauge 4	Verification Run (with Traer Rainfall) 5
Daily Flow Residuals[d] (cms)					
Mean		0.013	0.079	0.0889[c]	−0.0685
Standard Deviation		0.3004	0.4202	0.4347[c]	1.069
Maximum		4.87	7.53	7.67	13.6
Minimum		−2.97	−1.9	−2.04	−10.5

[a] 95% of flows are within 10%, except in winter, when records are poor (errors >15%).
[b] Calibration/verification periods.
[c] Based on 3-year statistics (10/75 to 9/78).
[d] Simulated minus recorded values.

The mean flow residual corresponds to the difference between the simulated and recorded mean flows (i.e., 0.186 − 0.172 = 0.014), whereas the standard deviation, maximum and minimum residuals reflect differences in timing of the daily flows.

Column 3 (Table I) shows the result of using the Traer rainfall record outside the watershed, in place of the four Iowa State University (ISU) gauges. Although the annual rainfall at Traer for the calibration period was less than 10% greater than the Four Mile Creek watershed average, the total runoff (391.7 mm) and mean daily flow (0.252 cms) were 35% greater than the values simulated with the detailed rainfall data. These differences demonstrate the errors introduced by nonrepresentative rainfall timing and intensity at Traer in the resulting simulated runoff.

To reduce model run costs, average slope and overland flow length values were then used to reduce the number of PLS from 12 to 3; these two parameters were the only differences among the PLS for a single land use category retained from the final calibration (column 2). Comparison of columns 3 and 4 shows that this change had an insignificant effect on the overall simulation; however, impacts on individual storms would be more significant. Thus, column 4 indicates the total effects of errors in the simulation introduced primarily by using the Traer rainfall and three PLS.

The results of the verification run in Column 5 show that the annual runoff volumes are generally within 10–20% of the observed volumes, except for the 1973 water year that accounts for more than 60% of the difference in the total runoff volumes for the verification period. The mean simulated flow of 0.425 cms agrees well with the observed value of 0.4933 cms, and the difference (0.0685 cms) represents only 14% of the observed mean flow. These differences fall well within the errors associated with the use of the Traer rainfall gauge, as reflected in the absolute value of the mean residual (0.0889 cms, column 4) and its percentage of the observed mean flow (18%). Consequently, this demonstrates that the hydrologic calibration at Four Mile Creek is verified.

The goal of these procedures was to establish a valid basis for model verification. In this example, the use of different rain gauges for calibration and verification does not provide a fair test of model verification. Consequently, the error introduced by the use of the Traer gauge provided a tolerance within which model verification could be assessed.

Another example of possible input errors that are especially critical for chemical fate modeling is errors associated with input loadings of chemicals. For a watershed modeling effort, these errors would be associated with chemical input with rainfall or dry deposition; chemical land application rates, methods and timing; and point discharges of chemicals. For nutrients, input with rainfall and dry deposition can be a significant source. For pesticides and other, more exotic, chemicals, these sources can usually be ignored unless drift from aerial applications in neighboring airsheds is significant. In the Four Mile Creek study, monthly estimates of nutrient input with rainfall were included in the simulations to perform a nutrient balance of the watershed. However, pesticide content of the rainfall was assumed to be negligible.

Assumptions on chemical application rates, methods and timing were shown to have a major impact on concentrations and loads measured at the watershed outlet. Application rates determined by an inventory of farmers can include substantial error, since many farmers may not follow label recommendations. Surface vs subsurface application methods affect the susceptibility to removal by surface runoff, erosion and/or subsurface flow components. Timing of applications is especially critical for fast-decaying pesticides, since the majority of the pesticide loss by runoff occurs in the first few storm events following application. For a complex watershed like Four Mile Creek, with more than 60 individual farmers, no model can accurately represent the time distribution of chemical applications to individual land areas within the watershed. Consequently, assumptions on the amount and timing of application, e.g., before or after a specific event, have a major impact on how much chemical loss occurs during the event.

For an aquatic model of chemical fate and transport, the input loadings associated with both point and nonpoint sources must be considered. Point loads from industrial or municipal discharges can show significant daily, weekly or seasonal fluctations. Nonpoint loads determined either from data or nonpoint-loading models are so highly variable that significant errors are likely. In all these cases, errors in input to a model must be considered to provide a valid assessment of model capabilities through the verification process.

System Representation Errors

System representation errors refer to difference in the processes and the time and space scales represented in the model vs those that determine the response of the natural system. In essence, these errors are the major ones of concern when one asks "How good is the model?" Whenever comparing model output with observed data in an attempt to evaluate model capabilities, the analyst must have an understanding of the major natural processes, and human impacts, that influence the observed data. Differences between model output and observed data can then be analyzed in light of the limitations of the model algorithm used to represent a particularly critical process and to ensure that all such critical processes are modeled to some appropriate level of detail. For example, a lake model that simulates dissolved oxygen (DO), without simulating biological components, cannot be expected to reproduce observed DO values during algal bloom conditions. Similarly, for chemical transport modeling, a compound that even partially adsorbs to sediment particles cannot be represented accurately without modeling sediment transport and sediment/chemical interactions. A model without algorithms for these processes may be quite appropriate for completely soluble (nonadsorbing) compounds, but it would not be applicable to many compounds that undergo adsorption and desorption.

Although the existence or absence of a particular process can often be determined from observed data, an assessment of how well an algorithm represents the process is often difficult to make because of observation errors, natural variations in field data, and lack of sufficient data on individual component processes. In such circumstances, model validity must be inferred or, possibly, based on comparisons with laboratory data obtained under controlled conditions. Laboratory data often provide the basis for developing an algorithm, since field data are so much more difficult and expensive to collect and interpret.

Figures 3 and 4 show the daily flow and sediment simulation results for the Iowa study for the 1978 and 1976 summer periods, respectively. Although the flow was calibrated for the entire period, April 1976 to September 1978, sediment was calibrated on data from April to June 1978 (Figure 2) and verified on the April to June 1976 data (Figure 3). For the summer periods critical to pesticide simulation, the flow representation is quite good. Since both the peaks and recessions of the daily hydrographs are well simulated, it can be inferred that the relative distribution between surface and subsurface flow is well represented. However, it should be clear that this is an inference from the model results, since the observed data are not divided into components.

Sediment calibration was concentrated on the April to June 1978 period, when flow simulation was quite accurate. Since sediment transport, deposition and scour are so highly dependent on flow, it is critical to choose sediment calibration periods when the flow simulation is representative. Although the sediment calibration (Figure 2) is quite good, the verification results are significantly lower than observed values, partially because of undersimulation of flow during the verification period. Since the goal of this work was simulation of alachlor, which moves primarily in solution, the sediment results were judged to be acceptable. However, additional calibration was recommended, especially for more highly

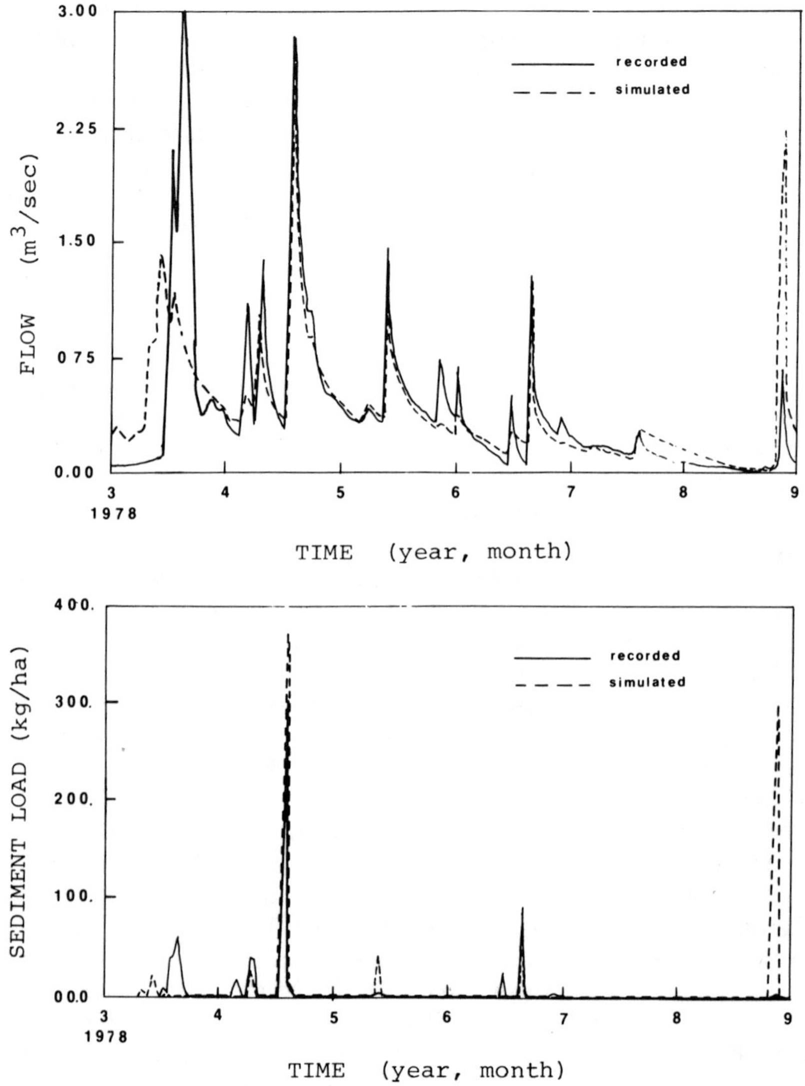

Figure 3. Calibration results for flow (top) and sediment loading (bottom) at Traer, IA.

adsorbed compounds. The observed data included sediment concentrations at the watershed outlet, particle-size distribution of the sediment and infrequent samples of bed composition; thus, as with the flow simulation, the ability of the model to simulate intermediate processes, such

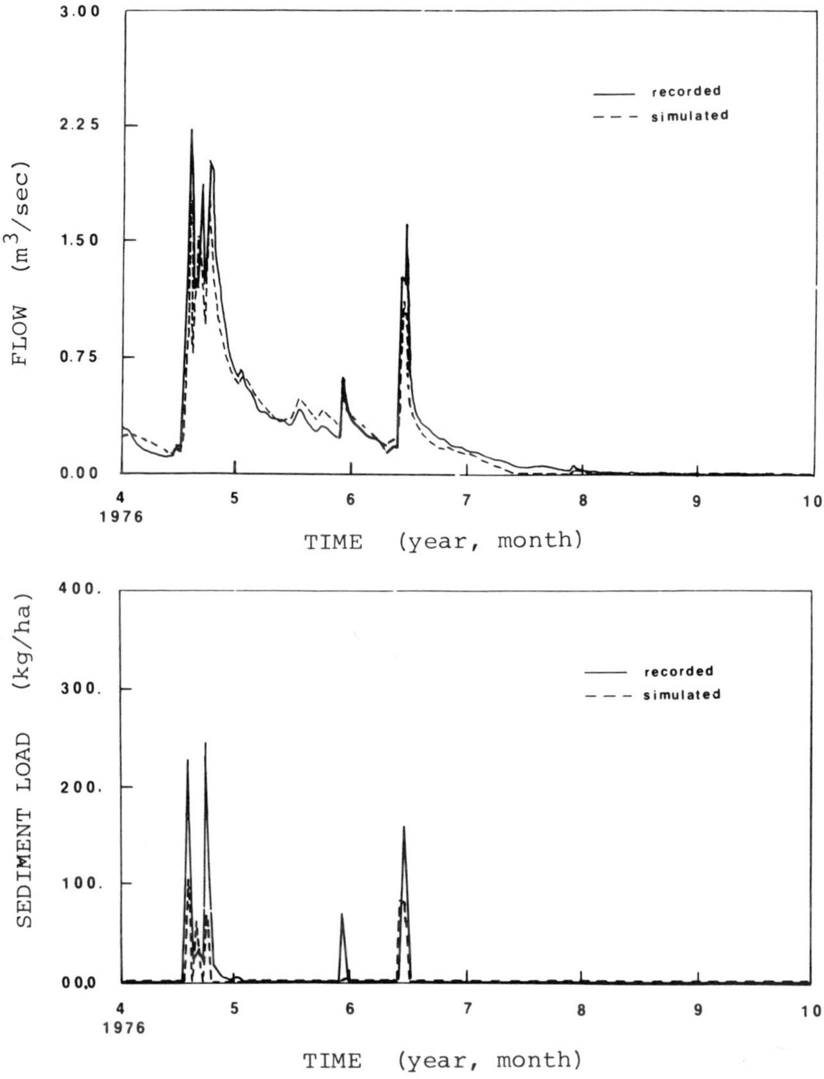

Figure 4. Verification results for flow (top) and sediment loading (bottom) at Traer, IA.

as bed scour and deposition, had to be inferred, since specific data on the bed behavior were not available.

Similarly, the simulation of the alachlor, shown in Figure 5, indicates that the model represents the "spiked" behavior of the natural system, with alachlor concentrations occurring only during and immediately following late spring and summer storm events. Although the discrepancies for particular events can be attributed to data errors (to be discussed below), flow simulation and assumptions on alachlor applications, the model does represent the general characteristics of the watershed. The release of deposited alachlor from the channel bed following a storm event is another example of a process representation that must be inferred

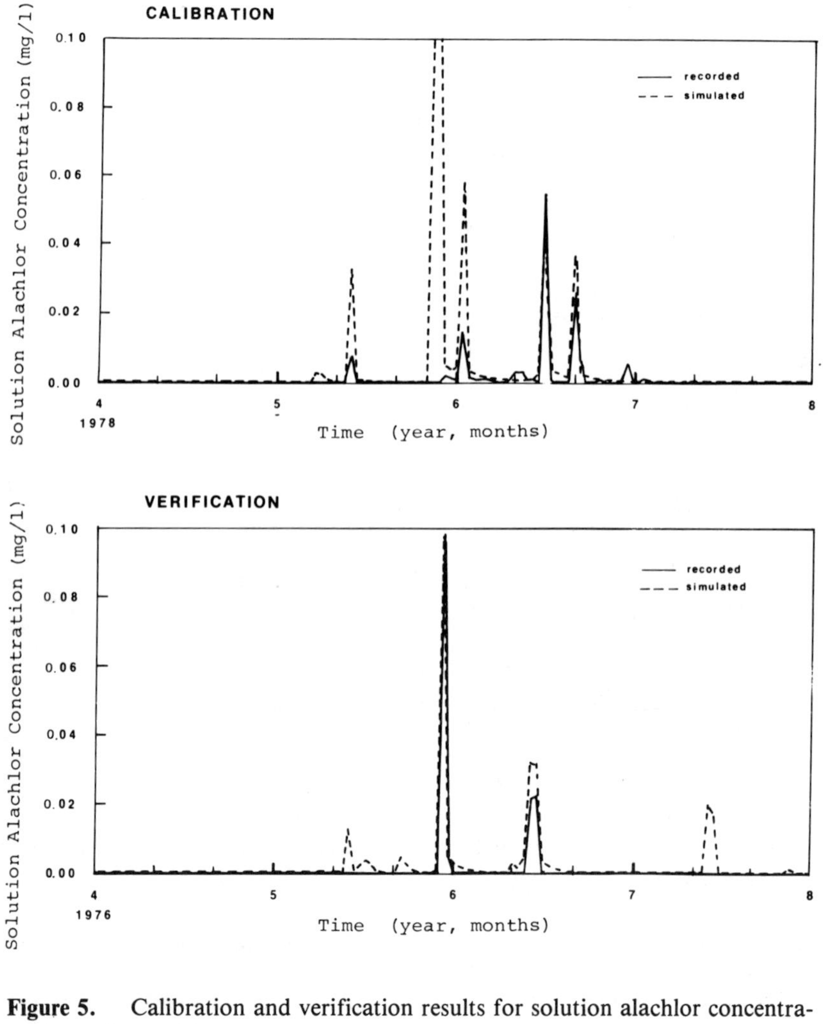

Figure 5. Calibration and verification results for solution alachlor concentration at Traer, IA.

from the outlet data, since bed samples were not taken and analyzed. Figure 5 shows that the simulation is slightly higher than observed values in the days following an event, indicating that possibly the model is adsorbing and subsequently releasing too much alachlor from the channel bed. Further parameter adjustments might correct these differences.

The above discussion shows how every individual process simulated by a chemical fate and transport model cannot specifically be evaluated from field data available from even the most extensive data-collection programs. Although valid inferences can often be made, validation of many intermediate processes may require data from laboratory experiments as the only alternative for an accurate evaluation.

Appropriate time and space scales depend on the specific response times of the natural system and processes being modeled, and they are further constrained by the economics of model operation. In natural free-flowing streams, the appropriate time interval of simulation (based on system flow-through time) would likely be less than that for large reservoirs and impoundments. Obviously if diurnal variations are important for specific processes, a daily time step would be too large to represent the system behavior. Since most currently used chemicals, especially pesticides, are not highly persistent in natural environments and undergo relatively fast reaction processes, time steps measured in hours and generally less than one day would likely be appropriate. For the Four Mile Creek watershed, covering 20 mi^2, an hourly time step for simulation was judged to be adequate; this could have been varied between 30 min and 2 hr without a significant difference in model accuracy. Alternately, the 3000-mi^2 watershed of the Iowa River Basin above Coralville Reservoir will be simulated on a 4- or 6-hr time step because of the large inertia in the system.

Time and space scales are interdependent. Thus, longer time steps are usually associated with a larger spatial definition of the system. Although models are available for simulating an individual farmer's field, this same spatial detail would not be appropriate for modeling the entire encompassing watershed. Even in the Four Mile Creek watershed, modeling the behavior of individual field-size areas was not appropriate to the goal of evaluating downstream water quality. Many surface processes important to field-size areas are greatly attenuated by the time the flow and constituents arrive at the watershed outlet. Also, subsurface processes that have very little impact on flow and pollutant runoff from fields are critical to modeling the entire watershed accurately.

Although "distributed" parameter modeling approaches, which superimpose a grid system on the watershed, attempt to provide greater spatial definition than the "lumped" parameter approach [e.g. HSPF,

Agriculture Runoff Model (ARM)], these models are almost exclusively single-event models that cannot practically be used to evaluate system behavior under a wide range of conditions. Moreover, for large watersheds, a certain degree of averaging or lumping of spatial characteristics is required even for the distributed models to perform simulations within reasonable time and cost constraints. Clearly, time and space scales and the degree of process representation required depend on the purpose or goal of the modeling effort and the types of questions to be answered.

Parameter Errors

Parameter errors are derived primarily from the inability to accurately measure and predict many of the parameters that characterize the natural system and, for chemical fate modeling, the relevant chemical processes under field conditions. Errors are associated both with parameter values obtained from actual measurements and from calibration. Because of natural variations in topography, soil characteristics, crop cover densities, etc., a single parameter value or a time-variable parameter function (e.g., crop cover) will always have some associated error. The goal is to obtain measured parameter values that, to the extent possible, represent mean or average conditions over the watershed segment. If the specific parameter values demonstrate significant variations, then separate land segments (as discussed previously) may be needed to develop regions with relatively uniform characteristics. Such is the case for different crop types and vastly different soil characteristics. This is also related to the spatial scales required to represent the watershed adequately.

In some cases, parameters may not be sufficiently sensitive to require a separate land segment definition solely on the basis of one parameter. For example, column 3 in Table I demonstrates that using average land slope and overland flow length values in place of the specific values for each segment shown in Figure 2 did not significantly affect the calibration results.

On the other hand, land cover, evapotranspiration rates, tillage practices, chemical application rates, etc., associated with each crop type required definition of parameters for each crop segment. Clearly, the characteristics of the crops and the resulting behavior of the cropped areas were sufficiently different to justify separate land segments. Parameters for which measured values are not clearly defined or readily available are often determined through calibration with observed data. Traditionally, calibration has been associated with hydrologic parameters (e.g., infiltration rates, moisture storage and groundwater recession constraints) and, to a lesser extent, sediment transport parameters (e.g., erodibility coefficients, and scour and deposition rates) because the required flow and sediment data are often available. Although initial parameter values can always be estimated, calibration is usually recommended to account for local and spatial variations.

Chemical parameters (e.g., partition coefficients, decay rates, and temperature and moisture effects) are not usually considered as calibration parameters because they can be measured in a laboratory, and calibration is usually not possible because the data do not exist. However, most scientists will agree that extrapolation of laboratory parameter measurements to field conditions is a risky assumption. If observed chemical data are available, refinement of initial chemical parameters through calibration is recommended.

Errors in calibration-derived parameter values are often a function of how much calibration was performed or errors in system inputs and/or outputs. In many modeling efforts, conscientious model users will often overrun the calibration budget because of the natural tendency to continue to make calibration runs in an effort to minimize discrepancies between simulated and observed values. Parameter errors associated with calibration are more often a result of missing and/or erroneous data either as system inputs or outputs. For example, the values in Table I show that the flow calibration was generally higher than observed data, although the verification run produced simulated values lower than observed. This was partially the result of performing calibration on a period of record that included drought conditions. Thus the calibration parameters were slightly biased toward low and moderate flow conditions, resulting in an undersimulation of high flows.

Also, in a previous modeling study on Four Mile Creek [3], the rainfall record at Traer was used for calibration in the absence of any rainfall measurements within the watershed. This calibration produced hydrologic parameter values significantly different from those obtained in the HSPF application, with the difference largely attributed to nonrepresentative rainfall.

Errors in system output measurements can also produce calibration errors, since the model user will be attempting to calibrate against inaccurate or missing data. Errors associated with system outputs are discussed below.

Output Errors

Output errors are analogous to input errors; they can lead to biased parameter values or erroneous conclusions on the ability of the model to represent the natural system. As noted earlier, whenever a measurement is made, the possibility of an error is introduced. Table I noted that the streamflow data can be 10–15% or more in error; this, in effect, provides a tolerance range within which simulated values can be judged to be representative of the observed data. It can also provide a guide for terminating calibration efforts.

Output errors can be especially insidious, since the natural tendency of most model users is to accept the observed data values as the "truth" on which the adequacy and ability of the model will be judged. Model users should develop a healthy, informed skepticism of the observed data, especially when major, unexplained differences between observed and simulated values exist. For example, Figure 5 shows a simulated alachlor concentration value of greater than 0.1 mg/l on May 27, 1978, whereas the observed value is zero. Since May is the month of alachlor applications, it is logical to expect significant alachlor concentrations for even minor runoff events during this period. Coincidentally, a storm of comparable rainfall and flow occurred on May 29, 1976 (bottom graph in Figure 5), producing a maximum alachlor concentration of above 0.1 mg/l, which was equal to the simulated value. Further investigation showed that problems in communications led to missing samples on both May 27 and June 1, 1978; the relative timing of samples and simulated alachlor concentrations for this period are shown in Figure 6. In comparing the May storms of 1978 and 1976, the simulated concentration values in Figure 6 are clearly more representative of what actually occurred than are the observed values. This is not meant to be a criticism of the sampling program, but an indication of how errors in observed data can exist and affect the model validation process.

In addition to outright errors in observed data, the model user needs to assess how representative the data are, considering sampling methods and locations, as well as subsequent manipulation of the data, for example, to estimate chemical loads based on the flow and concentration values. Observed data are measurements at a particular point on a stream, but the model simulation represents an average concentration in an assumed, completely mixed section of the stream. Also, the method of sample collection (e.g., depth-integrated, cross-sectionally integrated or grab samples) should be considered in terms of obtaining a representative sample.

Comparisons between observed data and model predictions must be made on a consistent basis, i.e., apples with apples and oranges with oranges. Since models provide a continuous time series, any type of statistic can be produced, such as daily maxima, minima, averages or medians. However, observed data are usually collected at infrequent

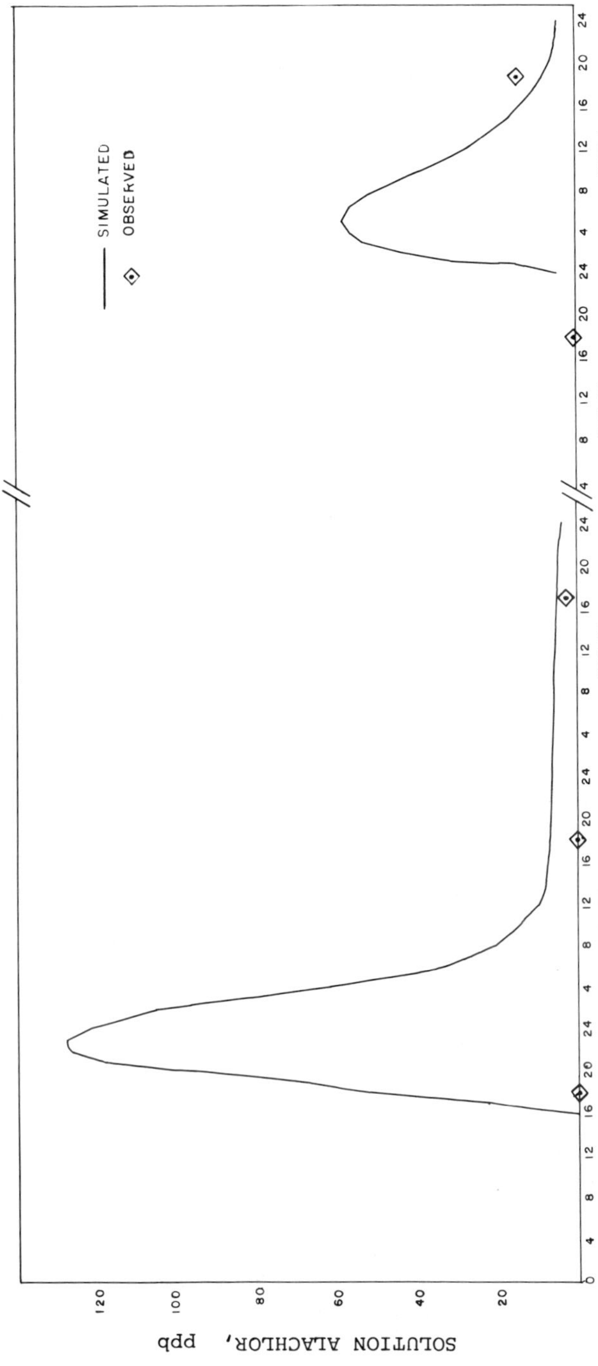

Figure 6. Comparison of detailed hourly simulation of solution alachlor concentration with observed data.

intervals, so only certain statistics can be produced. In Figure 6, the plotted values are daily maxima. The observed data were collected daily, or every other day during constant flow periods, with multiple samples during storm events. Thus, for days with storm events the comparisons are consistent; maximum observed and simulated values are plotted, although it is uncertain that the observations are the absolute maximum values that occurred. For nonstorm days, the observed value is the single daily observation taken usually at about 6 PM. Consequently, we might expect that the simulated value would be somewhat higher, especially in the days immediately following an event, since concentrations would be decreasing during this period.

Validation of chemical fate and transport models is often performed by comparing both simulated and observed concentration values and total chemical loadings obtained from multiplying the flow and the concentration values. Whereas the model supplies flow and concentration values in each time step, the calculated observed loads are usually based on values interpolated between actual flow and sample measurements. From Figure 6, for the Four Mile Creek watershed, we can see that unless observed concentrations are available on 2- to 4-hr increments with the closer intervals near the peak concentrations, it would appear that the observed chemical loads would be underestimated by these calculations. In other situations, the opposite might be true. The model user needs to be aware of how observed chemical loads are calculated to assess the veracity of the values.

Although this section has emphasized possible errors in the observed data, the above considerations should not be used as a crutch to support an invalid or inaccurate model. In most circumstances the observed data are the best indicators of system behavior. Combining model simulations with an informed skepticism of both the model and the observed data can lead to a better overall understanding of chemical fate and transport in natural systems.

CONCLUSION

The goal of this chapter has been to present concepts and considerations that should be examined whenever observed data and model predictions are compared in the process of validation of chemical fate and transport models. Although the various possible categories of errors to be examined have been listed as input, model representation, parameter and output errors, in reality there is considerable overlap. Input and output errors can lead to biased parameter values; model input and parameters are dependent on the particular model representation. Hopefully, the concepts presented here will initiate additional dialog and discussions that will in turn help to refine and improve model validation procedures.

REFERENCES

1. Johanson, R. C., J. C. Imhoff, H. H. Davis, J. L. Kittle and A. S. Donigian. "User's Manual for the Hydrologic Simulation Program—FORTRAN (HSPF): Release 7.0," U.S. EPA Environmental Research Laboratory, Athens, GA (1981).
2. Donigian, A. S., J. C. Imhoff and B. R. Bicknell. "Modeling Water Quality and the Effects of Agriculture Best Management Practices in Four Mile Creek, Iowa," U.S. EPA Draft Report on Contract No. 68-03-2895 (1981).
3. Leytham, K. M., and R. C. Johanson. "Watershed Erosion and Sediment Transport Model," U.S. EPA Report 600/3-79-028 (1979).

CHAPTER 18

FIELD VALIDATION OF EXPOSURE ANALYSIS MODELING SYSTEM (EXAMS) IN A FLOWING STREAM

Larry M. Games

Environmental Safety Department
The Procter and Gamble Company
Cincinnati, Ohio 45217

The need to be able to make realistic predictions of potential exposure to synthetic organic chemicals in the aquatic environment for use in hazard assessment is becoming increasingly evident. These predictions require that the use and disposal patterns of the specific chemical be defined clearly. They then require a quantitative estimate of (1) the distribution and transformation properties of the chemical; and (2) those properties of the environment that are important in controlling the fate of the chemical. Finally, a systematic framework for integrating these factors is needed, so that estimates of environmental exposure can be obtained. This last step, the model, has received increased attention in the last few years.

The need for models defining aqueous environmental exposure encompasses a broad range of philosophical viewpoints. From a purely academic view, the ability to create a model that very accurately reflects the behavior of chemicals in a complex ecosystem provides a practical assessment of our understanding of the interaction between the chemical and the environment. From a purely practical viewpoint, predictive models provide a framework for organization of input data which, for environmental exposure models, tells us if we are likely to affect aquatic populations or endanger human health by the proposed use of a chemical.

Evaluation of an academically oriented model is fairly straightforward, if not simple, since congruence of predicted and observed results is the measure of success. General-purpose, decision-oriented models are more difficult to assess because of the nature of the requirements. The more general models that have been proposed to answer more practical needs [1–3] are intended to approximate the environment sufficiently to provide adequate exposure assessments, not precise fits between measured and predicted concentrations. Some of the simplifying assumptions that are made in these models may preclude any attempt to obtain better-than-general agreement between measured and predicted exposures.

Nevertheless, extensive testing of these evaluative models is necessary if they are going to be used in decision-making processes. This testing is probably best accomplished by field validation studies. Model ecosystems, while they are extremely useful in studying specific parameters, may not be sufficient to model fate accurately enough in all respects for direct extrapolation to complex environments. In general, sediment/water dynamics are necessarily simplified, container effects are difficult to eliminate and important hydrodynamic properties are impossible to model. If models are sufficiently accurate for decision-making purposes, they should provide reasonable predictions of concentration in any specific ecosystem, and this can best be tested by comparing their predictions against carefully collected monitoring data from a natural environment.

Since general models are intended to be evaluative in nature, the question invariably arises as to what constitutes sufficient validation of the particular model under consideration. That is, when is the model sufficiently tested and validated so that its use in decision-making is justified? Two terms, validation and calibration, are often used to describe testing of a site-specific model for a given aqueous system (lake, estuary, river, bay, etc.). Calibration is the use of historical data to set values for unmeasured or difficult-to-obtain environmental parameters. If these calibrated parameters can then be used successfully to predict the concentration of a given chemical at that specific site for a separate time, then the model is considered to have undergone successful field validation. Unfortunately, validation and calibration of an evaluative model represent something of a dilemma. These models cannot be made site-specific and fulfill their intended function, and calibration over the whole range of potential environmental situations would be impossible. Calibration may be appropriate to determine the range of some parameters, but if the range is so wide that vastly different predictions can be made, understanding of that particular parameter becomes more important than understanding of the chemical under consideration. This is especially important if the parameter has no specific biological, physical or chemical basis that can be studied independently. Field validation of these evaluative models, then, requires not only comparison of measured and predicted concentrations, but also critical consideration of how much calibration has gone into providing a good fit. In general, the more calibration required for successful field validation, the less applicable the model is for use in a general evaluative sense, unless the calibration parameters are well defined and a range has been established based on extensive independent investigation.

Regardless of their limitations, if these evaluative models are eventually going to be used in decision-making processes, some determination of their applicability and robustness will be required. One logical approach to this is to compare model predictions to measured concentrations in a well-defined environment, and to include a critical examination of how well the assumptions of the model are met in that environment. Of critical importance is an assessment of how much calibration of the model is required to obtain a reasonable fit between measured and predicted exposure concentrations. Because of the complexity of the aqueous environment and the numerous processes that can occur, field validation studies should, in the beginning, be done in a way that limits, as much as possible, the variables involved. In this chapter, an attempted field validation of an evaluative model, the U.S. Environmental Protection Agency (EPA) EXAMS model [2], will be described. The environment chosen is a small stream in an isolated area with a single point-source input of domestic sewage. The test chemical is an anionic surfactant, linear alkylbenzene sulfonate (LAS), which is a common component of granular detergents. This chapter will focus on the data required to use the model and a sensitivity analysis for several laboratory-measured variables plus a single-model calibration term.

TEST CHEMICAL

LAS is an anionic surfactant used commercially in detergents. Sales in the United States were 640×10^6 lb in 1978 [4]. As shown in Figure 1, it is not a single chemical entity, but is instead a mixture of various alkyl homologs and phenyl isomers. The average chain length of commercial LAS is approximately $C_{11.8}$ and phenyl isomer distribution is variable, depending on the manufacturing process. Analysis of LAS in the various samples is accomplished by a modified version [5] of the desulfonation-gas chromatography (GC) method [6]. The method is designed such that

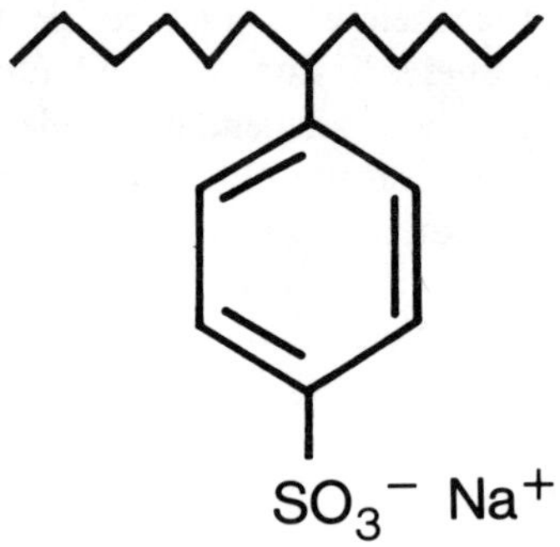

6Ø, C_{12}LAS

Typical Distribution for C_{12}LAS

	2Ø	3Ø	4Ø	5Ø	6Ø	7Ø
C_{10}	4.48	2.68	2.18	1.51		
C_{11}	11.22	7.49	6.46	5.59	4.75	
C_{12}	10.8	7.17	6.46	6.0	7.22	
C_{13}	2.89	2.14	1.88	2.2	2.03	1.02
C_{14}	0.84	0.55	0.49	0.59	0.72	0.62

Figure 1. Structure of LAS and typical distribution of C_{12} LAS. LAS is used as the sodium salt in detergents and is completely unprotonated in water at any environmental pH.

total LAS concentration and percent distribution of the individual members of the mixture are obtained.

MONITORING SITE

Monitoring data were obtained in Rapid Creek as it flows downstream for about 100 km from Rapid City, South Dakota, to the Cheyenne River. Rapid Creek is a mountain stream whose flow is regulated by the Pactola Dam 32 km upstream from Rapid City. The single known point-source input along the length of the creek is Rapid City itself. During the year, about 20% of the creek flow is directed for agricultural purposes, but this activity ceases during the winter (September to December), when the present monitoring program was carried out. During those months, the flow of the creek is relatively constant, about 1 m^3/sec below Rapid City. Reported flows are based on U.S. Geological Survey (USGS) gauging stations along the creek [7]. Detailed descriptions of the site are available from a study of phosphorus exchange in the stream [8].

Sampling stations in the stream are located 0.8, 7, 11.7, 25.3, 48 and 87.2 km downstream from the treatment plant. These are designated sites 1–6 in the downstream direction and are described in some detail in an earlier publication [8]. Grab water samples from midchannel and sediment samples from near-shore locations were collected at each site during sampling trips in September 1978, September 1979, November 1979 and October 1980. The samples collected during October 1980 were taken in a time sequence based on flow velocity, so that a single volume element of water could be followed downstream. The October 1980 results are used in modeling studies presented here, but concentrations in the water and sediment were relatively constant over the three-year time period.

TREATMENT PLANT

LAS in Rapid Creek comes entirely from the trickling filter sewage treatment plant (STP) serving Rapid City. Less than 10 μg LAS/l water and no LAS in sediment were detected in the creek above the treatment plant [9]. The STP processes almost entirely domestic sewage with a 6% input of dairy and meat-packing plant wastes. Effluent to the creek is 7–8 mgd, which results in about a 3:1 dilution factor in the stream during September to December. Influent and effluent concentrations of LAS in the treatment plant were 3.7 mg/l ($n=4$) and 0.51 ± 0.12 mg/l ($n=5$), respectively, over the period 1978–1980 as measured by analysis of five-day flow composited samples. Essentially no redistribution of homologs and phenyl isomers occurs during treatment.

EXAMS MODEL

The EXAMS model has been thoroughly described [1] and is available in batch and interactive versions from the EPA laboratory at Athens, Georgia. EXAMS requires a description of the environment, a quantitative estimate of the distribution and rates of transformation of the chemical, and a steady-state input of the chemical to the environment. From these data, EXAMS is designed to calculate the steady-state concentration of the chemical in the various compartments of the environment under consideration. In EXAMS the environment is held constant, but the distribution and transformation processes affecting the chemical are modified according to the nature of the environment. For example, the rate of hydrolysis is adjusted for pH of the various compartments. Further details are available in the manual written to describe operation of the program [2].

Figure 2 shows the technique used to compartmentalize Rapid Creek for use in EXAMS. There are five segments, each containing sediment and water compartments. Each segment ends at a sampling point, and because of stream-mixing considerations, the input (mass of chemical) to the segment modeled is the LAS concentration in the water at site 1. The geometry of each segment is also shown in Figure 2. Velocity measurements suggest that, as expected, the geometry of the creek is not constant over the whole length modeled, and average depth is varied as an artificial input parameter to reflect changes in residence time observed. Visual decrease in depth as the creek flows downstream was also observed, but no attempt was made to profile the creek depth in any detail. A constant sediment-mixing depth was assumed.

MODEL INPUT FOR TEST CHEMICAL

The important physical, chemical and biological parameters for LAS are shown in Table I for the C_{10}–C_{14} homologs. These tests were done with representative phenyl isomer mixtures of each chain length. Adsorption to bacteria was measured with activated sludge solids, and this value

Table I. Properties of LAS Homologs[a]

		Adsorption Partition Coefficient		Biodegradation Rate Constant at 24°C (/hr)	
	Solubility (mg/l)	Sediment	Bacteria	Aqueous	Sediment-Adsorbed
C_{10}LAS	80–160	41	220		
C_{11}LAS	40–100	100	1000		
C_{12}LAS	35–90	330	3070	0.021[(b)]	0.040[(b)]
C_{13}LAS	17–35	990	9330		
C_{14}LAS	0.2–0.9	2950	ND[c]		

[a] Each represents a phenyl isomer mixture approximating commercial mixtures, and variations of properties are much less for phenyl isomers.
[b] From Larson and Payne [9].
[c] Not determined.

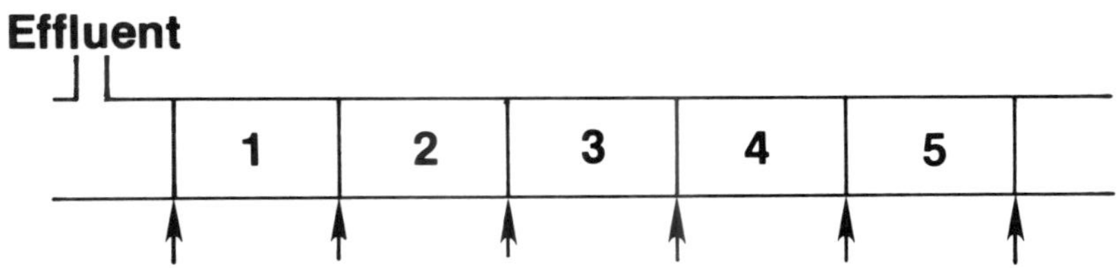

Sampling Site	1	2	3	4	5	6
Distance from effluent (km)	0.8	7	11.7	25.3	48	87.2

Segment Geometry	1	2	3	4	5
Volume (m^3)	2.6×10^4	2.1×10^4	6.4×10^4	8.2×10^4	1.3×10^5
Water Depth (m)	.56	.53	.51	.47	.43
Width (m)	7.6	7.6	7.6	7.6	7.6
Sediment Mixing Depth (m)	.03	.03	.03	.03	.03
Residence Time at 1 m^3/sec. in hours	7	5.7	18.4	25.7	45.6

Figure 2. Structure of the EXAMS model for Rapid Creek and location of sampling sites. The geometry of each model segment is also shown.

is assumed to represent adsorption to biomass in general. Reported K values are simple partition coefficients. The only identified transformation mechanism for LAS is biodegradation, and the rate constants shown were measured in Rapid Creek water and Rapid Creek water/sediment mixtures at locations downstream from the Rapid City treatment plant [9]. The rate constants are for ultimate degradation of ring-labeled ^{14}C-C_{12} LAS. In the modeling work, the aqueous biodegradation rate is used for stream and sediment interstitial water, and the sediment rate is applied to material adsorbed to sediment and suspended solids. The biodegradation rate constants in segment 5 are decreased in the water and sediment by factors of 10 and 4, respectively. These decreased rates correspond to observed rates upstream from the Rapid City treatment plant outfall [9], and they were used because the creek in segment 5 appears to have recovered completely from the impact of the sewage input.

Since LAS is a mixture of many alkyl chain length homologs and phenyl isomers, a theoretically correct approach to inputting fate data for EXAMS would include parameters for each species. The model would then be applied with input concentrations of each component based on the percent distribution of the LAS mixture. To simplify the modeling process for illustrative purposes, input parameters here have been based on the properties of C_{12} LAS. Commercial preparations are usually close to $C_{11.8}$ LAS, and the effluent to Rapid Creek averaged $C_{12.1}$ LAS, so that the use of C_{12} LAS properties as average appears reasonable. In addition, although some redistribution of components occurs in the creek, the average composition stays in the range $C_{11.5}$–C_{13} LAS.

The only important factors in assessing the fate of LAS in a moving stream are adsorption and biodegradation. The use of LAS as a test chemical precludes testing of the EXAMS model in its ability to deal with several other transformation mechanisms, as well as volatilization losses. In some respects this decreases the rigor of the test, but it also allows for easier evaluation of the parts of the model actually used. There are considerably fewer variables that need to be considered.

MODEL INPUT PARAMETERS FOR RAPID CREEK

EXAMS in general requires a detailed description of the environment under consideration, providing input for parameters such as pH, oxidant concentration, chlorophyll content and wind speed. Many of these parameters are included to enable calculation of first-order transformation rates from inputted second-order rate constants. Most of the these parameters are not critical here, since first-order biodegradation rate constants were measured directly, and no other transformation processes are considered.

Aside from geometry, advection and residence time considerations, only a few other environmental variables are important for LAS. These are acclimation and temperature as they affect biodegradation rates and the phenomena occurring at the sediment/water interface. Acclimation effects were included in the first-order biodegradation rate constants used since measurements in Rapid Creek water were made. Temperature effects have been included by assuming a Q_{10} of 2 based on previous work [10], and temperature ranged from 7° to 4°C in the stream during the monitoring period that was modeled. In addition to the adsorption coefficient, the magnitude of the dispersion coefficient between the water and sediment is also important in determining how much of a chemical enters the sediment. All mixing processes at the sediment/water interface in EXAMS are summed and expressed within a single dispersion coefficient. This coefficient is in some respects a calibration term which expresses the rate of mixing of sediment and water at the interface, and its magnitude is expected to vary over the range 10^{-4}–10^{-6} m^2/hr [2]. Several values were investigated here, but the best individual fit was obtained at 5×10^{-5} m^2/hr.

No sediment bed transport was assumed to occur in the creek. As support for this assumption, suspended solids analysis of the water at each site shows that entrained solids remain constant. The suspended solids levels were between 10 and 17 mg/l, a low enough concentration that very little LAS transport by suspended sediment advection occurred during the period studied.

MONITORING RESULTS

Table II shows total LAS concentrations measured at the six sites in Rapid Creek during November 1979 and October 1980. Inputs to the creek from the treatment plant were relatively constant in 1978–1981 (0.51 ± 0.12 mg/l), and this is reflected in the similar aqueous concentrations. Sediment levels vary somewhat more. This probably reflects many factors, including difficulty in obtaining representative grab sediment samples, occurrence of nonsteady-state transport phenomena (scouring, etc.) and general lack of understanding of mixing processes at the sediment/water interface. For example, intermittent scouring of the creek during transient high flow has been observed [8]. Also, the extremely high LAS concentration at site 1 in the October 1980 samples correlates with a visible change at that location. A new bank composed of a black

Table II. Concentrations of Total LAS Measured in Rapid Creek Water and Sediments

Site	November 1979	October 1980
Water (μg/l)		
1	a	270
2	190	150
3	120	120
4	80	80
5	b	40
6	a	10
Sediment (mg/kg)		
1	44.6	275
2	3.2	9.1
3	2.1	8.4
4	10.1	2.7
5	b	1.4
6	a	1.5

[a] Analytical interferences prevented quantification.
[b] Samples not collected.

ooze (anaerobic, 2- to 50-μm particles) was being formed, and this coincided with an extended period of low flow. It is reasonable to surmise that either a larger than normal interval between washout by high flow had occurred or there may have been a recent release of sludge solids from the treatment plant. How these intermittent events may affect the modeling process will be discussed in the next section. They will have some effect since the EXAMS model assumes steady-state conditions.

MODEL RESULTS

Predicted steady-state concentrations of LAS in water and sediment based on EXAMS are plotted with measured concentration for October 1980 in Figure 3. Qualitatively, agreement is quite good, even in the last segment (site 6) after nearly five half-lives of LAS removal. It is important to recall that sediment bed transport is assumed not to occur, so that the unusually high concentration in the sediment at site 1 (275 ppm) does not enter into the comparison. Only LAS in the water at site 1 is used in the input calculation. Modeling results are for sites 2–6 with an input equal to the aqueous concentration measured at site 1. Accepting these assumptions, EXAMS does predict concentrations in the stream with

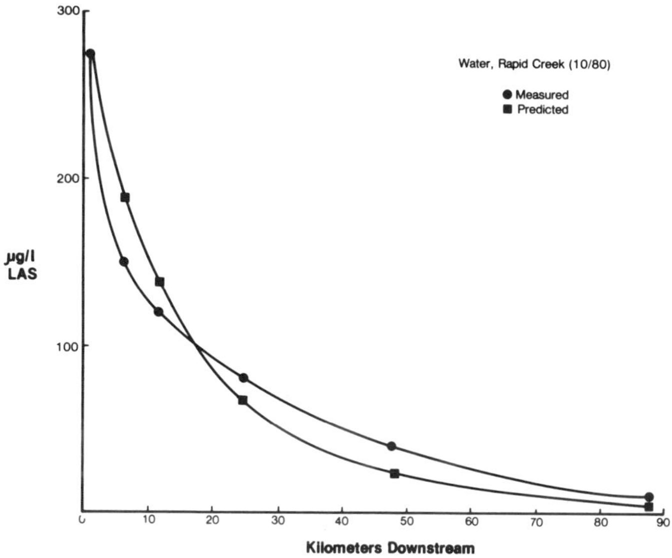

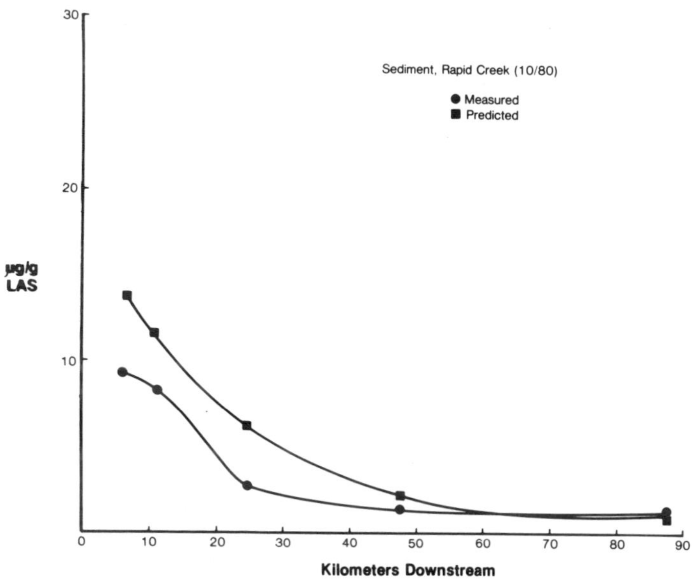

Figure 3. Predicted concentrations of LAS at each site were calculated by EXAMS using input data for C_{12} LAS shown in Table I. Input data for Rapid Creek geometry are shown in Figure 2 and a dispersion coefficient of 5×10^{-5} m^2/hr was assumed. Measured concentrations of LAS are from Table II, October 1980.

apparent accuracy, although the slopes of the predicted and measured concentrations are somewhat different.

Although the agreement between predicted and measured concentrations is excellent, this agreement was only obtained by assigning a relatively arbitrary value to the dispersion coefficient at the sediment/water interface. The value chosen (5×10^{-5} m^2/hr) is in the range expected [2], but little or no rationale for this particular value can be provided. Figure 4 depicts the range of predicted concentrations if this coefficient is allowed to vary from 10^{-4} to 10^{-5} m^2/hr. The effect is relatively large, especially in the water, and suggests that some effort is required to define better the types of situations governing the actual value. In practical terms, higher dispersion coefficients correspond to more churning of the sediment as a result of biological activity or irregularities caused by storms, high flow, etc. [2]. Since this term is fairly important and is seldom measured, it acquires the characteristics of a calibration parameter. In fact, allowing the dispersion coefficient to vary in each segment can lead to almost perfect agreement in predicted and measured aqueous concentrations with no decrease in accuracy for the sediment compartments. Further analysis of the sensitivity to other input parameters of the model results is required to determine if the dispersion coefficient is a controlling factor in successful field validation.

SENSITIVITY ANALYSIS

The sensitivity of the EXAMS LAS model to errors in measurement of creek flowrate, biodegradation rate constants and adsorption coefficient, was investigated by varying each over a range estimated as the maximum potential error. These results are shown in Figures 5 through 8. Biodegradation rate constants were varied by an order of magnitude, flowrate was varied by 10% and the sediment adsorption coefficient was varied ± 100 from a nominal value of 330. Under these conditions, the sediment biodegradation rate constant was the only parameter that caused a change in the predicted concentration approaching that caused by the variation of the dispersion coefficient. This is not totally surprising, since sediment biodegradation and the dispersion coefficient are both involved in determining the sediment concentrations of LAS.

Unfortunately, the model calculations are most sensitive to the least-understood parameters that are provided as input: the dispersion coefficient and sediment biodegradation rate constant. There is no clear evidence to define the extent of biodegradation of organic compounds that can occur while they are adsorbed to naturally occurring solids. In defining transformation processes for LAS, it was assumed that biodeg-

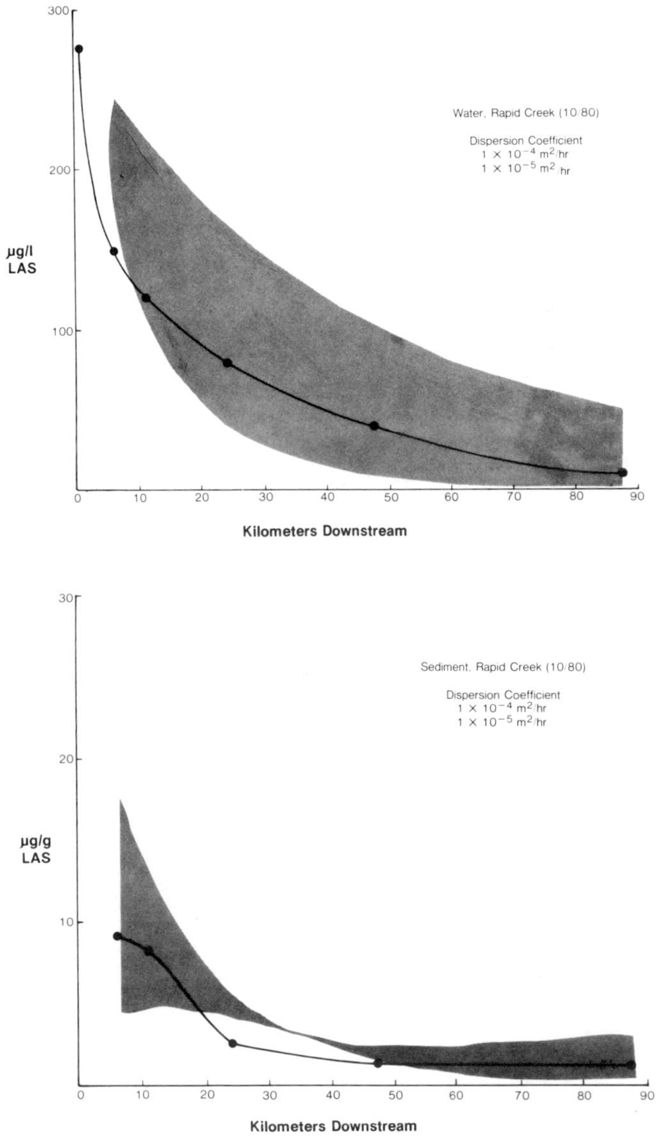

Figure 4. All parameters remained constant, except the dispersion coefficient, which was varied as indicated to obtain the range of predicted concentrations shown by the shading. The crossover shown in the predicted range for sediment concentrations represents a point where the two extreme dispersion coefficient values predict the same concentration. The line represents measured concentrations.

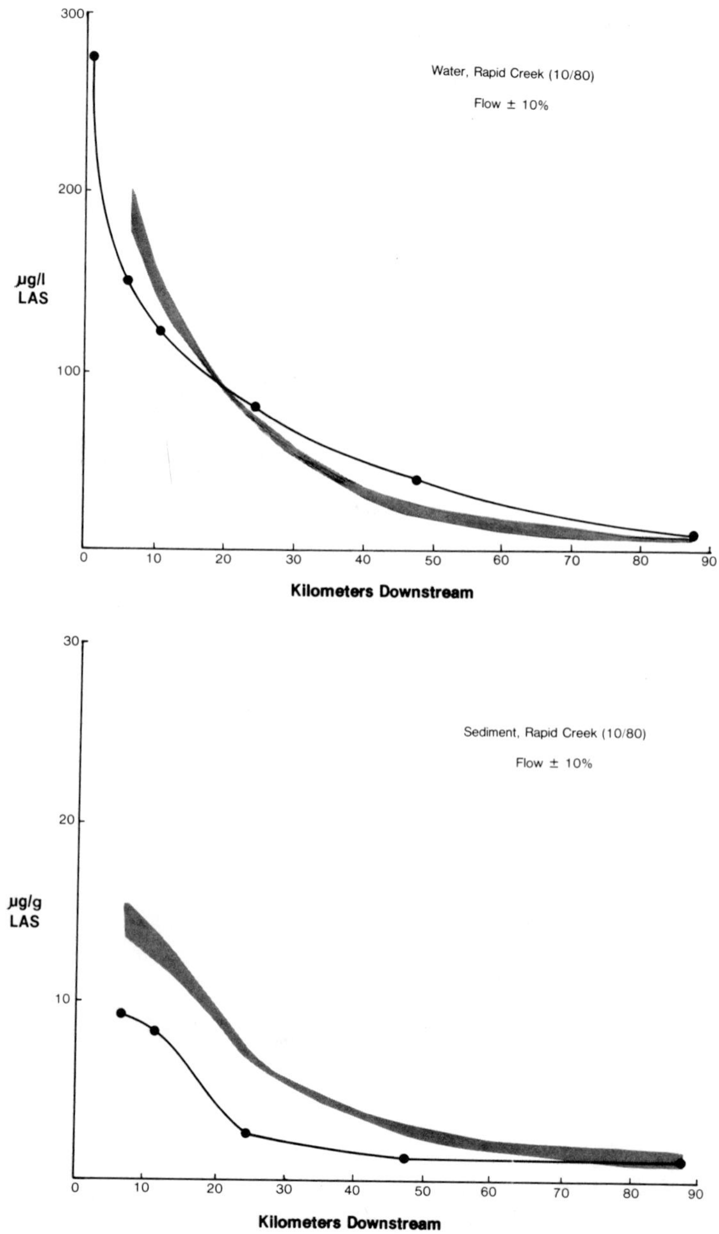

Figure 5. All inputs are as in Figure 3, except the flowrate, which was varied as shown. The shading represents the range of predictions obtained and the line represents measured concentrations of LAS.

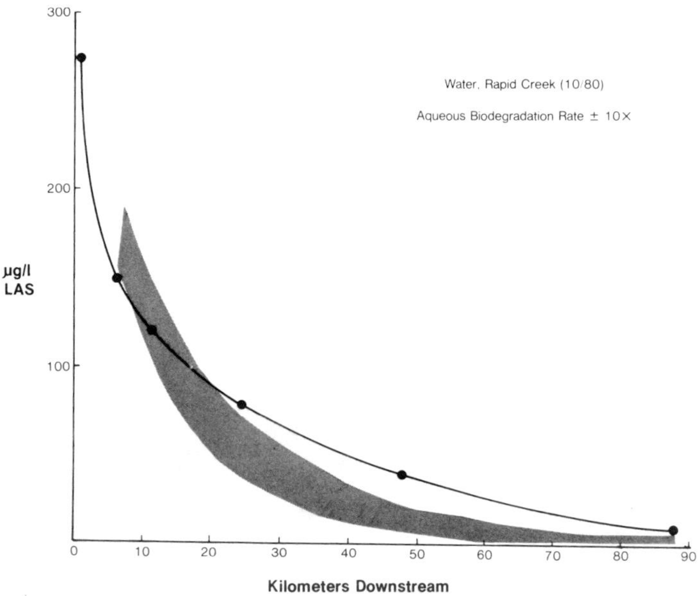

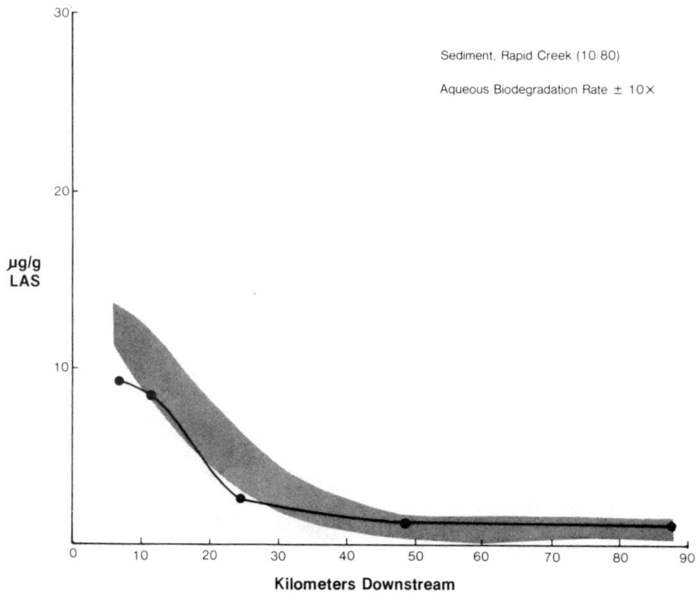

Figure 6. All inputs are as in Figure 3, except the aqueous biodegradation rate, which was varied as shown. The shading represents the range of predictions obtained and the line represents measured concentrations of LAS.

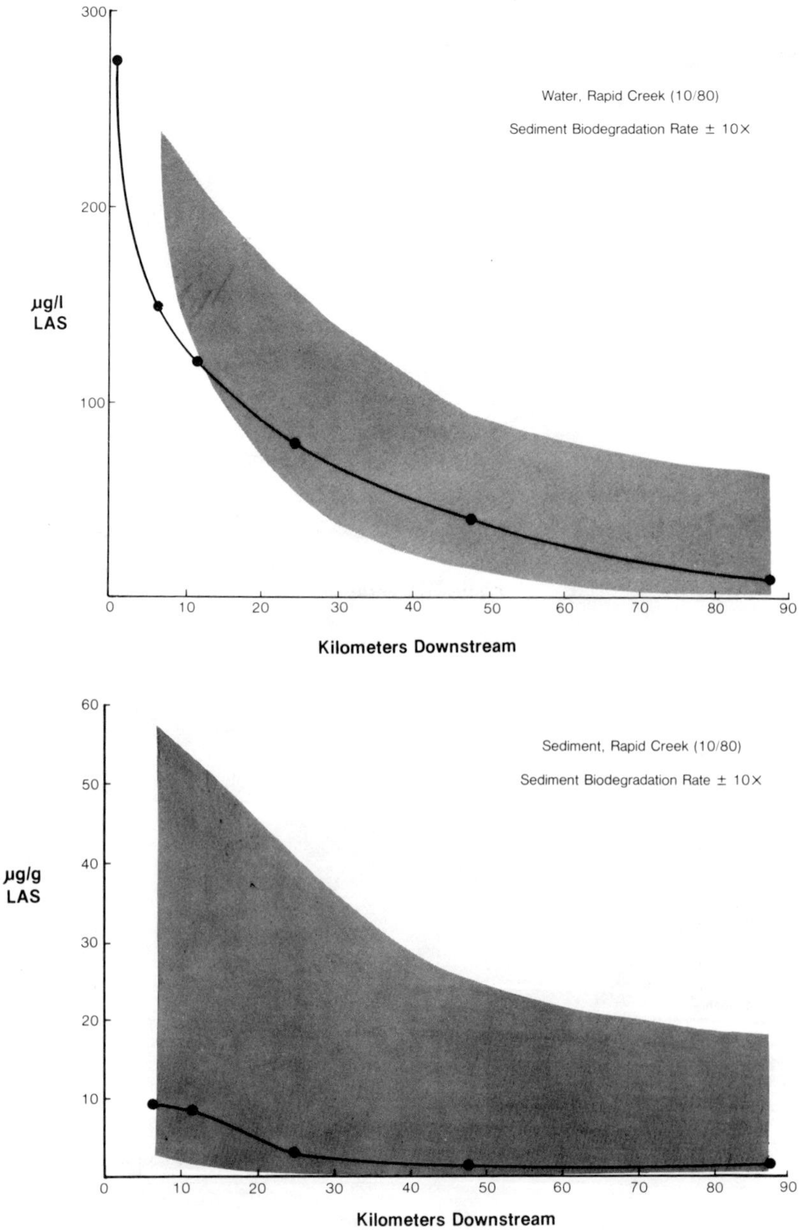

Figure 7. All inputs are as in Figure 3, except the sediment biodegradation rate, which was varied as shown. The shading represents the range of predictions obtained and the line represents measured concentrations of LAS.

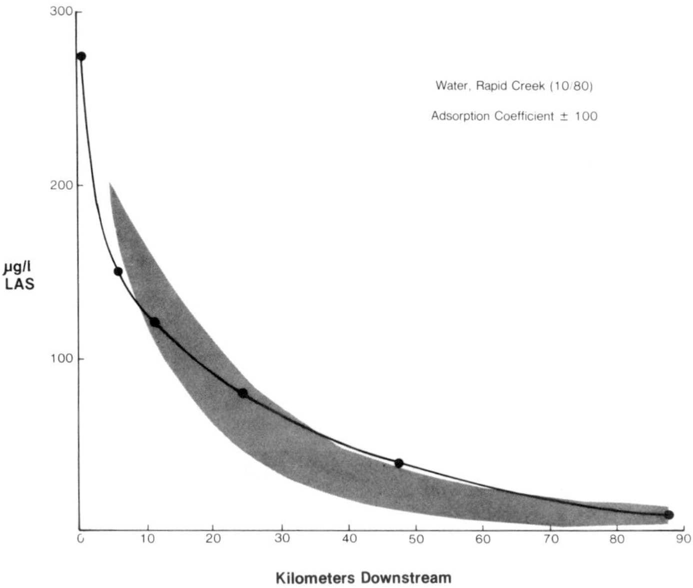

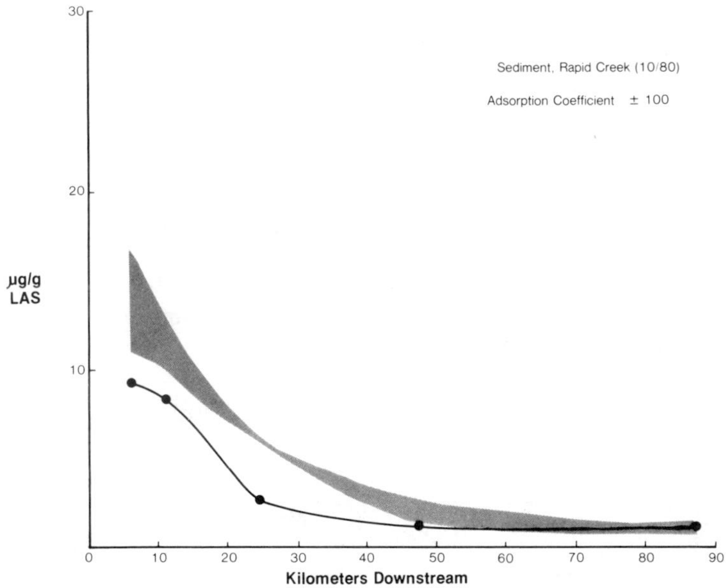

Figure 8. All inputs are as in Figure 3, except the adsorption coefficient, which was varied as shown. The shading represents the range of predictions obtained and the line represents measured concentrations of LAS.

radation occurred in the adsorbed state and at a rate faster than in solution. Some support for this comes from previous work [9], but that report did not define whether the enhanced biodegradation observed when sediments were added was due to an increased bacterial population in the water or because of adsorption. Dispersion coefficients in natural systems are equally difficult to assess, since their measurement is extremely difficult and time-consuming and they are subject to fluctuation due to transient phenomena. Figures 9 and 10 show that even additive errors in measurement of adsorption and aqueous biodegradation have substantially less effect on the fit between measured and predicted concentrations than a single error in measurement of the sediment biodegradation (Figure 7). Although it is unlikely that the aqueous biodegradation rate constant is in error by an order of magnitude, it is not difficult to reason that the sediment biodegradation constant may be inaccurate to that extent. Figure 7 suggests that, because of this uncertainty, wide error limits would need to be placed on predicted LAS concentrations in the sediment and water of Rapid Creek when using EXAMS, and probably any evaluative model, in a safety assessment. Further work to define biodegradation processes in sediment is needed to improve the predictive ability of the models for materials similar to those of LAS.

The stringency of a field validation test for an environmental fate model increases as the residence time of the chemical in the environment increases. At short times relative to transformation rates, dilution factors will predominate and control observed concentrations. As the number of transformation half-lives increases, errors in either the rates themselves or in the assumptions made to calculate the rates will be magnified. However, the concentration of the chemical continuously decreases during this process, so that potential errors in the analytical method used to measure observed concentrations also increase. It is important to recognize that the requirements of model validation must be balanced against the accuracy of the analytical measurements to be made and the errors involved at lower concentrations. For the EXAMS LAS model, site 6 represented the longest residence time (almost 5 days) considered and about the lowest routinely measureable concentration in water (10 ppb) and sediment (1.5 ppm). In this study, sites with intermediate residence times show a percent difference between measured and predicted concentrations that is similar to site 6 (Figure 3). This suggests that analytical limitations have not yet become important.

It is apparent that, given the proper assumptions, EXAMS can successfully predict the concentrations of LAS resulting from its steady-state input to a flowing stream. Whether this constitutes validation of the parts of the model used in this study is less clear. The input parameters

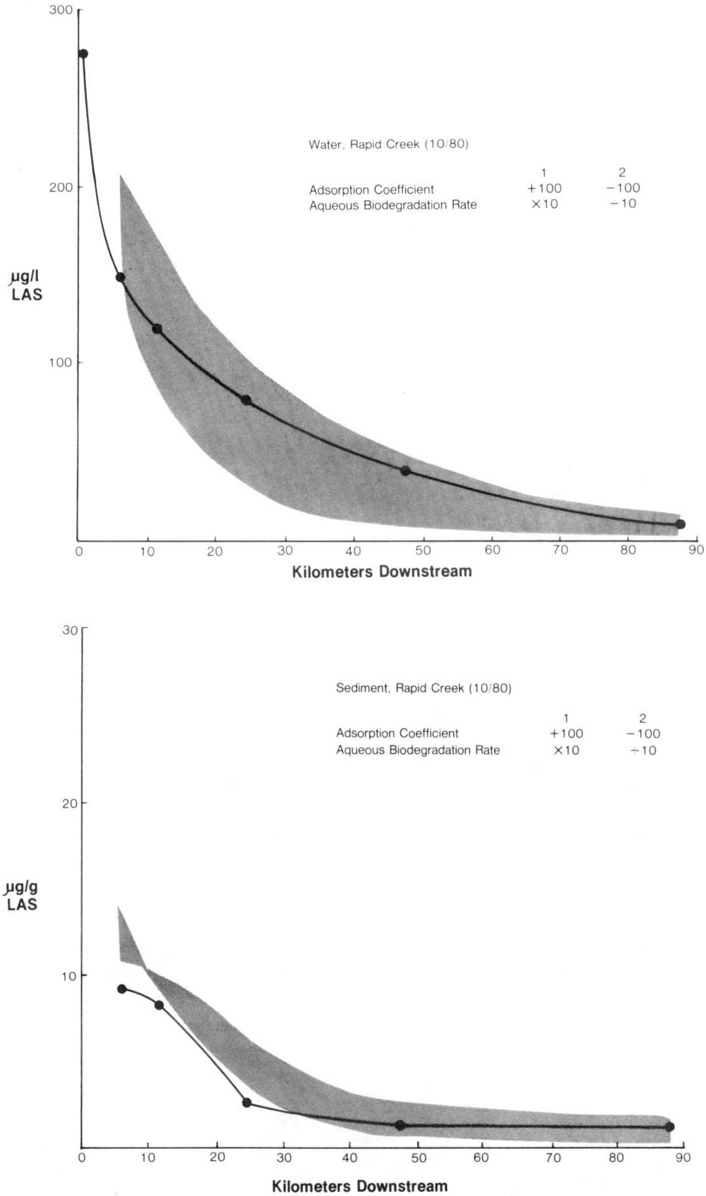

Figure 9. All inputs are as in Figure 3, except aqueous biodegradation and adsorption coefficient, which were varied as shown. The shading represents the range of predictions obtained and the line represents measured concentrations of LAS.

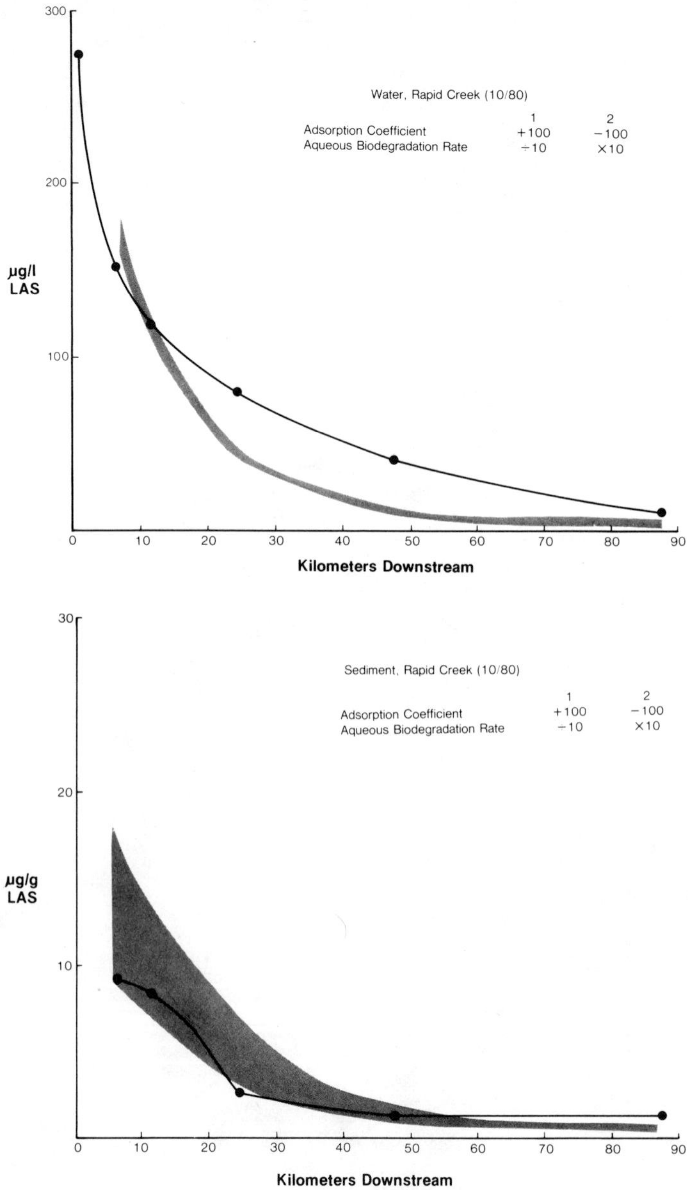

Figure 10. All inputs are as in Figure 3, except the aqueous biodegradation and adsorption coefficient, which were varied as shown. The shading represents the range of predictions obtained and the line represents measured concentrations of LAS.

that are least understood have the most impact on the predicted concentrations, and the current problems are centered around the least definable processes: sediment/water interface interactions and the bioavailability of adsorbed species. It is perhaps surprising that these processes can be so controlling for a highly soluble material like LAS with a relatively low adsorption coefficient. These processes are likely to be as important, or perhaps even more important, for many synthetic organics with moderate adsorption properties. To the extent that EXAMS has successfully defined the important processes for determining the fate of LAS, its utility as an evaluative tool is confirmed. Whether it can be successfully field-validated in all aspects awaits a further definition of the accuracy of the less well defined inputs.

SUMMARY

Field validation of evaluative models for predicting exposure of synthetic organic chemicals in the aqueous environment is necessary before their use in decision-making processes. One of these models, EXAMS, has been applied successfully to the prediction of concentrations of an anionic surfactant, LAS, in a flowing stream. However, a careful analysis of the results reveals that the success of the model is closely tied to a proper choice of the two least-understood parameters in the model. For LAS, and probably for many synthetic organic chemicals, improvements in modeling results will require a better understanding of sediment biodegradation and a better definition of sediment/water interface dynamics. These improvements should lead to less uncertainty in predicted concentrations and better estimates of environmental concentration for use in safety assessment schemes.

REFERENCES

1. Mackay, D. "Finding Fugacity Feasible," *Environ. Sci. Technol.* 13:1218–1223 (1979).
2. Burns, L. A., D. M. Cline and R. R. Lassiter. "Exposure Analysis and Modeling System (EXAMS): User Manual and System Documentation," U.S. EPA Environmental Research Laboratory, Athens, GA (1981).
3. O'Connor, D. J., and J. P. St. John. "An Assessment of Modeling the Fate of Chemicals in the Aquatic Environment," Chapter 1, this volume.
4. *Chemical Economics Handbook* (Menlo Park, CA: SRI International).
5. Osburn, Q. W., Procter and Gamble Co. Personal communication.
6. Sullivan, W. T., and R. D. Swisher. "MBAS and LAS Surfactants in the Illinois River," *Environ. Sci. Technol.* 3:481–483 (1969).

7. "Water Resources Data for South Dakota, Water Year 1980," USGS Water-Data Report SD–78–1, National Technical Information Service, Springfield, VA (1981).
8. Harms, L. L., P. H. Vidal and J. E. McDermott. "Release and Sorption of Phosphorus by Sediments in a Moving Water Course," Report PB272–715 National Technical Information Service, Springfield, VA (1977).
9. Larson, R. J., and A. G. Payne. "Fate of the Benzene Ring of Linear Alkylbenzene Sulfonate in Natural Waters," *Appl. Environ. Microbiol.* 41:621–627 (1981).
10. Larson, R. J. "Role of Biodegradation in Predicting Environmental Fate," in *Biotransformation and Fate of Chemicals in the Aquatic Environment,* A. W. Maki, K. L. Dickson and J. Cairns, Jr., Eds. (Washington, DC: American Society for Microbiology, 1980), pp. 67–86.

CHAPTER 19

ASSESSMENT OF TRANSPORT AND FATE OF TOXIC MATERIALS IN AN EXPERIMENTAL STREAM ECOSYSTEM

William E. Cooper and R. Jean Stout

Michigan State University
East Lansing, Michigan 48824

Federal and state regulations require scientists to perform an anticipatory form of risk assessment. The National Environmental Protection Act (NEPA) and Toxic Substances Control Act (TSCA) are prime examples of legislative mandates to implement ecosystem predictions of the transport, fate and effects of toxic substances. The limited knowledge of systemic processes and the large commitments of resources necessary to parameterize aggregate population models have forced consideration of simple surrogates. Unitized microcosms and simplistic analytical models have been given the most attention. Microcosms assume that reductions in diversity, scale and environmental stochasticity will not generate significant bias. The simplistic models usually are based on first-order processes, deterministic and devoid of ecological feedbacks, and do not consider spatial heterogeneity. These abstractions are necessary to obtain biological and mathematical tractability, given time and resource constraints.

Much money and time has been spent in developing a number of these predictive tools. Before we develop a new generation of abstractions, these physical and mathematical models should be tested for validity, precision and utility. It is time to test these models against controlled field experiments with large-scale systems. It was with this intent that we initiated a 3-year research program at the Monticello Experimental

Stream Laboratory operated by the U.S. Environmental Protection Agency (EPA) in Minnesota. Using *p*-cresol as a test toxicant, we tested (1) laboratory-to-field transfer of standard acute toxicity tests for the minimum data set recommended by EPA; (2) use of microcosms to represent the stream ecosystem; and (3) the utility of the Exposure Analysis Modeling System (EXAMS) in predicting transport and fate.

We utilized 24-, 48- and 96-hr exposure times to replicate temporal conditions commonly found in laboratory acute toxicity tests. The pulsed 144-hr exposure was performed to uncouple the impacts of secondary ecological responses from the primary effects of *p*-cresol.

FACILITY

In fall 1978 Michigan State University was awarded a contract by EPA to design and implement an experiment involving five of the eight 500-m artificial stream channels at the Monticello Ecological Research Station, Monticello, MN, to study the transport, fate and effects of toxic stress on large-scale experimental stream ecosystems.

That interdisciplinary-designed project uses two independent, isolated, recycling stream ecosystems containing a water volume of 1.5–1.8 million liters each, which are physically, chemically and biologically typical of warm-water, grassland, third- to fifth-order streams. The 1000- × 3-m channels are driven by recirculation pumps that propel the stream water up to 11,000 liter/m, which can generate mean velocities up to 12.2 cm/sec (riffle velocities range up to 25 cm/sec). The earthen channels have alternating riffles and 30-m-long pools.

Control and monitoring of the channel flows and water chemistry (pH, dissolved oxygen (DO) and temperature), local meteorological conditions and toxicant injection are executed by a dedicated onsite computer system.

An organic chemistry laboratory was established onsite and geared to our prototype toxicant, *p*-cresol. Thus, we have established at the Monticello site a set of large experimental stream ecosystems and facilities for analyzing toxicant concentrations in all ecosystem components.

EXPERIMENTAL OPERATIONS

24-hr Exposure

p-Cresol was added at 8 ppm from 1:30 PM on July 19 to 1:00 PM on July 20 (Figure 1). The retention time for the channels was approxi-

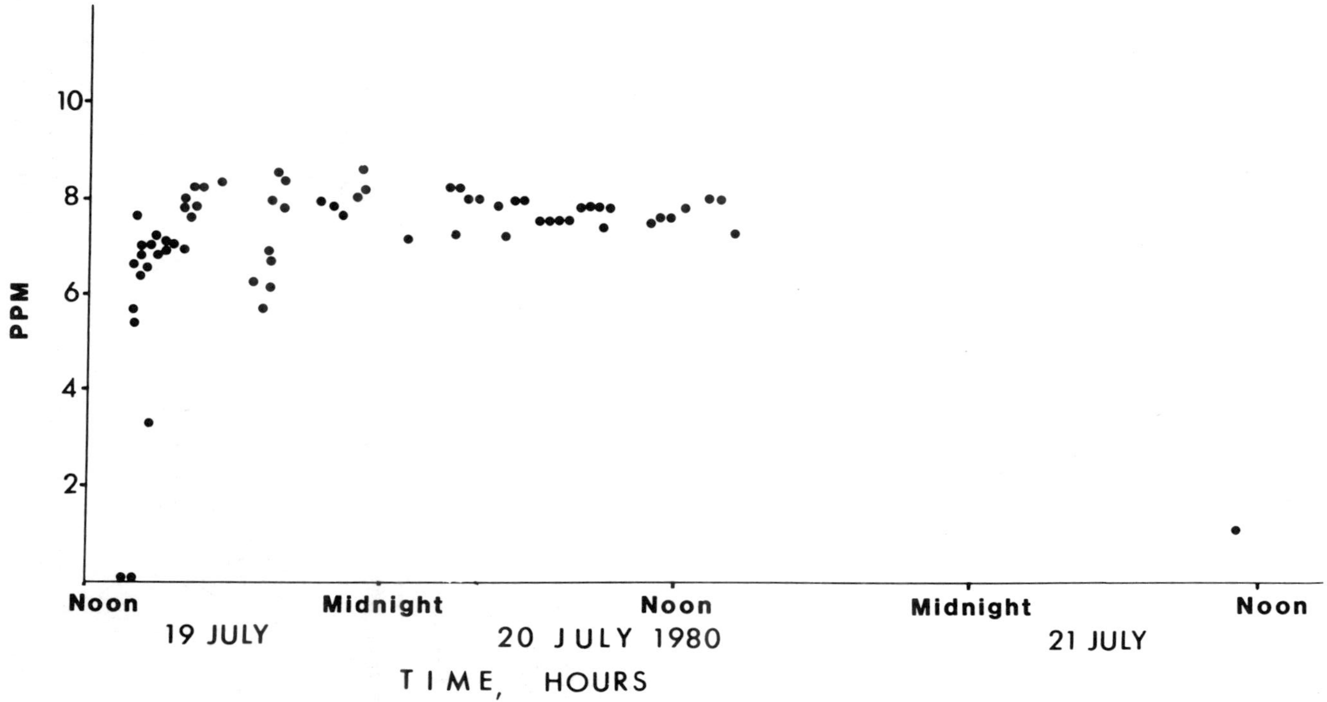

Figure 1. Ambient concentrations of *p*-cresol at head end of experimental channel during the 24-hr dose.

mately 6 hr at a flowrate of 400 liter/min. By 10:00 PM, an equilibrium profile was established with a slight oscillation of concentration rippling through the system.

The flow-through phenol analyzer and the gas chromatography (GC) analyses produced similar estimates. The estimated half-life was approximately 64 hr for the first 24 hr of the dosing.

There was no serious reduction in DO and no indication of mortality in invertebrates or fish. Data from gut analyses of the four species of resident fish indicated an inhibition of the feeding process. The bioaccumulations of *p*-cresol in fish are given in Figure 2. The average total body load was 18 ppm, which represents a bioaccumulation factor of 2.3. The average ambient water levels varied from 7.5 to 8.0 ppm until approximately 11:00 PM on July 20. All four species showed the same accumulation and release pattern. By 11:00 AM on July 21, the ambient concentration was reduced to 1 ppm and the fish contained 2–5 ppm; 24 hr later, only trace amounts were measurable in the fish.

Bacterial samples demonstrated a change in microbial flora associated with the *p*-cresol application. The number, size and color of bacterial plaques shifted dramatically during the 24-hr period of injection. Initial attempts to demonstrate significant increases in *p*-cresol degradation by these new bacteria were not definitive.

Figure 3 presents the bacterial density estimates. The four sampling stations in channel B were distributed uniformly along the stream. Bacterial counts increased consistently in channel B, but no increase was observed in the control stream. The increased density was maintained for at least 5 days after dosing ceased.

48-hr Exposure

The 48-hr dose occurred between 10:00 AM on August 9, and 10:00 AM on August 11. Ambient concentrations were maintained around 8 ppm (Figure 4). Again, the *p*-cresol appeared to be very stable during the 6-hr retention time in the recycling mode.

Figure 5 presents the ambient concentration at the tail end of the stream just before the water entered the sump to be recycled. The 6.25-hr retention time is illustrated by the fact that dosing started at 10:00 AM and the first wave of *p*-cresol reached the last sampling station at approximately 3:30 PM.

By comparing Figures 4 and 5, an estimated average of 1.7 ppm phenol was lost during the 6.25-hr retention time. This would produce an estimated 14.7-hr half-life, assuming an exponential drop in concentration

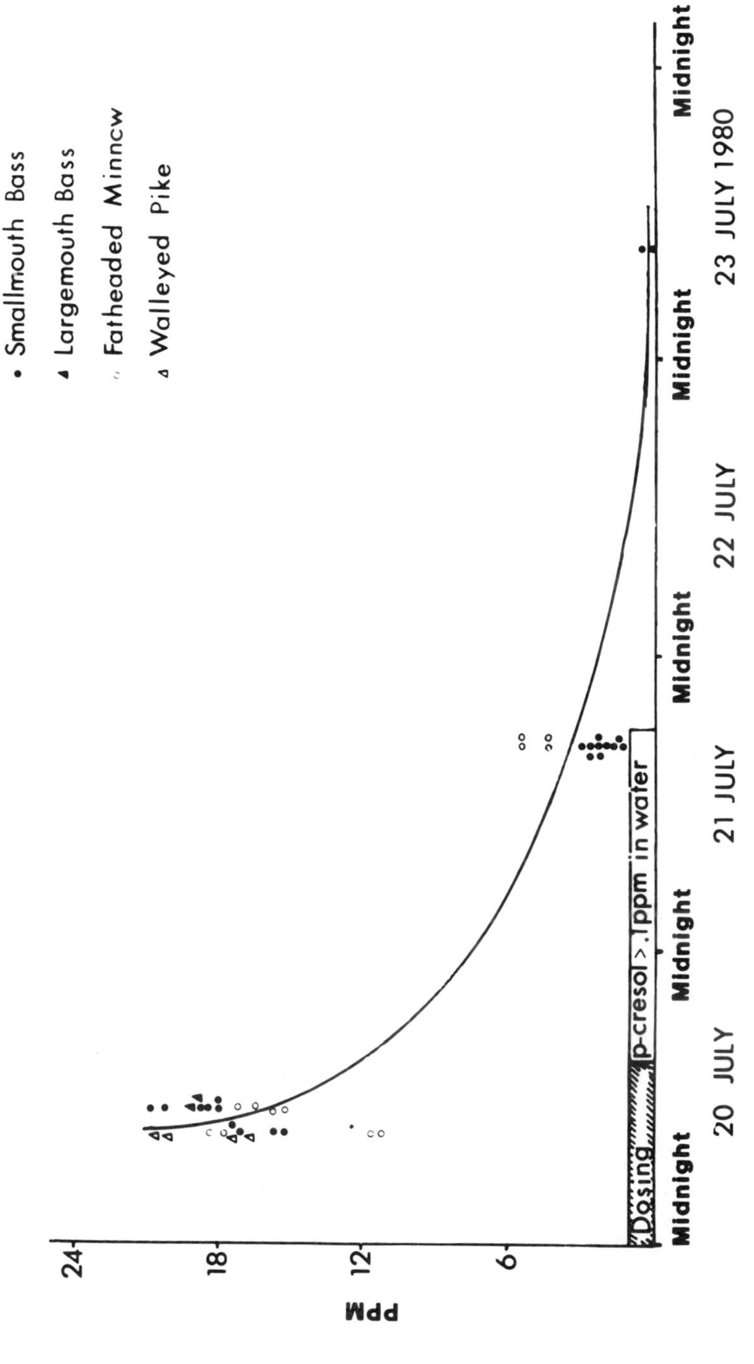

Figure 2. Bioaccumulation of *p*-cresol in fish as a function of a 24-hr dose of *p*-cresol at 8 ppm.

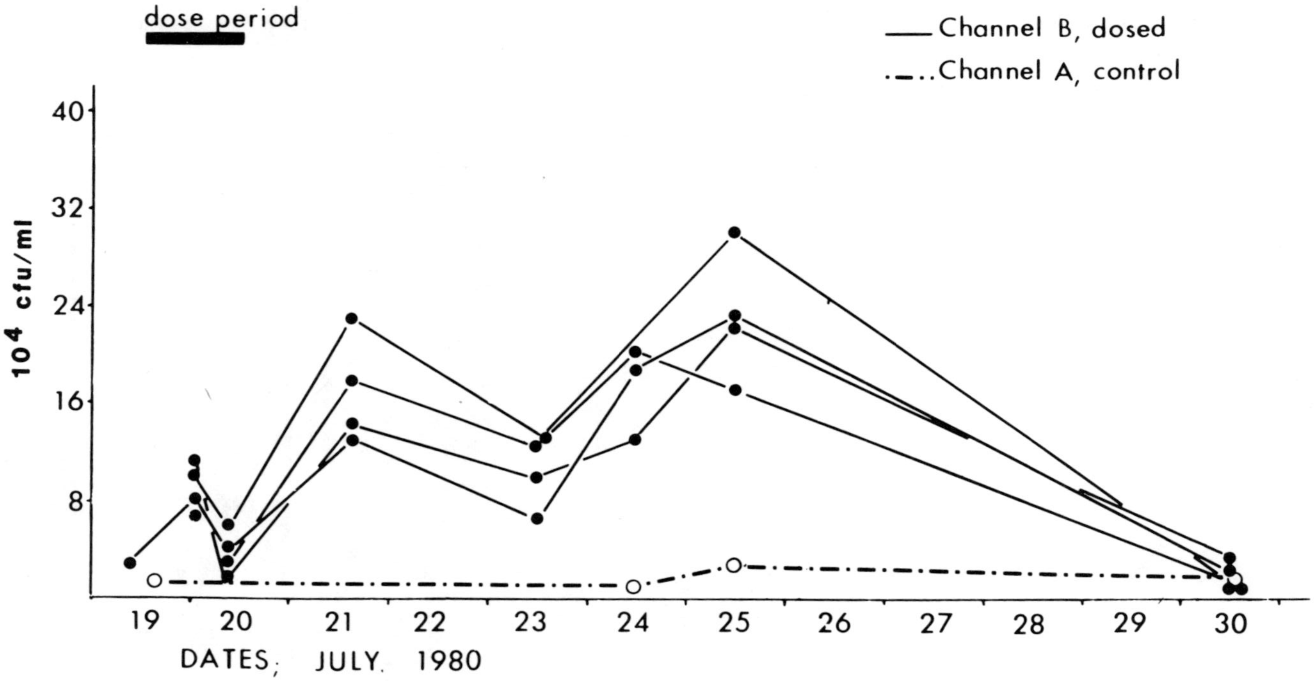

Figure 3. Comparison of baterial density counts (10^4 cfu/ml) in control and experimental channels during a 24-hr dose of *p*-cresol at 8 ppm.

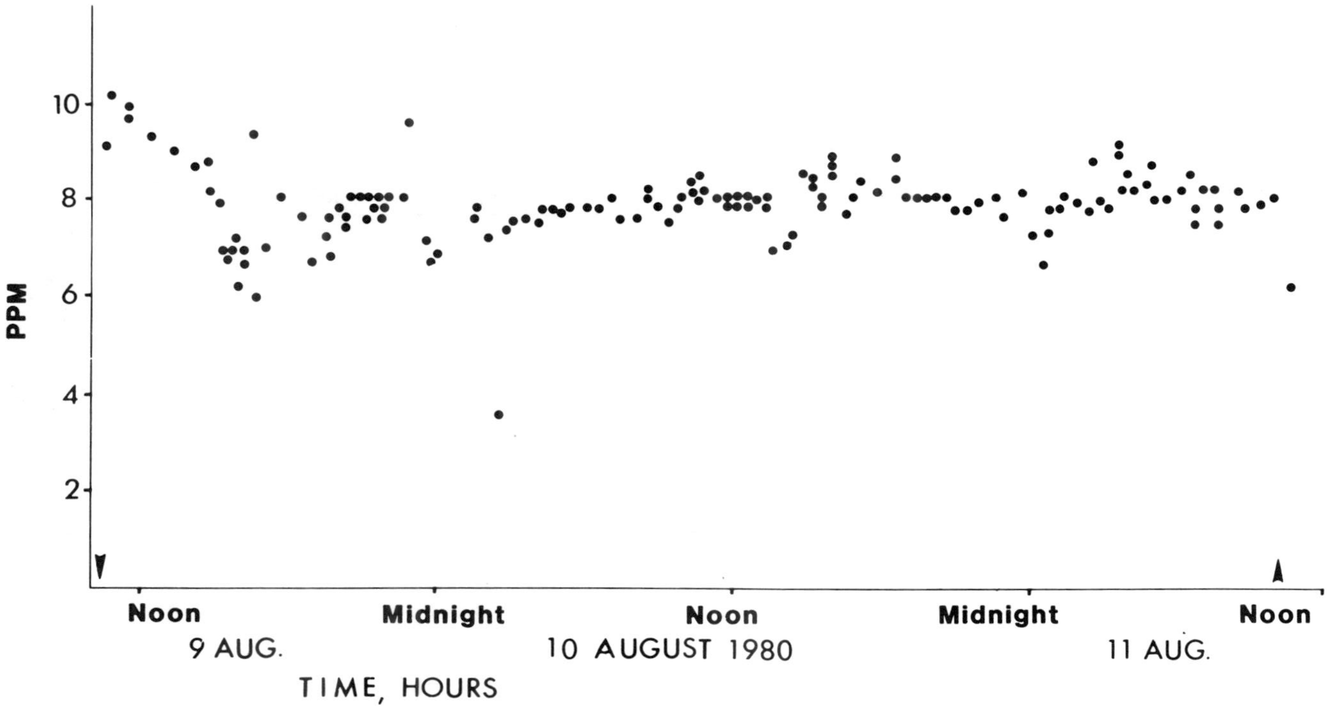

Figure 4. Ambient concentrations of *p*-cresol at head end of experimental channel during 48-hr dose of *p*-cresol at 8 ppm.

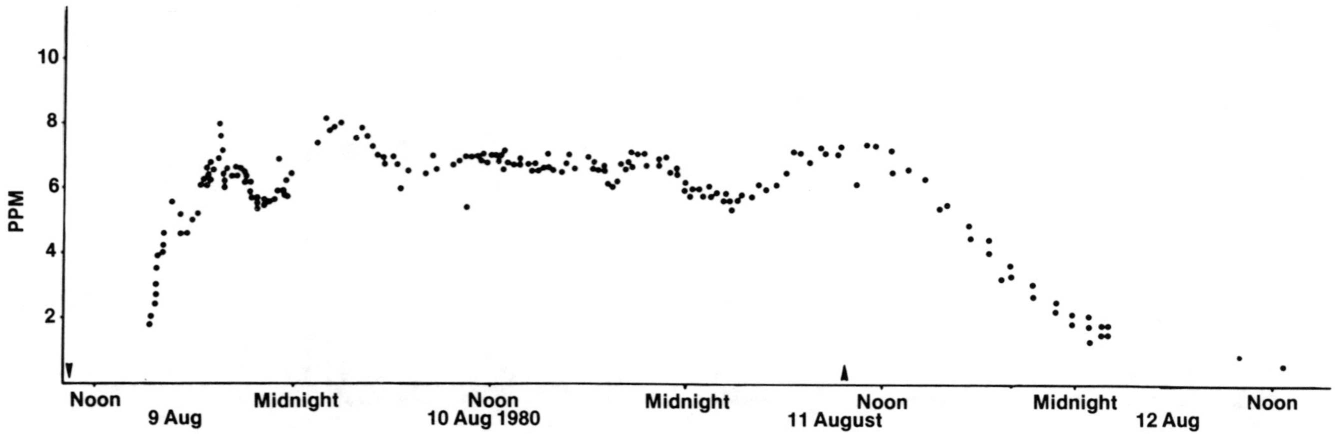

Figure 5. Ambient concentrations of *p*-cresol at tail end of experimental channel during 48-hr dose of *p*-cresol at 8 ppm.

over the stream. The half-life during the decrease from 6 to 2 ppm was approximately 5.3 hr. Obviously, the degradation rate increased significantly after the dosing terminated.

Examination of the differences in total phenol measurements between tail end of the stream and head end of the stream indicates that between hour 14 and hour 30 the differences steadily increased, giving a shift in interpolated half-life from 69 to 23 hr. A later decrease in concentration at the tail end of the stream implies a further decrease in half-life from 23 to 12 hr. This is much closer to the half-life immediately after dosing (5.3 hr). Thus, adaptation by microbial populations to *p*-cresol, and, therefore, generation of faster biodegradation rates, occurred in the stream.

During the dosing experiments, the temperature showed normal diurnal variation from 22 to 27°C, and the pH varied from 7.0 to 8.5. However, DO showed a marked change during and after dosing and was obviously related to the effect *p*-cresol had on the system, as shown in Figure 6. The DO profile of the control stream showed a normal diurnal cyclicity between 4 and 12 ppm. It is hypothesized that the mechanism of effect is probably the increased respiration of organisms (plants, animals and microbes) and not the inhibition of photosynthesis. The DO curves for the control and dosed streams are approximately parallel, with the major difference being the minimal levels reached at 5:00–6:00 AM; i.e., the rate of photosynthesis was similar and the pattern of the diurnal cycle was not disturbed. Chemical oxidation of *p*-cresol is possible, yet the DO increased during daylight at the same rate as the control stream, while the ambient levels of phenols were 7–8 ppm. DO was reduced to low levels (1–1.5 ppm) during the night, even though the phenol levels dropped to less than 2 ppm. The nighttime DO sags are due to the high concentration of *p*-cresol the previous day.

During the period of low DO on both August 11 and 12, certain fish and invertebrates experienced massive mortality. Walleyed pike died in appreciable numbers. Fish removed before death and placed in clean water recovered. Smallmouth bass were gasping for air at the surface, but none died. Largemouth bass were not visibly stressed, but their feeding was inhibited. Fathead minnows were not visibly stressed and, in fact, deposited eggs on the surface of fish traps.

Large numbers of *Stenonema* mayflies, amphipods and some isopods were observed dead on the surface of the algal mats. Field observations indicated a severe impact on macroinvertebrates.

The results of bacterial sampling are presented in Figure 7. Bacterial densities showed a sharp increase in experimental channel B, but there was no similar increase in control channel A. The timing of the increase in relation to the dosing interval was similar to that observed during the

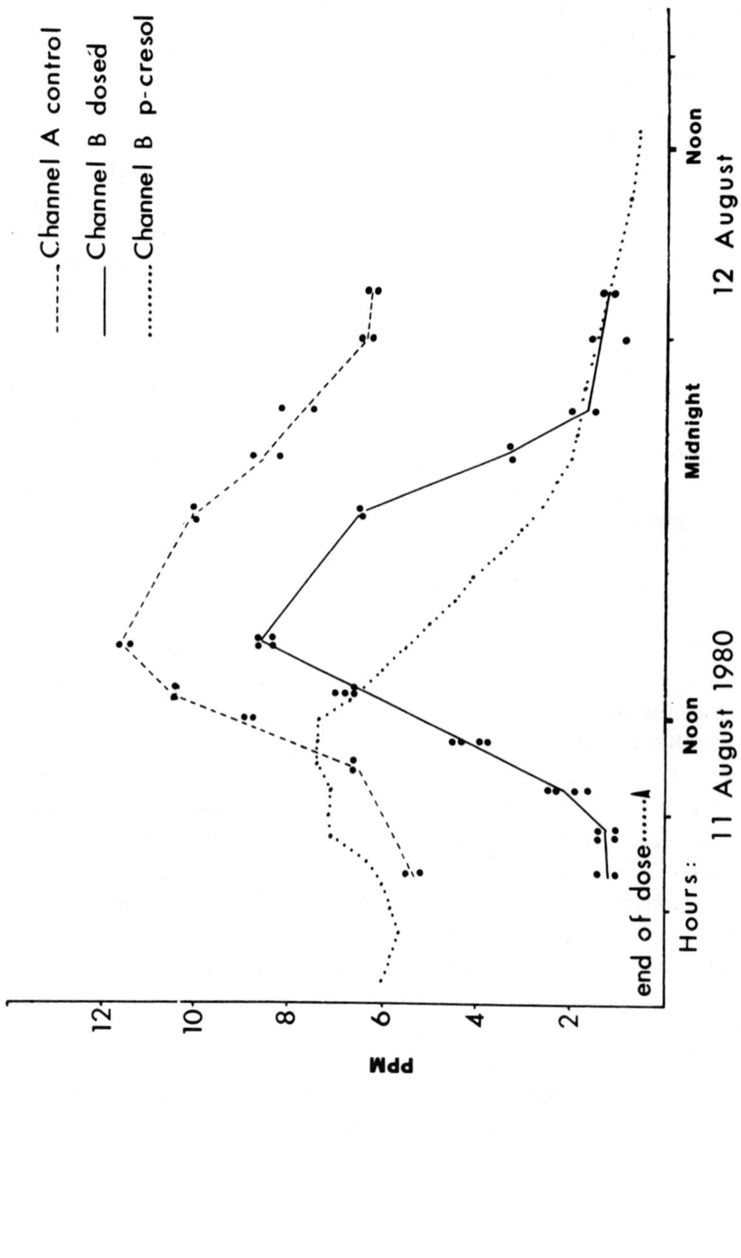

Figure 6. DO concentrations (Winkler method, in ppm) in experimental and control channels during a 48-hr dose in channel B.

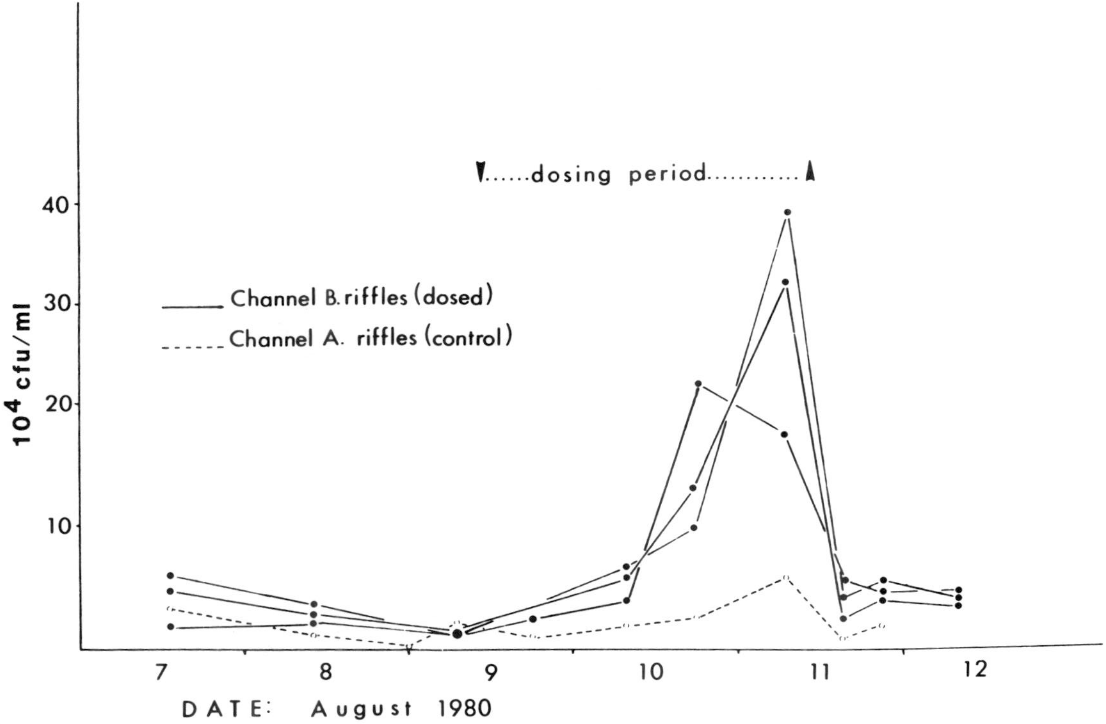

Figure 7. Comparison of bacterial density counts (10^4 cfu/ml) in control and experimental channels during a 48-hr dose of *p*-cresol at 8 ppm.

24-hr application (Figure 3). The maximum densities obtained are not very different, but the duration of the bacterial pulse was significantly different. Bacterial densities declined very rapidly after 48 hr of exposure. The *p*-cresol concentration declined much more slowly than the bacterial densities.

BIOACCUMULATION

Table I gives the results of the bioaccumulation of *p*-cresol by five species of fish that were sampled during and after the 48-hr dose. At 24 hours into the 48-hr dose, walleyed pike and largemouth bass showed levels of *p*-cresol similar to those found near the end of the 24-hr dose. *p*-Cresol concentrations for those species and for bluegills and fatheads remained high throughout the dosing. At 26 hours after dosing stopped, *p*-cresol levels in the fish dropped. The fish accumulated the toxicant rapidly and then quickly eliminated it.

Bluegills collected 20 hr after cessation of the 48-hr dose were dissected for organ analyses. The eye (without lenses) had the lowest concentration (2.8 ppm) and muscle tissue the next lowest (4.7 ppm). Gill tissue contained 7.3 ppm. The highest bioaccumulation values were in the livers (76.1 ppm) and intestines (96.7 ppm). High accumulations in the liver suggest sequestering of *p*-cresol from blood. The high values in the intestines could have resulted from the fish eating soon after dosing ceased. Fish gut analyses showed that bluegills stopped feeding during dosing, but began feeding within hours after dosing stopped.

Various biotic components of the channel were anlayzed for *p*-cresol concentration. Table II shows that only one component of those tested, a crayfish, showed significant amounts of *p*-cresol. The crayfish, *Orconectes virilis,* bioaccumulated nearly three times more p-cresol than occurred in background concentrations in the channel. This crustacean is the largest and one of the most common invertebrates in the channel.

Selected fish and invertebrates from the sites were tested at the EPA laboratory at Duluth for laboratory response to *p*-cresol. Those data appear in Table III. The fish species show a distinct "all or none" pattern: fatheads, for the most part, died after 48 hr in 20 ppm *p*-cresol, but none died at concentrations of 10 ppm. Largemouth bass all died after 24 hr at 32 ppm, and none died at 16 ppm; smallmouth bass died after 24 hr at concentrations of 20 ppm, but all survived at 10 ppm. The damselfly, *Ischnura verticalis,* showed no response, even at concentrations of 40 ppm. The amphipod *Hyalella azeteca* showed a more gradual dose pattern. For short periods of time (24 hr) most of the animals survived,

Table I. Bioaccumulation[a] of *p*-Cresol in Fish During and After 48 hr of Dosing Channel B at 8 ppm *p*-Cresol

		Amount (ppm) of *p*-Cresol in Five Species of Fish				
Date	Time	Walleyed Pike	Largemouth Bass	Bluegill Sunfish	Fathead Minnow	Smallmouth Bass
8/10	10:00 AM		18.7	5.9		
			12.0	13.7		
	1:00 PM	17.5	18.6		12.4	
		18.9	16.8		17.8	
		14.4	16.7			
			22.7			
			13.4			
8/11	5:30 AM			14.9		
	7:45 AM	15.4				
		14.8				
	10:00 AM		15.9	24.5		
				29.0		
	11:30 PM			14.0		
8/12	10:00 AM		20.3	12.0		
	12:00 noon		4.8		0.9	
8/15	8:00 PM					0.5

[a]Gas chromatograph data.

Table II. Concentration[a] of *p*-Cresol in Sediments, Plants and Invertebrates After 48 hr of Dosing Channel B at 8 ppm *p*-Cresol

		Amount (ppm) of *p*-Cresol in Components			
Date	Time	Sediment	Algae *Spirogyra*	Dragonfly *Anax*	Crayfish *Orconectes*
8/10	10:00 AM		0.8		
			0.0		
	11:00 AM	0.9			
8/11	9:45 AM			0.8	23.9
	10:00 AM	3.8	1.1		
		1.8	4.0[a]		
8/12	10:00 AM	0.7			
		0.0			
8/14	10:00 AM		0.0		

[a]Gas chromatograph data.

even at concentrations of 80 ppm, but only 50% survived after 72 hr at concentrations of 10 ppm. Death also occurred at concentrations of 5 ppm after 48 hr. The waterflea, *Daphnia magna,* was very sensitive to the toxicant at 2.5 ppm, even after 24 hr.

Table III. Dose-Response Data on *p*-Cresol with Selected Fish and Invertebrates from Monticello, Minnesota[a]

Organism (number/run)	Initial Concentration of *p*-Cresol (mg/1)	Time: 24-hour Survival (%)	48-hour Survival (%)	72-hour Survival (%)	96-hour Survival (%)
Fathead Minnow (10)[b]	40	60,30	20,0	0,0	0,0
	20	60,90	30,40	0,10	0,0
	10	100,100	100,100	100,100	100,100
	5	100,100	100,100	100,100	100,100
	2.5	100,100	100,100	100,100	100,100
	Control	100,100	100,100	100,100	100,100
Largemouth Bass (5)	32	0			
	16	100	100	100	100
	8	100	100	100	100
	4	100	100	100	100
	2	100	100	100	100
	Control	100	100	100	100
Smallmouth Bass (5)	40	0			
	20	0			
	10	100	100	100	100
	5	100	100	100	100
	2.5	100	100	100	100
	Control	100			
Largemouth Bass, Test 2 (5)	20	0			
	10	100	100	100	100
	5	100	100	100	100
	2.5	100	100	100	100
	1.25	100	100	100	100
Damselfly (10)	40		100		
	20		100		
	10		100		
	5		100		
	2.5		80		
	Control		70		
Amphipod, *Hyalella azteca* (8)[b]	80	50,37.5	0,0	0,0	0,0
	40	87.5,50	25,25	12.5,12.5	12.5,12.5
	20	87.5,100	37.5,100	37.5,37.5	12.5,25
	10	100,100	75,87.5	50,50	27.5,50
	5	100,100	87.5,97.5	87.5,50	75,37.5
	Control	100,100	100,100	100,100	100,100

Table III. Dose-Response Data on *p*-Cresol with Selected Fish and Invertebrates from Monticello, Minnesota[a]

Organism (number/run)	Initial Concentration of *p*-Cresol (mg/1)	Time 24-hour Survival (%)	48-hour Survival (%)	72-hour Survival (%)	96-hour Survival (%)
Waterflea, *Daphnia magna* (5)[b]	40	0,0	0,0		
	20	0,0	0,0		
	10	0,0	0,0		
	5	60,80	60,40		
	2.5	80,80	60,20		
	Control	100,100	100,100		

[a] Flow in chambers ranged 21–60 ml/min; initial DO values ranged 5.5–7.9 ppm; final DO values ranged 5.5–8.4 ppm; pH ranged 7.2–7.5; alkalinity = 40; hardness = 46, temperature ranged 24–26°C; measured concentration of *p*-cresol was done for each run. Those data are not presented here, but are part of the Mount and Norberg report [1], and are available on request.

[b] Two replicates.

96-hr Exposure

The 96-hr dose occurred between 10:00 AM on June 29 and 10:00 AM on July 3, 1981. Figure 8 presents the ambient concentrations at the head end (YE) and the tail end (GR) of the channel. These measurements were obtained from a phenol analyzer modified for continuous operation and verified by GC. The magnitude of the difference between these curves at any point in time is directly related to the half-life, since the retention time was a constant 5 hr. The diurnal degradation rate cycle is apparent in Figure 9, which presents the rate of *p*-cresol addition required to maintain the desired 8 ppm at the head end. *p*-Cresol was very persistent during the first 48 hr. Decreasing half-lives cycling diurnally were very apparent over the last 48 hr of the 96-hr dose. There is no fixed value of a degradation rate that would characterize the process.

Water temperatures reflected a normal diurnal pattern, with no apparent effect by *p*-cresol input (Figure 10). However, pH showed a strong response to the *p*-cresol input. The diurnal cycle of pH was retarded in amplitude, and the minimum levels lowered by the end of the 96-hr exposure (Figure 11). The reduction in pH occurred at the time of maximum algae death and decomposition and maximum anaerobic conditions (Figure 12).

DO demonstrated the same dynamic response as was observed the pre-

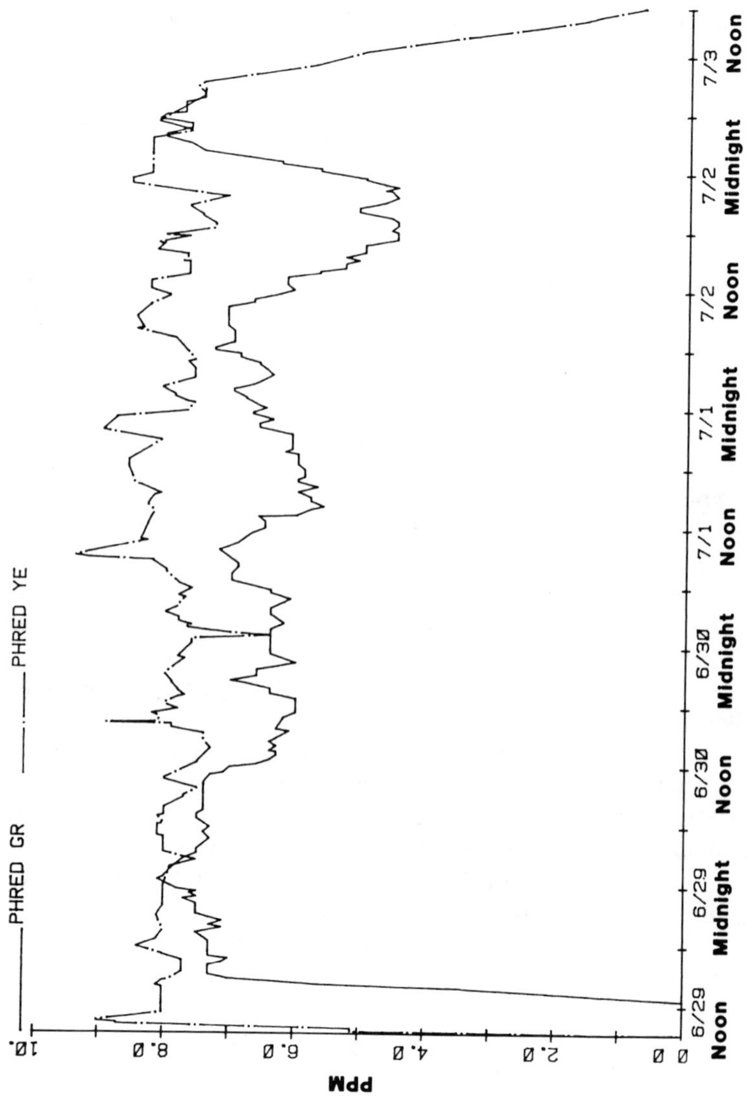

Figure 8. Ambient concentrations of *p*-cresol (ppm) during a 96-hr continuous dose.

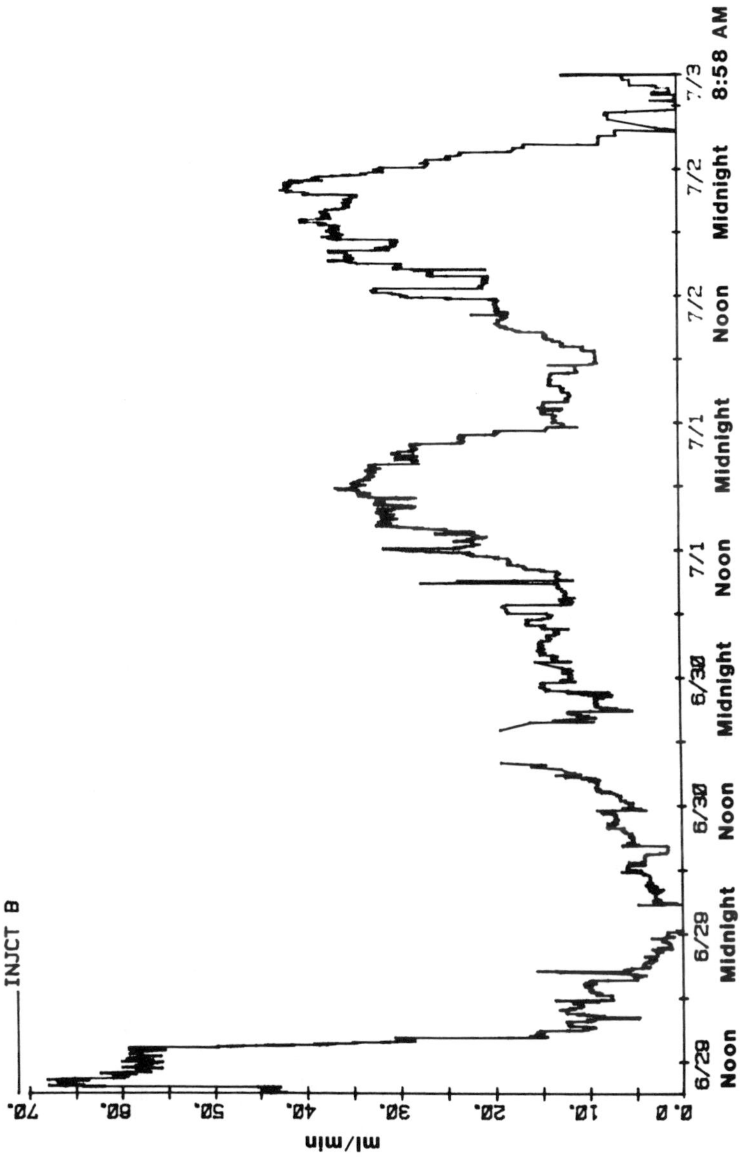

Figure 9. Amount of *p*-cresol added during a 96-hr continuous dose (ml/min).

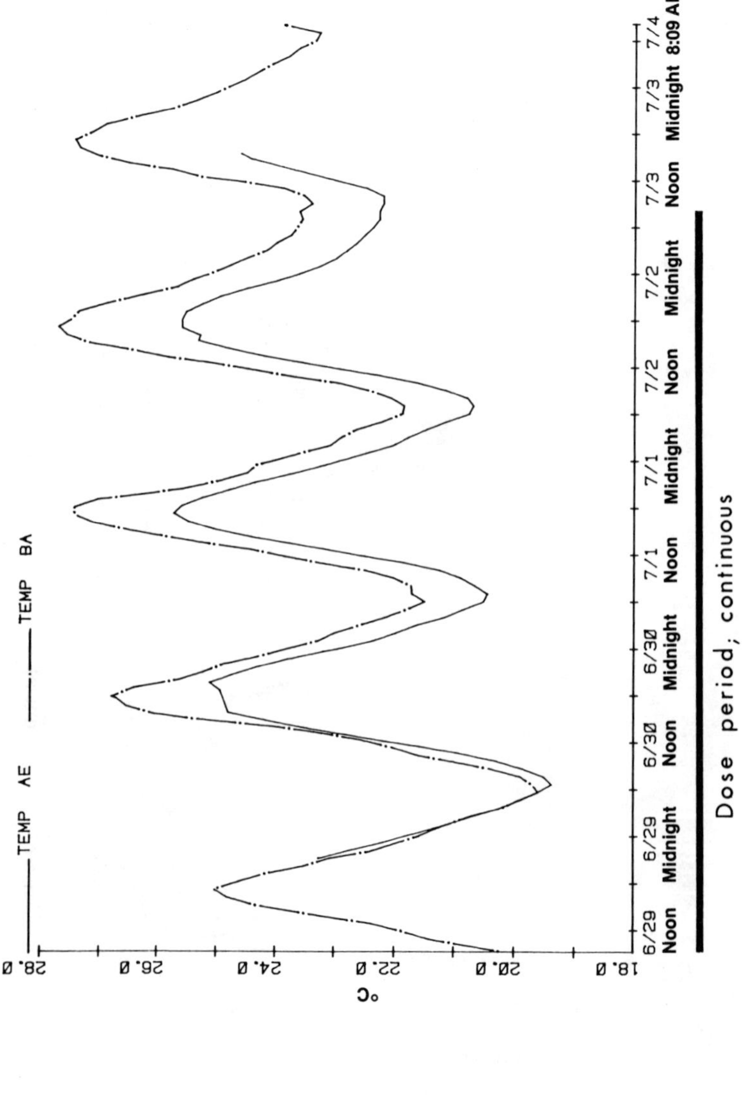

Figure 10. Water temperature (°C) in control channel (–) and experimental channel (– · – · –) during a 96-hr continuous dose.

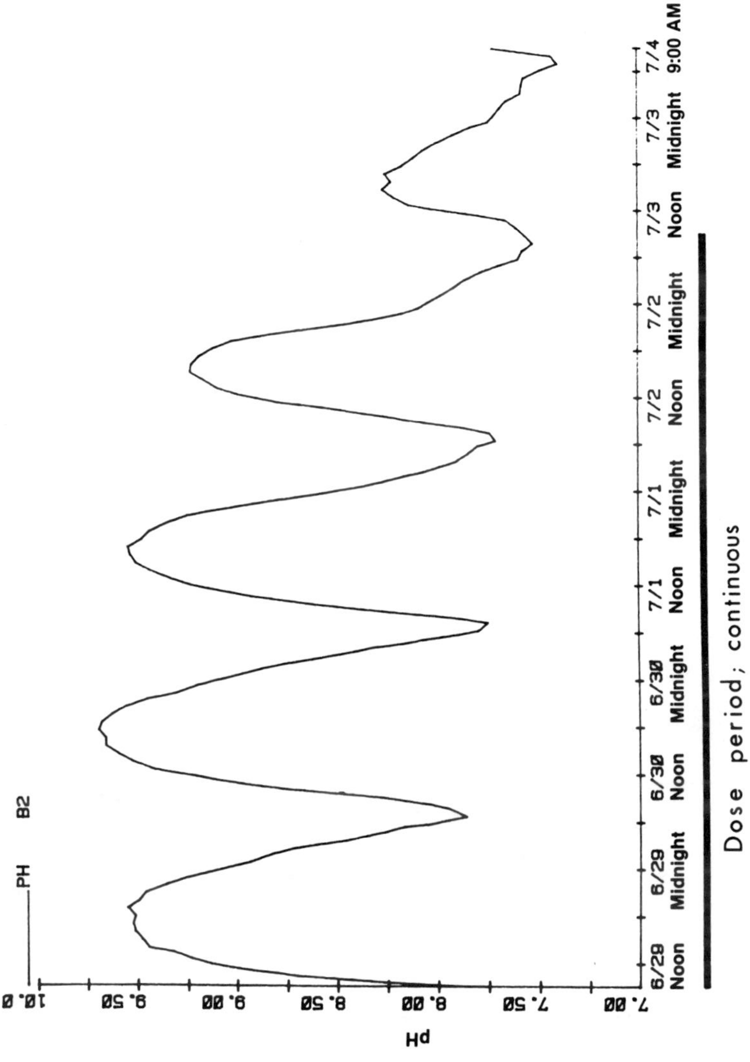

Figure 11. pH in experimental channel during a 96-hr continuous dose.

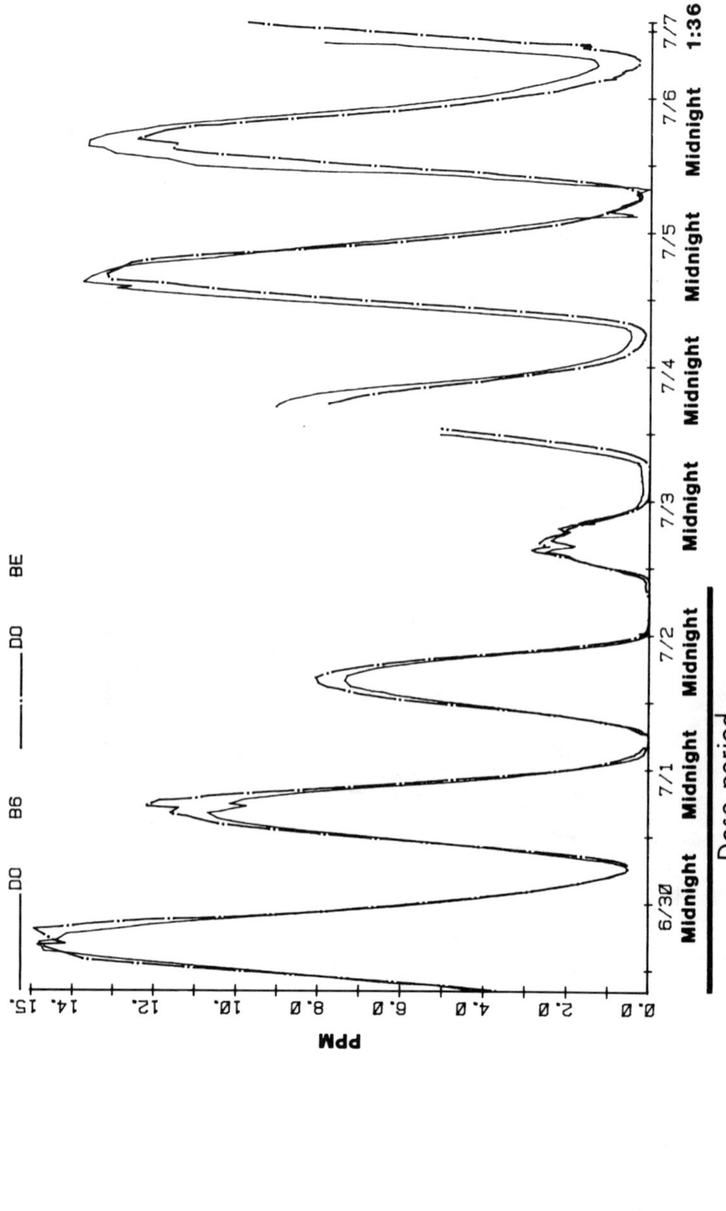

Figure 12. DO (in ppm) at two locations in experimental channel during 96-hr continuous dose.

vious year (Figure 6). At the end of the first 48 hr, anaerobic conditions developed, and significant mortality of fish and macroinvertebrates occurred. This anaerobic pattern intensified through the remainder of the dose. Extensive mortality of all animals and filamentous algae was readily apparent. Benthic and drift samples are currently being enumerated, but the sharp reduction in densities of organisms was qualitatively very obvious. The rate of DO shifts (Figure 13) can be used to compare exposure impacts on oxygen fixation rates of primary producers with impacts on respiratory demands by the entire biotic community. The maximum impact of *p*-cresol on the periodicity and amplitude of the rate of dissolved oxygen shifts occurred during the last 48 hr of the dose. Recovery took place within 3 days after the termination of dosing, even though the animal and algae populations took weeks to recover. The symmetrical reduction in fixation and consumption rates indicate that both processes were affected.

The bioaccumulation of *p*-cresol in fish and crayfish was slightly higher after 60 hr of exposure than those observed with the 48-hr dose (Tables I and IV). Data were not available for 96 hr, because most of these animals died under the anaerobic conditions. The distribution of *p*-cresol within the fish organ was similar to that observed for the 48-hr exposure (Table V). The liver, gut and intestine experienced the highest concentrations. Skin and gill levels were not much different from the ambient concentrations. This would indicate an oral pathway of uptake.

144-hr Pulsed Exposure

The pulsed exposure experiment was designed to separate the primary impact of *p*-cresol as a toxicant from the secondary ecological impacts resulting from anaerobic conditions. During the continuous 96-hr exposure, indicator species experienced mortality at much lower exposure thresholds than were predicted from the laboratory toxicity tests (Table III). Furthermore, the rate of bacterial degradation of *p*-cresol was highly correlated with DO concentrations. The pulsed dose was designed to generate the same integrated "concentration exposure time" measured as "ppm × hr" as was experienced in the 48-hr continuous dose. By pulsating the dose we attempted to stress the system without driving it into a sustained anaerobic state.

Figure 14 presents the ambient concentrations of *p*-cresol at the head end (YE) and tail end (GR) of the channel. Three applications designed to bring the whole channel to an 8 ppm initial condition after the initial 5 hr (the retention time) in addition to the degradation phase produced the

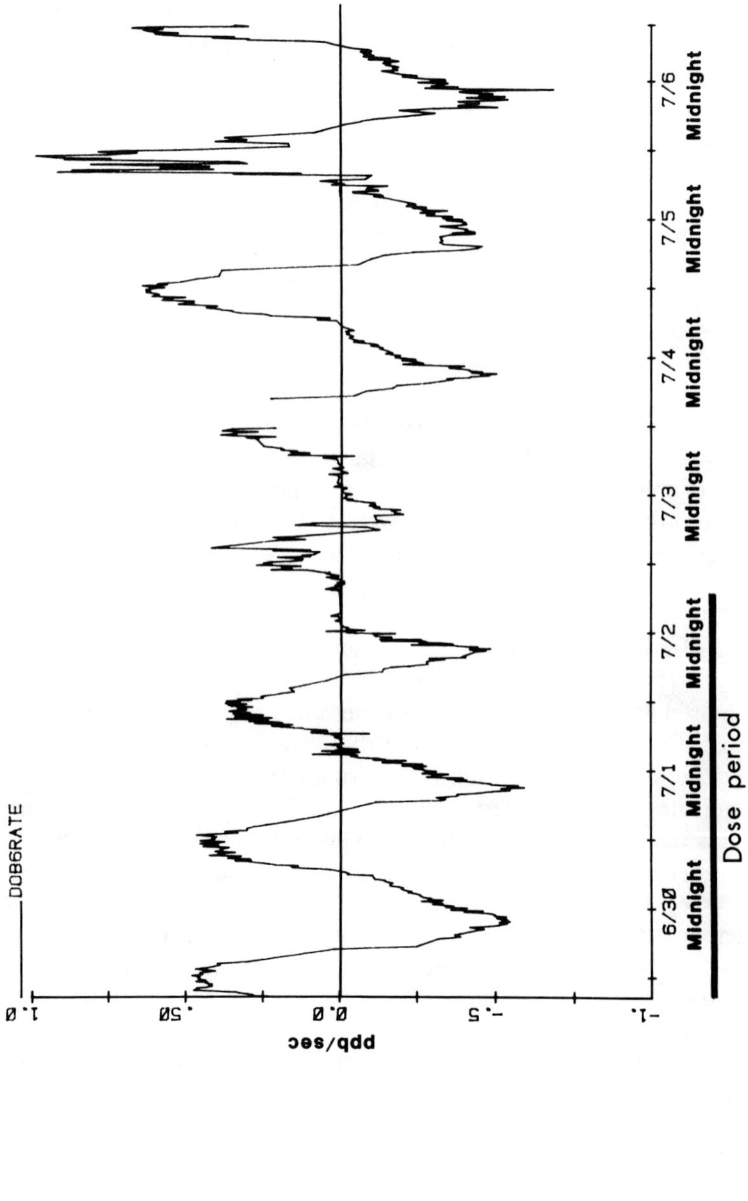

Figure 13. Rate of change in DO (ppb/sec) in experimental channel during a 96-hr continuous dose.

Table IV. Bioaccumulation: 96-hr Continuous Dose

Type	Hours After Dose Initiation	Bioaccumulation (ppm)
Walleyed Pike	44	17.5
	63	15.3[a]
Largemouth Bass	45	22.3[a]
	66	30.1
	66	25.4[a]
Smallmouth Bass	52	15.5
Fathead Minnow	25	7.7[a,b]
	51	10.4[a,b]
	66	16.7[a,b]
Bluegill	24	28.1[a,b]
	64	29.1[a,b]
Crayfish	25	13.6
	45	20.1
	65	23.2
	92	29.4
	119	16.9

[a]Caged animals.
[b]From Big Lake.

Table V. Organic Partitioning: 96-hr Continuous Dose

Type	Hours after Dose Initiation	Bioaccumulation (ppm)
Largemouth Bass		
Liver	27	18.0
Gut and Intestines		43.4
Muscles		8.95
Skin		7.6
Bluegills		
Liver and Gall Bladder	27/66	74.0/52.2
Gut and Intestines		103.7/89.1
Gills		12.5/9.8
Muscles		10.1/9.2
Skin		9.7/9.0

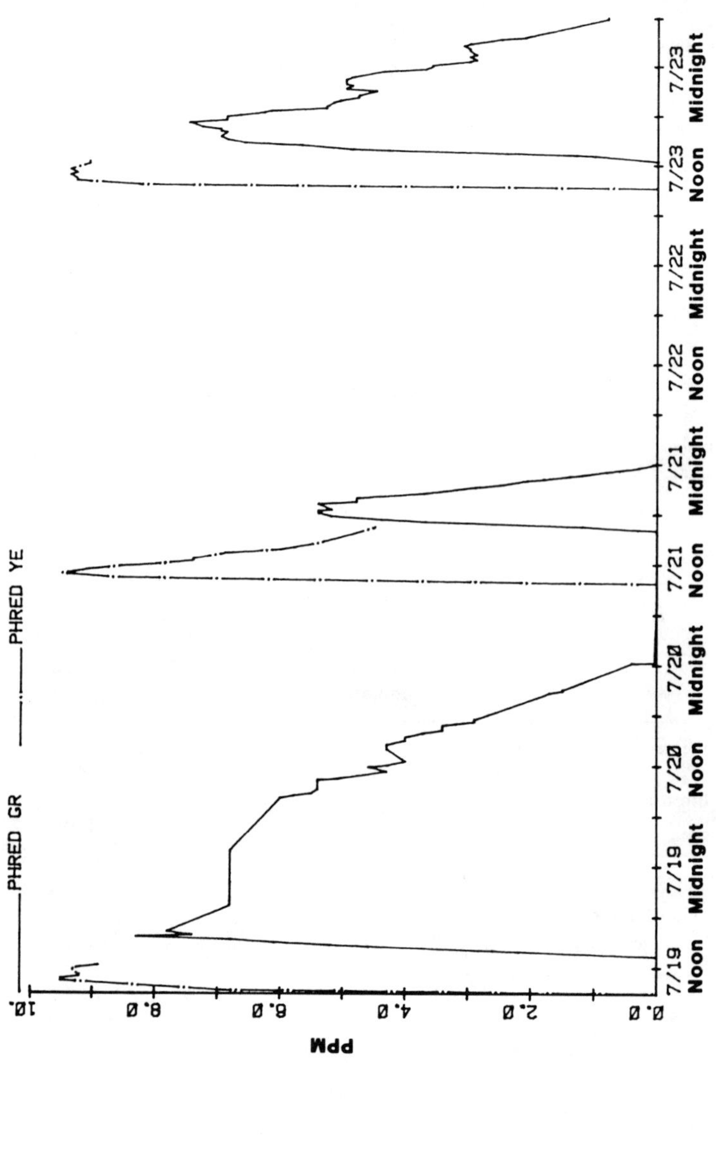

Figure 14. Ambient *p*-cresol concentrations (ppm) at the head (YE) and tail (GR) ends of experimental channel during a 144-hr pulsed dose.

required 356 ppm × hr needed to equal the integral under the curve in Figure 4.

Degradation rate increased from pulses 1 to 2. Adaptation of resident bacteria is the most likely explanation. The rate decreased, however, between pulses 2 and 3. Samples of benthic organisms, drift and bacteria are currently being enumerated. There are, however, some preliminary analyses available at this time that suggest mechanisms of interaction.

A normal diurnal temperature cycle was observed with a slight cold spell during the last part of the experiment (Figure 15). pH (Figure 16) demonstrated the same reduction as observed in the continuous 96-hr exposure (Figure 11). The effect was correlated with the accumulated exposure and not with strict chronological time.

DO levels (Figure 17) again demonstrated a marked response to *p*-cresol. The pulsed inputs took place between 10:00 AM and 3:00 PM on the 19th, 21st and 23rd of the month. The response to *p*-cresol was apparent within 24 hr and indicated an additive effect. The most severe reductions in amplitude and duration occurred 24 hr after the final application. The recovery phase took more than one week before the normal diurnal patterns were reestablished. Several sampling procedures were used to quantify the impacts on the filamentous algae. These data are currently being analyzed, but there is evidence to demonstrate that the algae were severely damaged. There were no indications from field observations, or from preliminary inspection of the benthic samples, that macroinvertebrates or fish experienced mass mortalities. The amphipod *Hyalella azteca* and the cladoceran *Ceriodaphnia verticulatus* should have experienced measurable mortalities. These predictions will be validated when the field samples are analyzed. We were successful in uncoupling the primary and secondary effects, and the preliminary results appear to be in agreement with the laboratory predictions of acute toxicities.

The major direct impact of *p*-cresol was on the filamentous algae. Both processes of community photosynthesis and respiration were affected. Figures 18 and 19 present the rates of change of DO for the control experimental streams. The rate of change became less, especially the rate change due to respiration (rate of change below horizontal bar). Also, the duration where there was no rate change increased in the experimental channel. The maximum impacts occurred from the 23rd to the 26th of the month.

The bioaccumulation of *p*-cresol is presented in Table VI. The pattern is more difficult to characterize, since the rates of uptake and depuration are rapid and the exposure is pulsed. It is obvious that the body levels did not approach those observed after the 48-hr continuous exposure.

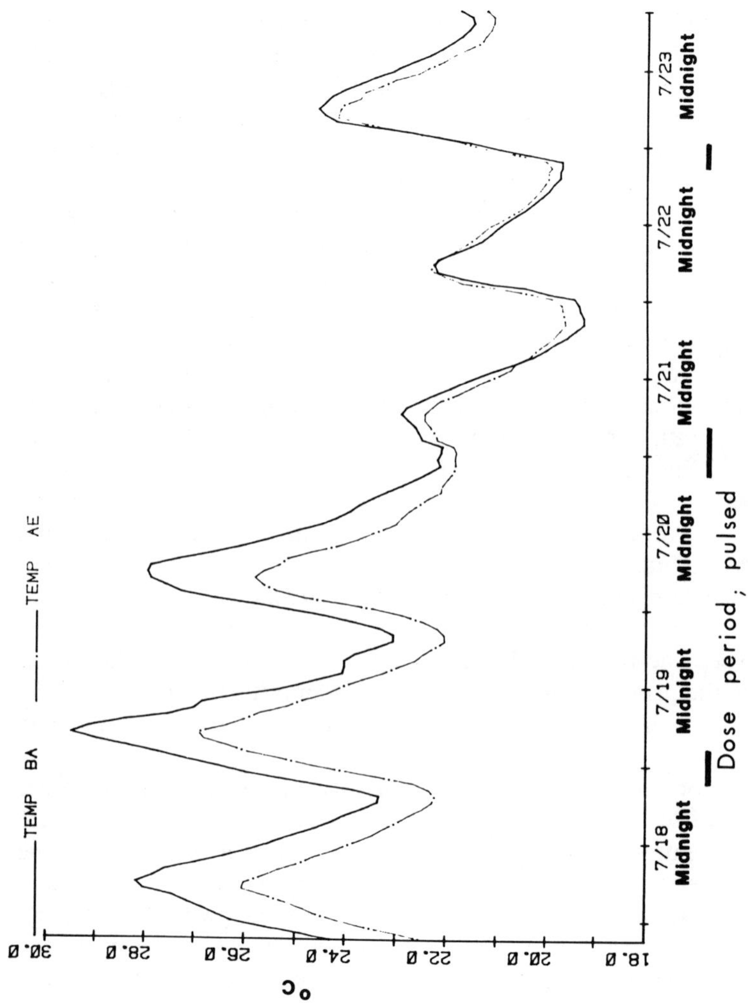

Figure 15. Water temperature (°C) in experimental channel (AE) and control channel (BA) during a 144-hr pulsed dose.

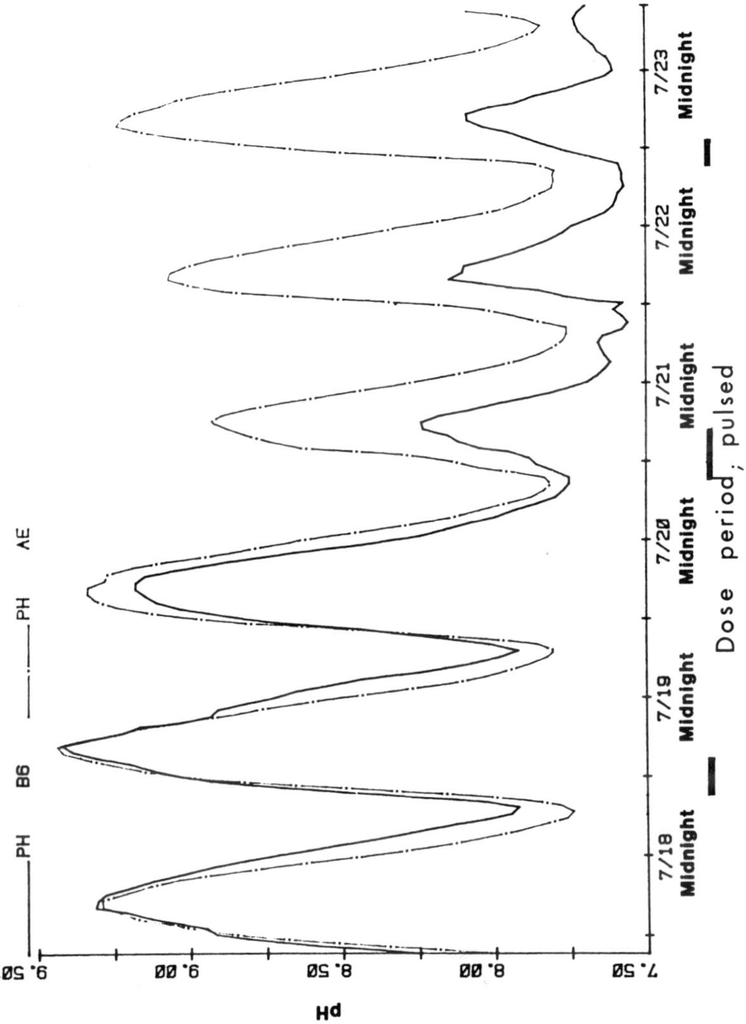

Figure 16. pH in experimental channel (BG) and control channel (AE) during a 144-hr pulsed dose.

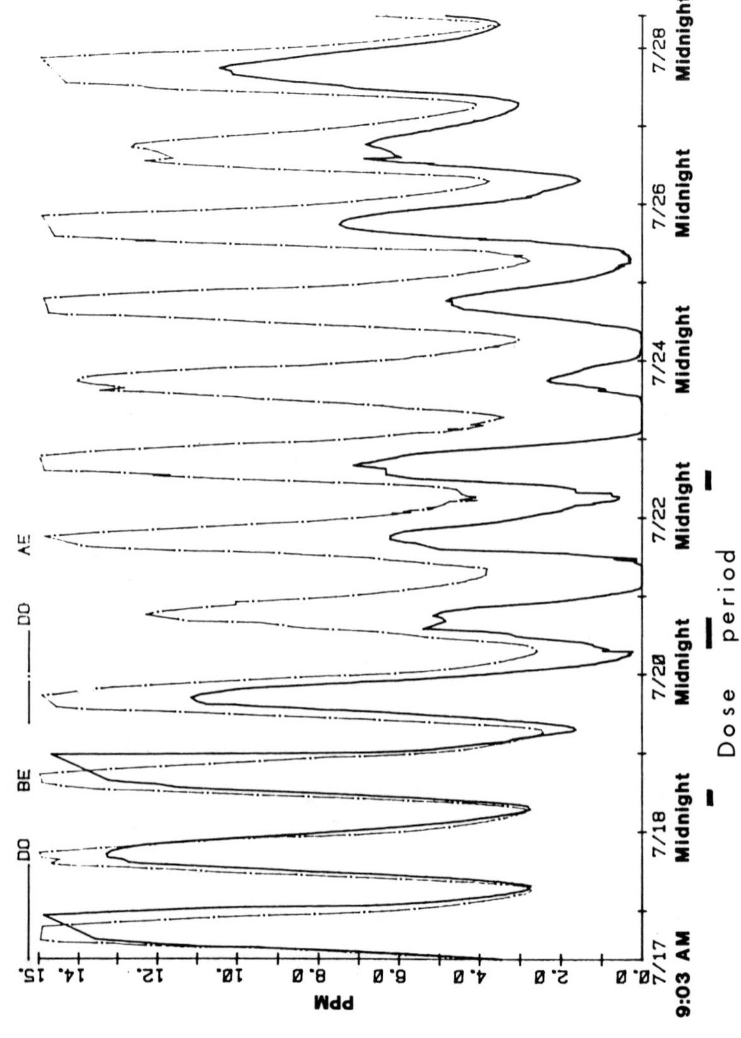

Figure 17. DO (ppm) in experimental channel (BE) and control channel (AE) during a 144-hr pulsed dose.

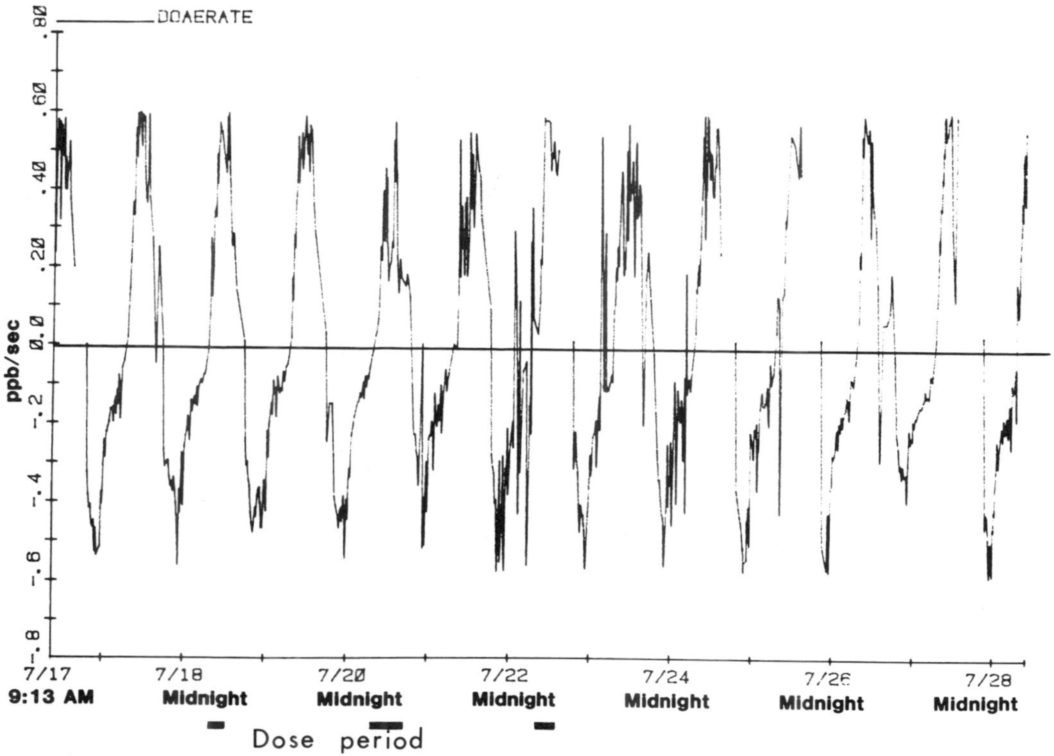

Figure 18. Rate of change in DO (ppb/sec) in the control channel during a 144-hr pulsed dose in experimental channel.

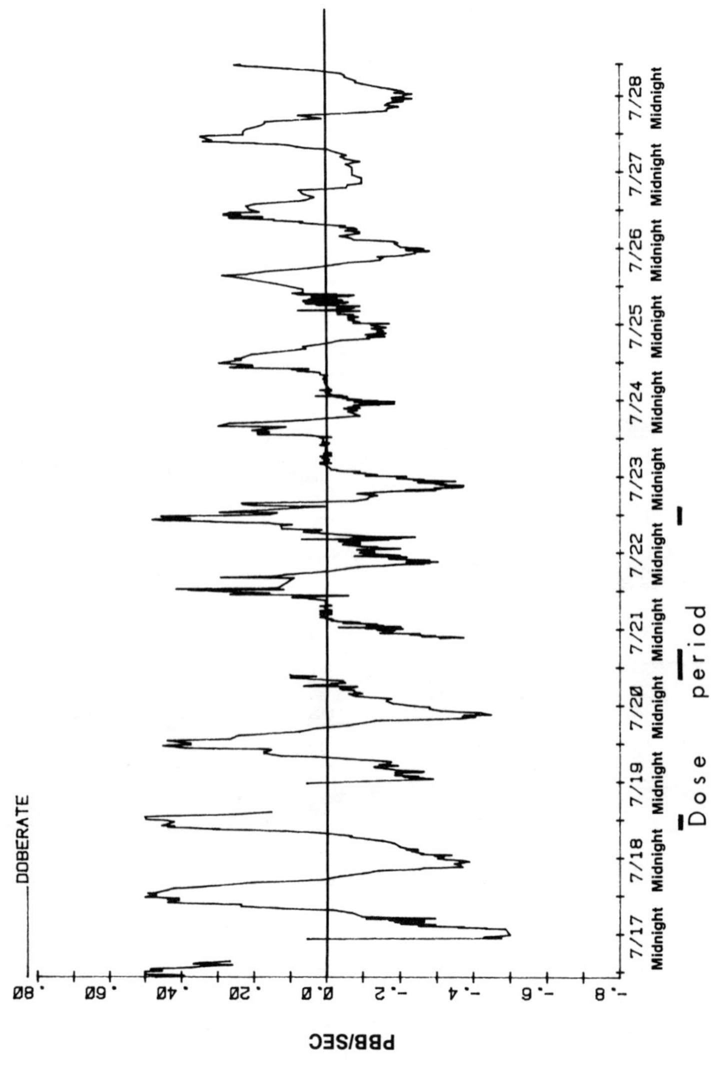

Figure 19. Rate of change in DO (ppb/sec) in the experimental channel during a 144-hr pulsed dose.

Table VI. Bioaccumulation: 144-hr Pulsed Dose

Dose Schedule	Largemouth Bass	Smallmouth Bass	Bluegill	Crayfish
Dose I (7/19)		7.3 (4:30 PM)	6.6 (3:45 PM)	4.4 (4:10 PM)
Interval I				
7/20	6.7 (10:00 AM)	8.3 (8:45 AM)	6.4 (8:45 AM)	5.7 (8:45 AM)
7/21	5.3 (7:30 AM)		5.2 (7:30 AM)	8.0 (7:30 AM)
Dose II (7/12)	13.0 (5:10 PM)	4.9 (7:45 PM)	1.4 (10:50 AM) 4.3 (5:10 PM)	3.2 (7:45 PM)
Interval II				
7/22	5.4 (6:30 AM)	1.6 (6:30 AM)	2.2 (6:30 AM)	4.2 (6:30 AM)
7/23	2.4 (8:45 AM)	0.0 (8:45 AM)		17.4 (6:45 AM)
Dose III (7/23)				
Interval III (7/24)			2.5 (9:30 AM)	4.0 (9:30 AM)

DISCUSSION

The major purpose of this chapter is to demonstrate the diversity of system responses that one might observe when stressing a large ecosystem with a toxicant. The direct, first-order impacts would involve mortality measurements on indicator species (the minimum data set for the EPA water quality criteria). In fact, for most small-scale flow-through test systems, first-order effects are all that one would expect to observe. The residual impacts are washed out with the discharge. If we had performed our experiments with a flow-through design utilizing Mississippi River water for make up, the external influence would have drastically dampened the system responses. A flow-through system is desirable if one is investigating the impact of the ecosystem on the transport and fate of the toxicant. A recycling system, however, is desirable if one is investigating the impact of the toxicant on the ecosystem.

p-Cresol caused significant shifts in DO, pH, bacterial densities and algal biomass. These caused diurnal oscillations in biodegradation of *p*-cresol. There were no steady-state conditions established for which constant rate functions would serve as adequate model parameters. Models utilizing degradation rates that are functions of state variables, and therefore are time varying, are required to predict the observed results.

Data are currently being analyzed for effects on photosynthesis and respiration of filamentous algae; leaf pack colonization and processing; physiological adaptation of bacteria for *p*-cresol degradation; drift of macroinvertebrates; population and community dynamics of riffle and pool communities; and the transfer dynamics of *p*-cresol in fish. When all of these data are available, we will have a unique data set to evaluate the utility of physical and mathematical models of the fate, transport and effects of organic toxicants.

REFERENCE

1. Mount, D. I. , and T. J. Norberg. Internal report, U.S. EPA Water Quality Laboratory, Duluth, MN.

CHAPTER 20

LEGAL VALIDATION OF ENVIRONMENTAL FATE MODELS

Frederick G. Ziegler
ERM-Southeast, Inc.
Brentwood, Tennessee 37027

Legal challenge is not a new topic for modelers. Admittedly, it is probably the most distasteful aspect of modeling, and one that has frequently caused the use of these excellent tools to be misunderstood or ignored in a science that desperately needs all the tools it can muster. Over the years, models have been used and misused in an effort to establish waste-load allocations and impacts on the environment associated with point- and nonpoint-source discharges. Their application has become almost an individual preference as a result of past positive or negative legal experiences. Criticism has developed over the years with regard to the use of models. However, under present economic conditions and with a more conservative view of the regulation of discharges, there appears to be evidence of a swing toward the use of an assessment of environmental impacts as opposed to the arbitrary application of treatment technology. Such a shift will certainly encourage aggressive application of environmental fate models in an effort to enhance our knowledge of the impacts on the environment and, obviously, the ultimate impact on ourselves.

Validation of environmental fate models in the legal sense immediately causes one to consider the defense of challenges that may occur as a result of technical/legal scrutiny of the tool. The technical and legal challenges associated with modeling are not easily separated. Both will be considered to present an overall picture of the potential legal challenges that will occur as environmental fate models are applied.

GENERAL REALITIES OF LEGAL VALIDATION

One of the most difficult facts for a modeler to appreciate is that in the legal world, models are perceived as absolute, whereas legal decisions are generally based on many factors, most of which are not quantifiable. In reality, a decision with regard to the application or acceptability of a model to predict specific environmental conditions is generally based on the "weight of evidence" and not solely on the specific modeling tool.

As an example, the reliance on nitrifier population density to assess the significance of nitrification in the receiving stream is a valid and very useful approach. However, nitrifier data are not generally used directly in dissolved oxygen (DO) modeling. The reliance on these and other data, not directly used in the modeling exercise and, perhaps not necessarily quantifiable in nature, may cause the uninitiated to doubt the modeler's credibility. We must also recognize that modeling, by its nature, may be the most complex investigative tool available to the environmental scientist. Complexity often frustrates the nontechnical person.

The term "weight of evidence" can easily be divided into two aspects:

1. those aspects of evidence that are most directly related to scientific studies and research commonly performed within a university setting and published to demonstrate its "truth and accuracy"; and
2. field justification of a specific modeling tool, including field calibration and validation and demonstration of the reproducibility of specific phenomena associated with an environment's specific problem.

One must not overlook the fact that published results of field surveys representing conditions similar to those of the specific area can also be of great value in model justification.

Obviously, as models become more sophisticated or are more intensively reviewed and evaluated, the cost of their development will increase. The increase in cost is more exponential than linear in nature. Admittedly, there are exceptions; adequate funds are often unavailable for rigorous, scientific confirmation of models in any setting other than laboratory and research environments. Even in the research environment, which is obviously the only place for the initial development of a modeling tool, funds may not be adequate and, where available, are frequently insufficient to permit the development of an accurate modeling tool.

It is not uncommon to find research programs without adequate financial backing to develop a complete and satisfactory tool. Often these circumstances may result in costly errors because of the application of improperly validated models. Private-sector funds are more likely to be adequate for specific application requirements. However, they are generally only available when large financial commitments for additional wastewater treatment technology or siting of new facilities is required.

The adequacy of a model to prevail in a legal, technical and/or political challenge is more often attributable to salesmanship than to pure technical merit. In reality, a modeler who fails to perceive the legal and political challenges that will occur as a result of his activities and fails to orient his technical justification to confront and diffuse these legal/political criticisms will lose his arguments and, therefore, the use of his modeling tool. Frequently it is difficult for purely scientific professionals to appreciate the less tangible legal and political world that often controls the decision-making processes.

There is hope, however. Many of us have been exposed over the last few years to the defense of modeling tools in legal, political and scientific atmospheres. One basic reality does protect the scientific modeler from complete frustration. That is, if the model can withstand the scrutiny of scientific and legal criticism, it cannot be ignored easily. The ability to convince the technical staff of a regulatory body, or for that matter a discharger, that they should accept the results of a comprehensive modeling analysis often precludes the political and legal bickering. As previously indicated, there is also hope because of an apparent shift towards environmental analysis as opposed to an arbitrary application of technology.

TECHNICAL CHALLENGE

The purely technical and legal challenge cannot be separated once the initial phases of evaluation and assessment of the adequacy of the model have begun. Two facets of the technical aspects of modeling need to be considered. The first of these concerns the development of models to separate phenomena and generally be used for either pure or applied research. The legal aspects of this use are primarily left for the discussion following the presentation.

The second facet of model usage can be generally collected into the termination of regulatory usage. From a national economic standpoint, this area of usage is by far the most significant. The regulatory usage includes development of plans and projections of the impacts of certain specific compounds on the environment, as well as the regulation of their use by predicting the impact on the environment of point- and nonpoint-source discharges.

One must separate model development into two parts to discuss adequately the potential legal and scientific challenges that will occur as a result of model usage.

Basic Model Development and Validation

Obviously the building block for any accurately calibrated and verified environmental fate model is the basic theoretical development of the equations and validation of the model in the laboratory. Only through careful separation can the basic tool and the interrelationships of the various specific phenomena be developed. In the past, the relatively simple nature of the equations representing the various phenomena was relatively well documented prior to their broad application. The more complex nature of models associated with the fate of specific chemicals and their effects on the biological community will cause, hopefully, a more thorough assessment and testing program to assure the adequacy of these predictive tools before they are applied to the real world. It is the obligation of the model developer and aquatic scientist to properly qualify his tool, recognizing the level of refinement and testing that has gone into the development of the model at the time of its publication. It is critical that a model developer think like the model user, who has not had privilege of the model developer's background and experience, especially with the more complex environmental fate models that are presently being developed. It must also be recognized that these far more complex modeling tools may inadvertently be applied by individuals not familiar with the conditions associated with their development, thus causing significant errors in the predictions. This could result in severe financial hardship or perhaps even the far more serious implications of human health damage.

Validation of these complex models for predicting the fate of compounds must be rigorous. As indicated, the intensity associated with the validation and testing of environmental fate models cannot exclude the realization that these tools will be used to assess implications on public health as opposed to the relatively simplified predictions associated with such phenomena as DO. The potential repercussions associated with errors in predicting that are traceable to the basic model development could be quite serious.

Calibration and Validation

Calibration provides the initial establishment of constants for a specific field use based on field sampling and analytical analysis. Validation provides the demonstration that the calibrated constants will reproduce another set of field data.

For example, analytical techniques are not adequate to measure dilute concentrations of chemicals. We have all had experiences where dramatic contradictions in analysis compared to known hydrological flow conditions or dispersion characteristics within receiving waters have caused the invalidation of models or at least caused increased skepticism with regard to their adequacy in projecting future impacts. It is not the purpose of this chapter to review the status of specific analytical protocols. However, one must always include in an analysis of calibration and validation the range of adequacy of these procedures when confronted by a variation in analytical results. It would be inappropriate to extend credibility to models that have not been assessed accurately with regard to calibration and validation, considering the range and variability of analytical results.

If one becomes nervous after considering the range of analytical results associated with measurement of low concentrations of specific compounds in the environment, one may panic when the adequacy of field data acquisition is considered. The costs associated with field data acquisition have advanced much more rapidly than the scientific procedures associated with the appropriate protocol to obtain samples.

Perhaps the easiest way to destroy the use of a particular modeling tool is to review the field sampling techniques. Once these observations have been completed, it is often easy to invalidate a model based on criticisms of field data acquisition, ignoring all the other potential liabilities associated with the model. Criticism of inadequate field investigations associated with calibration and validation is a most effective way to limit model credibility in a court of law. By the same token, description and explanation of extensive field investigations for calibration and verification, thereby indicating a thoroughness in real-world analysis of modeling tool before its application, is a strong selling point for model adequacy. In fact, some model developers and validators are inexperienced field personnel who may not be aware of the limitations regarding the analytical accuracy of testing procedures or the crudeness associated with many field data acquisition procedures.

I have always believed it to be inappropriate for model developers to publish "user's manuals" without extensive discussions of the appropriate field survey protocol for calibration and validation. Accurate calibration and validation of environmental fate models are more critical than for the basic waste-load allocation tools associated with DO deficits and ultimate fate of oxygen-consuming materials within the environment. Admittedly, a modeler may not be experienced in all aspects of data acquisition. Nevertheless, by developing user's manuals in collaboration with scientists and engineers experienced in field data acquisition, a much more useful user's manual can be developed.

Data interpretation fulfills the Sherlock Holmes ambitions of modelers. However, it is shattering to be asked by an attorney in a court of law if one personally participated in the field survey work before developing the model and having to answer "no." It is therefore essential that model developers and users participate in field survey investigations so that they are exposed to the environmental conditions associated with the model, as well as with the logistical problems and complications of field survey work. It is also extremely useful for the initial model developer and validator to incorporate in the user's manual those aspects of data interpretation that might influence final development of a calibrated and validated model. There is often an apparent lack of communication between the original developer and the model user who is calibrating and validating the water quality model tool. We should all recognize that environmental fate models will not be as forgiving as the classical biochemical oxygen demand (BOD) models. The data-interpretation phase of model development is often ignored in the final documentation. This is a serious omission and could well result in legal invalidation of the model.

It is most effective to develop and maintain notes with regard to field observations during the calibration and validation of a water quality model. Being able to recount the findings and conclusions as they occurred during calibration and validation is most effective in "selling" the adequacy of the model to the appropriate regulatory or judicial body.

Summary

Model adequacy can be divided into two categories: (1) adequacy of initial model development, and (2) adequacy of specific application of the model. For example, in the past, DO models were relatively basic. They were also applied for many years. Because the establishment of the DO models occurred over a relatively long period of time, the specific equation development has been accepted. Future evaluation of environmental fate models may cause a substantial difference in the challenging of initial model development. The limited time over which environmental fate model development has occurred may not be adequate, or may at least not appear adequate. Additionally, the more sophisticated nature of environmental fate models will cause increased criticism and debate.

Adequacy with regard to a specific application of an environmental fate model dictates to the user a demonstration with a thorough and well designed field sampling investigation followed by proper interpretation of data, as well as calibration and validation. Attention to details of the specific parameters being modeled will be necessary to justify the results of the predictive tool. Realistic interpretation of the data and realistic discussions of adequacy regarding its intended purpose must occur within the modeling analysis.

DEVELOPMENT OF MODELING RESULTS AND PRESENTATION

At this time it is necessary for the modeler to sell his results to the appropriate legal or regulatory body. I have found that it is useful to identify and recognize limitations of the modeling tool before their disclosure in a negotiation or court of law. By recognizing these limitations and developing the appropriate responses during report preparation, a stronger defense will be created. It is also most important to prepare and provide the proper perspective regarding the accuracy of the model. In the past, initiation of adversary proceedings has resulted from reports prepared without the above information included. Today, it is realistic to anticipate a rigorous discussion of the limitations of the modeling tools.

DIRECT LEGAL CONSIDERATIONS

Many of the public health statutes, and therefore environmental regulations, anticipate the use of modeling to some extent. These laws include the Clean Water Act, Clean Air Act, Resource Conservation Recovery Act, Superfund, Toxic Substance Control Act, and appropriate state and regional regulations. Inclusion of a rigorous discussion of the limitations of the model is prudent.

To thoroughly appreciate legal validation, one must understand the legal profession's perspective. The objectives of an attorney in representing his client and those of a technical professional are often quite different. The difference is that an attorney is not necessarily limited by data in the development of his case in the defense of his client. However, a technical professional must always consider the existing data base and make his conclusions and recommendations based on the level of knowledge available to him at the time. This often frustrates attorneys, because they are looking for only an absolute answer in favor of their client. This may also become frustrating to the client, who also would prefer a simple, positive answer instead of one qualified by standard deviations and indications that certain phenomena had not been investigated adequately. Therefore, it becomes critical to a water quality management engineer or scientist to develop a rigorous defense of his position to be able to withstand the criticism, which can often be quite intimidating, by attorneys prior to negotiation or legal confrontation.

We must all recognize that water quality models primarily associated with DO have been developed and criticized to the extent that they became useless tools that were ignored during the final decisions associated with application of specific technology to wastewater treatment. Future uses of environmental fate models may become much more intensely argued as a result of their direct influence and implications with regard to aquatic and human health. Those of us working in the environmental fields are aware of the increased intensity and somewhat hysterical nature of engineering problems and solutions associated with hazardous waste and chemicals. These additional pressures have caused many environmental scientists to inquire about errors-in-omissions insurance, which until recently has been unnecessary for all except those in the design engineering business.

ACKNOWLEDGMENTS

The author recognizes the contributions of personal discussions with Jan Tai Kuo (General Software Corp., Landover, MD); P. Rosten (AWARE, Inc., Nashville, TN); D. Kleiver (Cleary, Gottlieb, Steen & Hamilton, Washington, DC); and M. McGee (U.S. Environmental Protection Agency (EPA), Atlanta, GA). Another source of information was a paper by J. L. Schnoor entitled "Assessment of the Exposure, Fate and Persistence of Toxic Organic Chemicals to Aquatic Ecosystems in Stream and Lake Environments," presented at the Water Quality Assessment of Toxic and Conventional Pollutants in Lake and Streams, sponsored by the EPA, June 23–25, 1981, Arlington, VA.

CHAPTER 21

SYNOPSIS OF DISCUSSION SESSION: VALIDATION AND APPLICATION TESTING

D. M. Di Toro
Session Chairman

A. S. Donigian, L. M. Games, S. Hern, R. R. Lassiter and Y. Matsuoka

The terms validation, verification, calibration and postaudit have led to confusion, since they have been used interchangeably or defined differently by various investigators. For the purposes of this workshop, validation is defined as the overall process of model testing. It is composed of a number of steps (Figure 1), which result in the capability of making predictions that are sufficiently accurate and precise for the intended use. These stages are different for empirical and theoretical models [1], and are defined as follows:

1. Model construction check: check the structure of the model algorithms. Computer runs are made to check performance over the model's applicable range.
2. Calibration (empirical part of the model): adjust the model parameters within the expected range, until the differences between model predictions and field measurements satisfy some criterion of goodness of fit, which preferably has a sound statistical basis. Fidelity test (theoretical part of the model): ensure that the model faithfully represents the theory and includes all of the important processes that occur in the prototype system. Sufficiently reliable data and the best available parameters are supplied to check the differences between model predictions and measurements.
3. Verification: compare model predictions to field measurements that were not used in the calibration or fidelity testing. Adequacy of the

model for the modeling purpose is ascertained. Field site and environmental conditions are often changed, but parameters determined at step 2 should not be altered unless they reflect known differences in environmental variables.

4. Postaudit: evaluate the decision based on the model prediction by comparing it with field measurements collected after implementation. Adequacy of the model can be established by the experience gained at this step.

Based on their structure and applicability, currently available environmental fate models can be divided into three classes:

1. Partitioning model: Mackay Level I and Level II models and the Mackay-Neely model [2];
2. Simplified fate model: Stanford Research International (SRI), Chemical Manufacturing Association—HydroQual Inc. (CMA—HydroQual) and Mackay Level III;
3. Detailed dynamic fate model: Sediment Radionuclide Transport Model (SERATRA), Hydrologic Simulation Program—FORTRAN (HSPF) and Unified Transport Model (UTM).

Each of these classes will be discussed in terms of their uses and limitations, as well as their state of validation.

CLASS 1: PARTITIONING MODELS

Use

Partitioning models use thermodynamic equilibria and first-order decay reactions to determine the distribution of a chemical in the environment. Mass transfer rates are assumed to be rapid relative to the first-order decay rates. Two levels of complexity have been considered: the first involving only equilibrium distribution and the second including, in addition, rate constants for important kinetic processes. Both levels, because of the equilibrium assumption, result in the same distribution of chemical (i.e., the same concentration ratios among the phases). The only additional information gained from the second level is a calculation of persistence (total chemical mass/input mass discharge rate).

Currently available partitioning models are designed to describe the thermodynamic tendency of organic chemicals to partition among phases. They can, for example, be used to direct research efforts for a new chemical to the proper compartment (i.e., phase) for measurements of the rates of transformation processes. At a preliminary stage, they

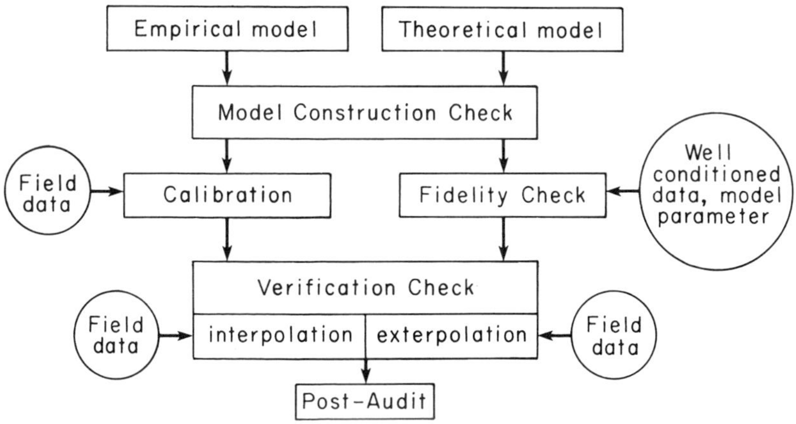

Figure 1. The process of model testing (validation).

can, in some cases, also direct testing for effects, although distribution alone cannot be used to assess risk to the ecosystem. The reason is that the absolute concentrations predicted by the model depend on the relative dimensions of the "unit world" [2], and these are somewhat arbitrary. Therefore, the predicted concentrations should not be used quantitatively in hazard assessment. In addition, the persistence of the chemical calculated by the Mackay Level II model also depends on the absolute concentration in the phases and therefore depends on the unit world dimensions. These limitations are well known to the authors of this model and are included here as guidance for potential users.

Validation

A full validation, in the sense defined above, is clearly inappropriate for partitioning models. What is required is that they provide a reasonable assessment of the ratios of chemical concentrations in the phases of the unit world environment. Limited data [3] suggest that, for widely distributed chemicals (e.g., polychlorinated biphenyls, DDT) the partitioning models give an approximately correct distribution.

Limitation

Current use of these models is limited to obtaining a qualitative sense of the environmental distribution of a chemical. Although the class 1 models can be used to calculate persistence, there is no way to judge the usefulness of this procedure at this time because, in addition to the difficulties discussed above, the effects of mass transfers are neglected. To provide realistic and supportable estimations of environmental persistence, some definition and inclusion of mass transfer coefficients are required as in the Level III Mackay model. The models discussed next include these mechanisms.

CLASS 2: SIMPLIFIED FATE MODELS

Simplified fate models are the least complex formulations that can be used to calculate chemical concentrations. The level of resolution is such that detailed environments can be evaluated. However, in typical uses, much of the spatial and all of the temporal variation are ignored. Because of this simplification, it is feasible to consider the expected chemical fate over a broad range of environments, thus developing an environmental profile of the probable fate of the chemical. Uses can vary considerably, however. One can specify an environment, on the one hand, that is equivalent to the unit world; on the other hand, spatial resolution can be made more complex. This flexibility is an important feature that allows adaptation to questions of different resolution as they arise in assessments of chemical fate. Currently, however, there are various limitations to the way that these models can be used, even within the general assessment setting.

Limitations

As they are now implemented, simplified fate models do not easily provide allowance of a time-varying mass discharge rate. If this is important, dynamic models should be considered. Additionally, as a practical matter, it is often difficult to decide on the set of environmental conditions to use for broad-scale chemical fate assessments. Also, the models require absolutely that the mass discharge rate be specified. For situations where these estimates cannot be made, environmental chemical concentrations cannot be calculated.

Validation

As the first stage of model validation, model construction checks have been initiated for the simplified fate models. To date, checks between the Exposure Analysis Modeling System (EXAMS) and the CMA—Hydro-

Qual model have been made for the simple lake and spatially uniform stream case. The models give the same results if the parameters that describe the kinetic, transfer and transport processes are properly interpreted, so that the specified flux rates are identical in both models. Testing is continuing for more complex environments that are spatially variable and involve more detailed segmentation.

Fidelity and Calibration

Simplified fate models, as currently constructed, are mixtures of empirical and theoretical formulations. They are all based on mass conservation equations. Kinetic formulations are included that are based on laboratory-derived rate parameters for most processes. However, certain empirical constants still remain that must be established by calibration to field data (e.g., the sediment/water transfer rate coefficients).

A few calibrations and utility checks have been performed to date. An extensive set of laboratory ecosystem experimental data using methyl parathion, accompanied with independent laboratory-derived kinetic rate coefficients, was gathered. Predicted concentrations from EXAMS were compared to observations and agreed to within $\pm 20\%$ [1]. The principal active mechanisms were kinetic decays and water column sorption. Sediment/water transfer processes were not significant for the particular experimental design.

Photolytic decay of DDE in an Indiana quarry [4,5] and *para*-dichlorobenzene in Lake Zurich [6] are in agreement with the theory included in EXAMS.

An initial calibration in a stream has been made for LAS in Rapid Creek, South Dakota [7]. The principal active mechanisms were water column and sediment biodegradation, for which independent laboratory measurements and sediment/water column bulk exchange were used. The exchange parameter was calibrated to the field data. The agreement between water column and limited sediment observations and calculations are within a factor ± 2 and ± 4, respectively. For an initial calibration, this result is encouraging. More detailed investigations of this stream are in progress.

When there is interest in environmental conditions that occur intermittently, such as runoff events, the class 3 models should be selected.

CLASS 3: DETAILED DYNAMIC FATE MODELS

The primary differences between these models and the simplified fate models are inclusion of dynamic watershed hydrologic and hydraulic transport mechanisms and chemical fate processes in soils. These models include simulation of watershed dynamics to represent the dynamic chemical loadings to the aquatic environment from applied chemicals (e.g., pesticides). Aquatic chemical fate processes are often represented in the same formulations as the simplified fate models.

Use

These models are designed to provide time-varying chemical concentrations for a specific chemical (or class of chemical characteristics) in an environment under defined physical and climatic conditions. These models are needed for environmental systems where steady-state assumptions may not be appropriate, such as the highly dynamic chemical loadings from agricultural and silvicultural use.

In any proposed application, a preliminary assessment should be performed either from analysis of the data or application of the simplified fate models to determine whether or not the dynamic models are required for the specific decisions to be made.

Validation

For models of this nature, the probability of undetected code errors increases because of the difficulty of detection and correction in all possible computational options in the model. Users should be on the lookout for counterintuitive results (which may indicate the presence of errors).

These models are also mixtures of empiricism and theory; thus, calibration and fidelity tests discussed above for the simplified fate models are also relevant here. Much of the empiricism in the dynamic fate models is related to modeling the dynamic transport mechanisms. Calibration of selected model parameters to determine appropriate values for site-specific field conditions is required to ensure model accuracy. Calibration is needed primarily for the transport parameters related to hydrologic and sediment transport and dispersion. For the majority of these parameters, well-defined ranges exist; sediment dispersion, mixing and bed exchange are notable exceptions.

The dynamic fate models have undergone testing and validation in a limited number of watersheds, such as the HSPF application to Four Mile Creek [8]. Other applications are currently underway. The hydrologic (e.g., Stanford Watershed Model), hydraulic and chemical runoff portions (e.g., Agriculture Runoff Model) of these models have undergone testing and applications in numerous watersheds across the country. However, the linkage of dynamic chemical runoff and instream chemical fate and transport models (e.g., HSPF) is a recent accomplishment.

FUTURE NEEDS FOR VALIDATION OF FATE MODELS

It is clear from the preceding review that the state of fate model validation is in its infancy. We recommend, therefore, that the highest priority for future experimental work is to validate the existing models. This can be done at three spatial scales: microcosms (i.e., controlled laboratory systems), macrocosms (i.e., controlled fabricated field systems) and natural aquatic systems.

For each of these spatial scales the experimental design is similar. It should include direct measurements of input mass discharge rate, chemical measurements in all compartments, i.e., soluble and particulate concentrations in both the water column and sediment phases, as well as in any biota in the system. In addition, separate, independent experimental determination of all the relevant kinetic and transfer process parameters must be available. This includes the kinetics discussed elsewhere in this publication and the physical transfers that occur in the particular system under investigation, i.e., liquid- and gas-phase transfer coefficients, sediment/water dissolved transfer rates, and aqueous and particulate mixing characteristics. Partition coefficients for the solids in the systems at the appropriate sorbent concentrations are particularly important. The best guide to the design of a validation study is a review of the input parameters and output concentrations of the model being validated.

A critical test of a model, in principle, requires all these determinations. However, it is still true that, at least for calibration purposes in field-scale investigations, there is a minimum data set that is still useful. At minimum, mass loading rate and total concentrations in the water column and sediment in time or space, together with geometry, physical characteristics and sorbent concentrations, can provide a starting point for calibration. Soluble and particulate concentrations are usually also essential as well. If these data sets are produced for many natural water systems and different chemicals and analyzed within the context of the model being tested, a great increase in model credibility will be developed.

For detailed dynamic fate models, it is usually impractical to produce a validation data set for a critical test in the manner described above. However, calibration data sets can also increase the credibility of these models.

Microcosms, in our view, are well suited for investigations that are designed to detect model inadequacies. That is, if a model fails in reproducing the chemical fate observed in controlled microcosm conditions, there is little reason to expect that the models are applicable to field conditions.

Macrocosms are well suited to investigate the fate in flowing waters since it is difficult to construct practical flowing-water laboratory systems. In addition, outdoor systems at a larger scale for both quiescent and flowing waters may develop more diverse environmental conditions and represent more varied transport and transfer processes. These experiments provide additional opportunities for the models to fail.

However, field-scale experiments and monitoring data sets are the most direct tests of model description and prediction capability. There is no problem with the physical scaling or biological realism potentially associated with microcosm and macrocosm experiments. Eventually models must be validated at field scale to assess quantitatively the precision associated with their predictions. Postaudit procedures will eventually decide the adequacy of predictions, and these studies are recommended.

It should be clear from this discussion that the degree to which the validation steps discussed above have been completed determines, at least qualitatively, the level of confidence associated with model calculations and prediction. Unfortunately, no adequate statistical theory has been applied to chemical fate models to determine quantitatively the confidence limits to be associated with model predictions. This is in large measure due to lack of adequate field data and, to a lesser extent, lack of relevant statistical methods. We recommend that these methods be developed in conjunction with field data evaluations, since the predictions of chemical fate made in models form an integral part of exposure assessment.

REFERENCES

1. Lassiter, R. R. "Testing Models of the Fate of Chemicals in Aquatic Environments," Chapter 16, this volume.
2. Neely, W. B., and D. Mackay. "An Evaluative Model for Estimating Environmental Fate," Chapter 8, this volume.
3. Klein, A. W., and F. Schmidt-Bleek. "Significance and Limitations of Environmental Compartmentalization Models in the Control of New Chemicals Based on the Organization for Economic and Cooperative Development Minimum Premarketing Set of Data," Chapter 5, this volume.

4. Zepp, R. G., N. L. Wolfe, L. V. Azarraga, R. H. Cox and C. W. Pape. "Photochemical Transformation of the DDT and Methoxychlor Degradation Products, DDE and DMDE, by Sunlight," *Arch. Environ. Comtam. Toxicol.* 6:305–314 (1977).
5. Zepp, R. G., and P. F. Schlotzhauer. "Effects of Equilibration Time on Photoreactivity of the Pollutant DDE Sorbed on Natural Sediments," *Chemosphere* 10:453–460 (1981).
6. Burns, L. A., Environmental Research Laboratory, U.S. EPA, Athens, GA. Personal communication.
7. Games, L. M. "Field Validation of Exposure Analysis Modeling Systems (EXAMS) in a Flowing Stream," Chapter 18, this volume.
8. Donigian, A. S. "Field Validation and Error Analysis of Chemical Fate Models," Chapter 17, this volume.

WORKSHOP SUMMARY AND CONCLUSIONS

C. M. Lee
Session Chairman
H. L. Bergman, W. P. Wood and J. H. Rodgers, Jr.

The potential of chemicals used in industrial and manufacturing processes, and consumer and agricultural products to create a hazard for human and environmental health has been given considerable attention in recent years. Indeed, legislation has been formulated and introduced to regulate the use of new chemicals [e.g., EEC Directive on the Discharge of Dangerous Substances (79/831/EEC) and the Toxic Substances Control Act (TSCA) of 1976]. Unfortunately, it is all too easy to ask searching questions about the impact of a chemical on the environment and a good deal less easy to provide the methodology and data that will give unequivocal answers. Much of the current approach to hazard assessment of chemicals is empirical and relies heavily on scientific judgment and the experience of the individual. This approach is not always easy to make compatible with the legal process. It is also true to say that the complexity of the questions asked and the expectations of the concerned community have in some cases outstripped the ability of science to provide the answers. Thus, considerable pressure remains to develop and refine the science of hazard assessment. For those who have had contact with it, mathematical modeling seems to hold a tantalizing prospect for objectively assimilating and manipulating the complex data generated during hazard assessment of a chemical. It also seems likely to enable us to move away from the present, rather simplistic, approach to take account of more complex processes and interactions in environmental compartments. The purpose of this workshop was, therefore, to examine this prospect against the following objectives:

1. to analyze the state-of-the-art of environmental fate modeling, considering both development and application;
2. to critically evaluate existing models for predicting environmental exposure concentrations of chemicals in various aquatic systems;
3. to critically examine the utility of various models as decision-making aids for their specific applications;
4. to position the role of environmental fate modeling for aquatic hazard assessment, considering both regulatory application and new product development; and
5. to develop recommendations for future research needs in environmental fate modeling.

CONTEXT

Previous workshops in this series [1–3] have presented a number of sequential assessment schemes whose general objective is to determine whether a chemical (1) is clearly acceptable for its proposed use, (2) is clearly unacceptable for its proposed use, or (3) needs further research before a decision can be made. These assessment schemes are generally based on a comparison of the exposure and effect concentrations of the chemical under review. The determination of its acceptability or otherwise is based on the size of the margin between two concentrations (Figure 1) [4]. Most schemes progress from simple to complex tests through a series of stages, each of which is followed by decision criteria. The intention inherent in this approach is to continually refine the estimates of exposure and effect concentrations until there is sufficient confidence to make a decision.

The exposure concentration of a chemical in the aquatic environment is determined by its use and disposal pattern and, subsequently, through its dispersion by a variety of physical and chemical transport mechanisms and its transformation through interactions with physical, chemical and biological entities in the open environment. In current hazard assessments, estimates of exposure concentrations are rarely based on more than a combination of production tonnage and dilution information. Dilution models alone, however, are notable to indicate potential problems associated with assimilation and transformation. In some schemes, biodegradation and hydrolysis information is used to indicate whether the actual concentration of the chemical in the environment is likely to be reduced substantially after its release [3]. Recent discussions in international groups such as the Organization for Economic and Cooperative Development (OECD) have recommended an increased data base for estimating the exposure concentration, including abiotic and biotic trans-

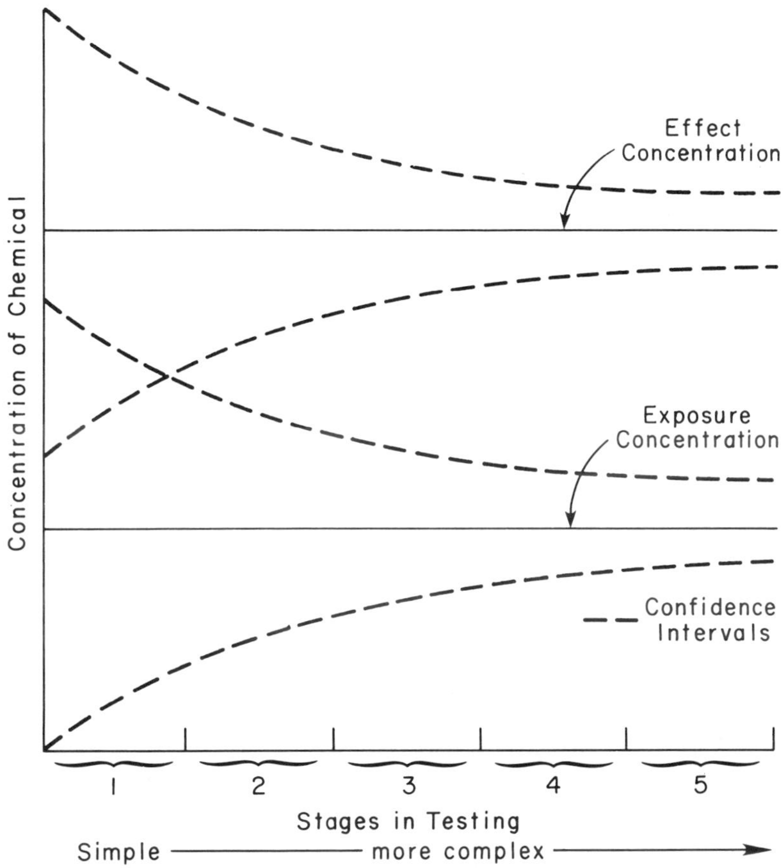

Figure 1. The principle of hazard evaluation (after Cairns et al. [1]).

formation and accumulation processes, and have extended the empirical schemes for utilizing and combining these data to refine estimates [5].

The current approach to determine effect concentrations of a chemical on the aquatic and terrestrial community involves gross simplification. It is clearly impossible to determine the effect of a chemical on every individual species; therefore, surrogates are chosen to represent important functional groups in the ecosystem. Extreme accuracy and precision in estimating the effect concentration are therefore not possible, and the confidence interval surrounding the estimate is large and usually measured in orders of magnitude.

The aim in this approach is always to assign a chemical to one of the above categories with the minimum expenditure of time, cost and effort.

When these assessments have been made, the conclusions are coupled with human safety assessments, cost/benefit analyses, etc., and the question of risk is addressed. Risk is considered to be the statistical likelihood or probability of the hazard being realized in the environment or human population.

Probably the greatest difficulty and challenge in assessing the potential hazard of a chemical is to account for the inherent complexity and variability in the environment. There is also a need to increase understanding of the relevance and utility of laboratory results in predicting events in the environment. In current hazard assessment schemes, decision criteria are based largely on empirical observation, and the data bases are created from tests that often are only the best option in the face of major compromises, which are usually enforced by practical, cost or effort considerations.

Estimating concentrations of chemicals in the aquatic environment using mathematical models will also demand compromises, which will be greater as the case departs from being site-specific to having widespread discharge. So long as this is recognized, it need not detract from the value of hazard assessment schemes. One objective of a hazard assessment scheme must be to select those chemicals that need a detailed examination. Another is to develop schemes that will lead to a correct judgment in the majority of cases. The problem is currently dealt with by making the decision criteria more or less conservative as the case demands. It is desirable to reduce the need for this conservatism to reduce the potential for failing safe chemicals. Environmental fate modeling may contribute to an improved estimate of the distribution of the chemical in environmental compartments. It may also qualitatively improve estimates of exposure concentrations, by taking account of a broader spectrum of physicochemical reactions and interactions, and quantitatively, by making more accurate integration of complex data and providing some associated confidence limits.

Despite the acknowledged limitations of current schemes, the recognition of an ordered approach and the delineation of the basic principles of hazard assessment achieved in earlier workshops [1–3] must be regarded as a quantum step forward. The task that now faces scientists in the field is to review and improve methods for estimating exposure and effect concentrations. As noted above, decisions on acceptability are made primarily on the basis of the margin between exposure and effect concentrations. It seems likely, on the basis of current methods and level of understanding, that the best chance of increasing the apparent safety margin is through a downward revision of the exposure concentration.

Nature and Scope of Contaminant Fate Models

There is a role for environmental fate models in the hazard evaluation process, in the initial and in the more detailed phases. Workshop participants were able to identify and have evaluated models that range from the very simple to the very complex, in terms of model architecture, data requirements and ease of application. No one model, however, has applicability to all stages of the assessment process, nor should a model built for one purpose be used for a different application without consulting the model developer. Workshop participants concluded that developers and users of fate models are distinct groups and should consult in advance to ensure that the model is designed to defined uses and to answer specific questions.

Some uses of environmental fate models discussed included the identification of potential problems requiring additional laboratory testing prior to introduction of a chemical into commerce, hazard assessment of specific uses of a substance, decisions on whether site-specific releases of a chemical will exceed environmental criteria or standards, decisions on the need to take remedial action to mitigate some perceived risk, and the evaluation of specific environmental processes.

At previous Pellston workshops, two kinds of models have been described and discussed with respect to their use in predicting the fate of chemicals in the environment: (1) physical (microcosms) models and (2) mathematical (rate) models. This book has emphasized mathematical environmental fate models. Those who are interested in physical models should consult the literature [3,6,7].

Environmental fate models are used to predict the behavior and expected exposure concentration of a chemical in the environment through use of some mathematical construct. Such models can be as simple as a logical relationship or single algebraic equation, or they can be so complex as to require solution with a computer. They may use either theoretical or empirical data and can be based on either equilibrium (steady-state) or dynamic (nonsteady-state) conditions. Because models are so diverse in their structure, the terminology applied to them can be somewhat confusing (e.g., theoretical vs empirical, equilibrium vs dynamic, stochastic vs parametric ... see also other papers in this volume by Yeh, by Lassiter, and by Di Toro et al.). To minimize confusion, we suggest adoption of the following classification of mathematical contaminant fate models (after Di Toro et al. [8]):

1. partitioning models, e.g., the Mackay-Neely unit world model [9];

2. simplified fate model, e.g., Stanford Research International (SRI), Exposure Analysis Modeling System (EXAMS), Chemical Manufacturing Association—HydroQual Inc. (CMA-HydroQual) and the Mackay Level III;
3. detailed dynamic fate model, e.g., Sediment Radionuclide Transport Model (SERATRA), Hydrologic Simulation Program—FORTRAN (HSPF) and the Unified Transport Model (UTM).

Within these three broad categories of approaches, most current models are developed for a specific environment or effluent type. The case studies session at this workshop considered runoff, groundwater, river, lake estuary and effluent models. All of these model classes or model types are needed, and each is normally used for a different purpose. For instance, partitioning models can be used to estimate quickly where a chemical will tend to go in the various environmental compartments if allowed to proceed to equilibrium. This helps direct further fate modeling and effects testing. A simplified fate model for movement of a chemical in a river can be used to estimate environmental concentration at various points downstream.

Despite the apparent differences among modeling strategies and model types, there is a remarkable similarity among available models with regard to the component processes of parameters that are incorporated. Examination of currently available ground- and surface-water models reveals that a base set of processes incorporating physical, chemical and biological phenomena appear to be common to all. The partitioning processes, such as sediment sorption/desorption, are relatively rapid and reach equilibrium with the aqueous chemical concentration in a short period of time. On the other hand, the kinetic (e.g., biotransformation and photolysis) and transport processes are time- and event-driven, and require more complex algorithms and data sets for successful application.

Implicit in all of these models is the assumption that the processes are decomposable (or can be dissected) and then recomposable. The models also assume that the chemical subject has no influence on the constituent processes. To date, most of the modeled processes have incorporated episodic or excursions that appear to be of considerable importance in nature in controlling the ultimate disposition of a chemical.

Models as a Tool in the Hazard Assessment Process

Environmental fate models can contribute to the consistency of the hazard assessment process and provide a framework for decision-making. Workshop participants discussed the similarities and differences in the existing models and the implications for their use. In general, all of the models discussed are quite similar in the environmental processes dealt with, but all processes are not equally well understood. Because of this, even though there are differences in the degree of detail by which these processes are described from one model to another, it is not yet clear how these differences will affect the correctness of the final conclusion. More distinct differences, however, exist among the models in terms of their spatial and temporal scales and the way in which transport is handled. The applicability of a specific model is often determined by its respective spatial and temporal scales.

Validation studies conducted to date indicate that the model predictions are consistent with observations; however, only a limited number of chemical classes and environments have been tested. Further work is continuing to test model components individually and in complex mixtures for an increasing range of chemical classes and under a range of environmental conditions. Sensitivity analyses of models' outputs to uncertainties in inputs are essential to any validation study, and increasing effort is being directed at establishing realistic ranges for many of the environmental data inputs required by the models. These ranges of environmental data inputs are to be compiled into data bases to facilitate the use of these models in hazard assessments. Only in this way can a representative range of environments and associated predicted concentrations be developed as part of the hazard assessment.

Mathematical models were considered to offer a number of particular advantages to the hazard assessment process. These major advantages are discussed within a decision-making context.

Organize Thinking and Decision Framework

To arrive at an environmental exposure concentration (EEC) for a chemical, one must consider more than simple dilution of an input in most cases. For this prediction, it is logical to consider the relevant processes that may dissipate the chemical in a given system. Models tend to provide a step-by-step decision framework from which an estimate of exposure concentration can be calculated. A model user may become aware of a process or processes that are important but would not have been considered without the use of a model.

Defensible or Accountable Decisions

An accountable decision may not be a totally defensible decision, but it can demonstrate how all of the options were considered. This does not mean that models can serve to replace well trained and experienced professionals, but models may greatly aid a newcomer to the profession.

Temporal Control of Decisions

Chemical fate models may reduce the variability in decisions that are made by an office (public or private) because of changes in personnel. The framework of complex models usually attempts to consider all relevant processes or parameters that may serve to reduce environmental concentrations of the chemical in question. The error in decisions introduced by the relative experience of a model user can never be eliminated, but it can be reduced. Ample and clear documentation by model developers is paramount and would go a long way toward reducing "model user's risk" (analogous to Type II statistical error). With using a model as a decision-making tool goes the responsibility for the user to feed back and develop any insights or suggestions that may arise through experiences concerning model construction or documentation. Models of chemical fate in multimedia environments may also be useful in experimental design when more information is required for decision-making. The critical points in space or time may be identified, and the resources for information-gathering can be distributed appropriately.

Information Reduction for Decision-Making

When faced with large numbers of inputs concerning both the chemical and receiving environment, a decision-maker or team of decision-makers may be hard-pressed to arrive at an agreeable simultaneous solution to multiple processes contributing to the reduction of the environmental concentration. Many of these decisions have further complications of limited time and resources. A model of chemical fate allows the examination of a variety of scenarios to assess the range of possibilities that are open to the distribution of a chemical after release and which will have a bearing on the ultimate decision on acceptability.

Quality Control or Assurance

Chemical fate models may have some utility in serving as a minimal quality control. Given a range of information about a particular chemical and candidate receiving environment, both experienced and inexperienced users ideally should arrive at the same decision. For the less-than-clear-cut decision, more information may be required (or more expertise may be brought to bear on the problem).

FUTURE RESEARCH NEEDS

Although there is a plethora of areas that require further research, the following were thought by the workshop participants to be the most crucial to advancement.

1. Sediment/chemical interactions: such processes and equilibrium phenomena as sediment-associated decay rates, sorption and sediment/water exchange need to be described.
2. Sediment transport dynamics: movement of a sorbed chemical with the bed load; sedimentation and resuspension are important factors.
3. Dynamics of mixtures as opposed to discrete compounds: does a mixture act as a discrete compound?
4. Biotransformation rates: are water column transformations adequately described by zero-, first- or second-order expression? What are the sediment-associated biodegradation rates?
5. Episodic events: models may require incorporation of episodic events that are important in both fate and effects of chemicals.
6. Validation and calibration: adequate field data are required for validation and calibration of chemical fate models.
7. Dynamic model development: there is a need for development of time- and spatially variable models in light of the limitations of the currently predominant equilibrium and steady-state models.
8. Need for harmonization of systematics and terminology: of seemingly small importance at first sight, but in a field where many people have no background in mathematical modeling this will be essential.
9. Importance of degradation products: degradation products may be of importance in the hazard evaluation process, but current models are keyed to primary degradation. This problem may be resolved via iteration using current models, but this should be clearly translated to the user.
10. Bioaccumulation: bioaccumulation may play a role in increasing the body burden of some chemicals in some organisms. More confidence in the predictability of this event could be gained through further research.
11. Coupling fate and effects models: although not a topic specifically covered by this workshop, it became clear that this will be essential in the future and that there is little current experience of the associated problems.

CONCLUSIONS

Hazard assessment of chemicals released to the aquatic environment is a function of exposure and effect concentration. Therefore, the ability to predict environmental exposure concentrations is essential for hazard assessment.

There are two groups interested in models: producers/developers and users. Communication is essential between these two groups, and is presently lacking, so that producers do not always appreciate the needs and objectives of users, and users do not always understand the implications of model construction, especially concerning data manipulation. The comparability of laboratory-derived data on process mechanisms with actual mechanisms in the environment is not well understood. Furthermore, there is a lack of good field data to validate models and to serve as data bases for assessment purposes.

The utility of a model depends on its being matched to the objective, which must be clearly stated. The limited number of validation studies has demonstrated the feasibility of predictive models for hazard assessment. Furthermore, the validation and improvement of models of proven utility should receive higher priority than the development of conceptually new models.

There is no doubt that this workshop brought into the hazard assessment arena for the first time a new and important scientific discipline. As suggested at the beginning of this chapter, environmental data modeling holds out the tantalizing prospect of providing the means to take hazard assessment another major step forward. Initially, however, it is expected that environmental fate models will raise more problems than they solve. However, in the process of resolving these problems, we will inevitably come to a better understanding of the reactions and interactions that govern the fate of chemicals released to the environment.

REFERENCES

1. Cairns, J., Jr., K. L. Dickson and A. W. Maki, Eds. *Estimating the Hazard of Chemical Substances to Aquatic Life,* ASTM Technical Publication 657 (Philadelphia: American Society for Testing and Materials, 1978).
2. Dickson, K. L., A. W. Maki and J. Cairns, Jr., Eds. *Analyzing the Hazard Evaluation Process* (Bethesda, MD: American Fisheries Society, 1979).
3. Maki, A. W., K. L. Dickson and J. Cairns, Jr., Eds. *Biotransformation and Fate of Chemicals in the Aquatic Environment* (Washington, DC: American Society for Microbiology, 1980).
4. Cairns, J., Jr., "Estimating Hazard," *BioScience* 30(2):101–107 (1980).
5. Klein, A. W., M. Harnisch, H. J. Poremski and F. Schmidt-Bleek. "OECD Chemicals Testing Programme, Physico-chemical Tests," *Chemosphere* 10(2):153 (1981).
6. Giesy, J. P., Jr., Ed. *Microcosms in Ecological Research,* U.S. Department of Energy, Symposium Series 52, CONF-781101 (Washington, DC: U.S. Technical Information Center, 1980).

7. Bourquin, A. W., and P. H. Pritchard. "Proceedings of the Workshop: Microbial Degradation of Pollutants in Marine Environments," U.S. EPA Report 600/9-79-012 (1979).
8. Di Toro, D. M., A. S. Donigian, L. Games, S. Hern, R. R. Lassiter and Y. Matsuoka. "Synopsis of Discussion Session: Validation and Application Testing," Chapter 21, this volume.
9. Neely, W. B., and D. Mackay. "An Evaluative Model for Estimating Environmental Fate," Chapter 8, this volume.

INDEX